Chemistry and Biochemistry of Herbage

Volume 2

Chemistry and Biochemistry of Herbage

Edited by

G. W. BUTLER

and

R. W. BAILEY

Department of Scientific and Industrial Research
Applied Biochemistry Division
Palmerston North, New Zealand

Volume 2
1973

ACADEMIC PRESS
LONDON AND NEW YORK
A Subsidiary of Harcourt Brace Jovanovich, Publishers

ACADEMIC PRESS INC. (LONDON) LTD.
24/28 Oval Road,
London NW1

United States Edition published by
ACADEMIC PRESS INC.
111 Fifth Avenue
New York, New York 10003

Library of Congress Catalog Card Number: 72 12264
ISBN: 0 12 148102 6

Printed in Great Britain by
William Clowes & Sons Limited
London, Colchester and Beccles

Contributors

R. W. BAILEY, *Applied Biochemistry Division, Department of Scientific and Industrial Research, Private Bag, Palmerston North, New Zealand.*

F. J. BERGERSEN, *Division of Plant Industry, C.S.I.R.O. Canberra, Australia.*

N. K. BOARDMAN, *Division of Plant Industry, C.S.I.R.O., Canberra, Australia.*

C. J. BRADY, *Plant Physiology Unit, C.S.I.R.O., Division of Food Research, School of Biological Sciences, Macquarie University, North Ryde, Sydney, Australia.*

G. W. BUTLER, *Applied Biochemistry Division, Department of Scientific and Industrial Research, Private Bag, Palmerston North, New Zealand.*

J. P. COOPER, *Welsh Plant Breeding Station, Plas Gogerddan, Aberystwyth, Wales.*

W. DIJKSHOORN, *Institute for Biological and Chemical Research on Field Crops and Herbage, Wageningen, The Netherlands.*

M. D. HATCH, *Division of Plant Industry, C.S.I.R.O., Canberra, Australia.*

E. F. HENZELL, *Division of Tropical Pastures, C.S.I.R.O., Cunningham Laboratory, Mill Road, St. Lucia, Queensland 4067, Australia.*

D. I. H. JONES, *Welsh Plant Breeding Station, Plas Gogerddan, Aberystwyth, Wales.*

J. F. LONERAGAN, *School of Biology, Murdoch University, Perth, Western Australia.*

R. W. LOWREY, *Department of Animal Science, University of Georgia, Athens, Ga., 30601, U.S.A.*

P. J. ROSS, *Division of Soils, C.S.I.R.O., Cunningham Laboratory, Mill Road, St. Lucia, Queensland 4067, Australia.*

R. W. SHEARD, *Department of Land Resource Science, Ontario Agricultural College, University of Guelph, Guelph, Ontario, Canada.*

S. R. WILKINSON, *Southern Branch, Soil and Water Conservation Research Division, Agricultural Research Service, U.S.D.A., Watkinsville, Ga., 30677, U.S.A.*

D. WILSON,* *Grasslands Division, Department of Scientific and Industrial Research, Palmerston North, New Zealand.*

* Present address: Welsh Plant Breeding Station, Plas Gogerddan, Aberystwyth, Wales.

Contents

CHAPTER 14

Vitamins

G. W. BUTLER

CHAPTER 15

Water in Herbage

R. W. BAILEY

CHAPTER 16

Biochemistry of Photosynthesis

M. D. HATCH and N. K. BOARDMAN

CHAPTER 17

Physiology of Light Utilization by Swards

D. WILSON

CHAPTER 18

Mineral Absorption and its Relation to the Mineral Composition of Herbage

J. F. LONERAGAN

CHAPTER 19

Mineral Biochemistry of Herbage

G. W. BUTLER and D. I. H. JONES

CHAPTER 20

Organic Acids, and their Role in Ion Uptake

W. DIJKSHOORN

CHAPTER 21

Symbiotic Nitrogen Fixation by Legumes

F. J. BERGERSEN

CHAPTER 22

The Nitrogen Cycle of Pasture Ecosystems

E. F. HENZELL and P. J. ROSS

CHAPTER 23

Cycling of Mineral Nutrients in Pasture Ecosystems

S. R. WILKINSON and R. W. LOWREY

CHAPTER 24

Changes Accompanying Growth and Senescence and Effect of Physiological Stress

C. J. BRADY

CHAPTER 25

Organic Reserves and Plant Regrowth

R. W. SHEARD

CHAPTER 26

Genetic Variation in Herbage Constituents

J. P. COOPER

Contents of Volumes 1 and 3

CHAPTER 14

Vitamins

G. W. BUTLER

Applied Biochemistry Division, Department of Scientific and Industrial Research, Palmerston North, New Zealand

I. INTRODUCTION

The subject of vitamins in herbage has limited significance for ruminant nutrition, because the intestinal microflora of ruminants supplies adequate amounts of many of these factors. Farm livestock other than ruminants which are feeding partly or wholly on grass have a greater dependence on the vitamins present in ingested herbage.

The topic is treated briefly here. The reader is particularly referred for further information to the detailed tabulations of vitamins in livestock feeds compiled by Aitken and Hankin (1970) and to extensive discussions of the comparative aspects of vitamins in man and domestic animals by Mitchell (1964) and Beaton and McHenry (1964). Up-to-date information on vitamin and coenzyme determination and enzymology has been compiled by McCormick and Wright (1970 and subsequent volumes). The biosynthesis of the vitamins is in general not considered here and the reader is referred to the foregoing reference and to Goodwin (1963). The biochemical roles of the fat-soluble vitamins are well discussed by DeLuca and Suttie (1969).

II. FAT-SOLUBLE VITAMINS AND VITAMIN C

A. *Vitamin A*

Vitamin A occurs naturally in the bodies of animals, principally as retinol (vitamin A_1) (1) and is stored in mammals principally in the liver. The chief sources of vitamin A for herbivorous livestock are

(1) Retinol (vitamin A_1)

carotenoids, especially β-carotene, in green feed (see Chapter 10, Vol. 1) and for ruminants it appears that 5 to 8 μg β-carotene in the diet is equivalent to 1 μg retinol at normal levels of intake.

As discussed in Chapter 10, Vol. 1, the β-carotene content of herbage is subject to very wide variation; Aitken and Harkin (1970) have summarized much data on levels in fresh, ensiled and dried green-feeds. The effects of drying on β-carotene content are discussed in Chapter 27, Vol. 3. In general it may be said that fresh green herbage will contain several hundred micrograms of carotene per g DM, with losses being likely with both ensiling and drying—but that very wide variation is possible as a consequence of physiological, environmental, genetic and processing factors.

"Vitamin A deficiency in grazing ruminants occurs only when animals have to exist on mature dry herbage for long periods as in drought, since the green plant provides abundant carotene and enables large amounts of retinol to be stored in the body. Animals grazing green pastures consume quantities of carotene far in excess of needs, but there is no record of hypervitaminosis A as a result; the mechanisms limiting conversion of carotene to retinol are still obscure", (McDonald, 1968). The same situation exists with other herbivores.

The daily intake of β-carotene required by cattle appears to be not greater than 70 μg/kg body-weight. This is in cold environments and the requirement may be only half this in warm environments (Mitchell, 1964).

B. Vitamin D (Calciferol)

When the plant sterol ergosterol (2) is irradiated, vitamin D_2 ergocalciferol (3) is formed; similarly the cholesterol oxidation product, 7-dehydrocholesterol, which is present in skin, is oxidized to vitamin D_3, cholecalciferol (4). Both forms of vitamin D are

(2) Ergosterol

(3) Vitamin D_2, ergocalciferol

(4) Vitamin D_3, cholecalciferol

active in preventing rickets in mammals—calcification of the bone matrix proceeds normally if there is adequate dietary vitamin D, but in vitamin D deficiency this does not occur. Chickens respond much more to vitamin D_3 than to vitamin D_2 for the prevention of rickets. Vitamin D has also been shown to mediate the absorption of calcium from the intestine.

The assessment of dietary vitamin D requirements is complicated by interactions between vitamin D requirement and calcium and phosphorus contents of the diet. The content of vitamin D in plant feeding stuffs is measured by rat bioassays and expressed as International Units, one International Unit being defined as $0 \cdot 025$ μg cholecalciferol. Calves are commonly considered to require 300 International Units per 100 pounds live weight per day. It appears that there will normally be ample vitamin D in herbage, provided there is adequate light—sun-cured hay is stated to be a very good source.

There is evidence for rachitogenic substance(s) in plant feeds which lower(s) the effectivity of vitamin D (Grant and O'Hara, 1957; Weits, 1958); the material appears to be in a carotene fraction, but has not been identified.

C. Vitamin E (Tocopherols)

Vitamin E, α-tocopherol (5), is one of seven tocopherols found naturally in plant oils and lipids of green leaves.

(5) α-Tocopherol, vitamin E

Of these, α-tocopherol has the widest distribution and greatest biological activity; leaves of grasses and legumes contain little or none of tocopherols other than α-tocopherol, the concentrations of which are usually in the range 80-250 μg/g DM (Aitken and Hankin, 1970). One milligram of synthetic DL-α-tocopherol acetate represents 1 International Unit of vitamin E.

The requirements of animals for vitamin E are greatly complicated by multiple interactions with other substances and by questions of

which criteria of inadequacy to use. Vitamin E deficiency is manifested in a variety of ways, with infertility and muscular dystrophy
being dominant symptoms; herbivorous animals are relatively sensitive to deficiency. Typically, Loosli (1949) found 0·23 to 0·37
mg/kg body weight required to protect lambs against muscular
dystrophy.

A property of tocopherol which seems important in relation to
deficiency symptoms is its ability to inhibit the auto-oxidation of
unsaturated fatty acids. Thus certain synthetic antioxidants prevent
manifestations of vitamin E deficiency. However, Green and Bunyan
(1969) have reviewed the evidence for vitamin E functioning solely
as an antioxidant—in particular, in the protection of unsaturated fats
against peroxidation—and have concluded that the relationship is too
complex to be interpreted solely in terms of fat peroxidizability and
an antioxidant function for vitamin E. The selenium status of
animals is also important in relation to deficiency conditions involving vitamin E, and a complex set of interactions exists where selenium
and vitamin E may be necessary together or singly to alleviate or
prevent specific conditions (see also Chapter 19). The differing
viewpoints on the mode of action of vitamin E are well discussed in
the book of DeLuca and Suttie (1969).

D. *Vitamin K*

Vitamin K_1, 2-methyl-3-phytyl-1,4-naphthoquinone (6), is present in green leafy material and is required by chickens for normal
blood clotting. With other animals and in man this vitamin is
supplied by intestinal bacteria; under conditions where the bacteria

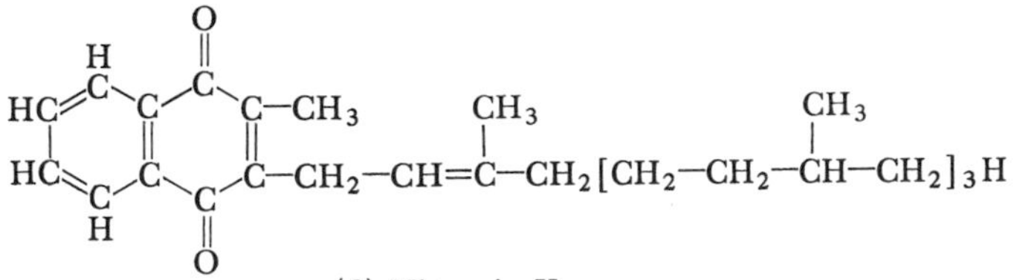

(6) Vitamin K_1

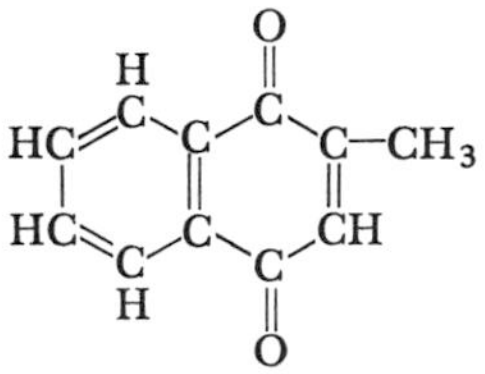

(7) Menadione

are inhibited or intestinal absorption of lipids is impaired, symptoms of vitamin K deficiency become apparent, with faulty blood clotting. Several related compounds also have vitamin K activity e.g. 2-methyl-1,4-naphthoquinone (menadione) (7).

Thornton and Moreng (1958) reported that the vitamin K_1 content of artificially dried alfalfa meal ranged from 21·5 to 30·4 μg/g DM, and of sun-dried alfalfa meal from 13·0 to 18·4 μg/g DM.

E. Vitamin C (Ascorbic Acid)

Vitamin C, 2,3-dienol-L-gulono-1,4-lactone (8) is synthesized by higher plants and by most species of animals except the guinea pig, primates, the fruit bat, the pipistrelle and some birds of the

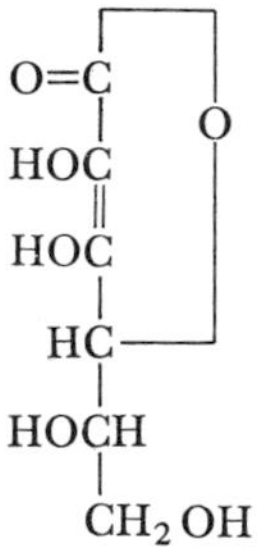

$$
\begin{array}{l}
\text{O=C} \\
\text{HOC} \quad \text{O} \\
\text{HOC} \\
\text{HC} \\
\text{HOCH} \\
\text{CH}_2\text{OH}
\end{array}
$$

(8) Ascorbic acid, vitamin C

Passeriformes order (Chatterjee, 1970). The pathways of biosynthesis have been elucidated in considerable detail (White et al., 1968; Loewus and Baig, 1970; Chatterjee, 1970). Vitamin C is not in general required by farm animals, although Aitken and Hankin (1970) report benefits to poultry, young pigs and calves. Fresh green herbage contains the vitamin in amounts which decrease with increased maturity, e.g. Aitken and Hankin (1970) cite figures for four common grasses ranging from 4250-5660 μg/g DM in young grass to 700-1100 μg/g DM in mature grass. Dried herbage will generally contain little or no ascorbic acid.

III. B VITAMINS

A. Thiamine (Vitamin B_1)

Thiamine (9) is easily converted to thiochrome (10) by mild oxidation and the blue fluorescence of thiochrome forms the basis of the usual method of estimation, after enzymic digestion to hydrolyse bound thiamine. The coenzyme form of thiamine, thiamine pyro-phosphate, is involved in the oxidative decarboxylation of α-keto-acids and in the transketolase reaction. All animals other than ruminants require a dietary supply of thiamine, deficiency resulting in beriberi in humans and characteristically in hemorrhagic brain lesions in humans, foxes and pigeons.

Concentrations of thiamine in herbage are generally reported to range from 2·5 to approximately 10 $\mu g/g$ DM, with ensiling and drying both appearing to lead to lower figures (Aitken and Hankin, 1970).

The lowering of thiamine absorption in animals is recognized in certain circumstances where a thiaminase is present in the ingested

(9) Thiamine chloride (vitamin B_1)

(10) Thiochrome

feed, e.g. in bracken (*Pteridium aquilinum*) and horsetails (*Equisetum* spp.) (Forsyth, 1968). Horses exhibit a simple thiamine deficiency, but the situation with sheep and cattle is more complex because of the synthetic capabilities of rumen microorganisms. Also, other toxic agents are present in *Pteridium aquilinum* (Evans, 1964).

B. Riboflavin

Riboflavin (11) is essential for all organisms because of its contribution to the synthesis of the enzyme cofactors flavin mono-nucleotide and flavin adenine dinucleotide, and occurs ubiquitously in all biological materials. It usually occurs as part of one of the two flavin coenzymes, and but rarely as the free vitamin. Riboflavin is most commonly estimated microbiologically with *Lactobacillus casei*, after acid hydrolysis or enzymic digestion with papain and

Riboflavin
(11) (6,7-dimethyl-9-(1'-D-ribityl)-isoalloxazine)

diastase. Rumen microorganisms synthesize adequate riboflavin for ruminants.

In general, riboflavin contents of fresh green herbage are relatively high, ranging from 5 to 25 μg/g DM and similar levels can be retained in dried herbage (Aitken and Hankin, 1970).

C. Nicotinic Acid (Niacin)

Nicotinic acid (12) occurs in biological tissues as the pyridine nucleotides and nicotinamide (13) occurs in the free form. Nicotinic acid can be released from unidentified substances in cereal and other

(12) Nicotinic acid (niacin)   (13) Nicotinamide

grains by alkali treatment, but these bound forms appear to be poorly available biologically. Such forms are not recognized in herbage. For ruminants, rumen microorganisms synthesize adequate amounts. L-Tryptophane can be converted to nicotinic acid in most animals, and this exerts a sparing effect on requirements of this vitamin.

Total nicotinic acid concentrations (including nicotinamide) in fresh and dried herbage range from 30 to 60 μg/g DM (Aitken and Hankin, 1970).

D. Pyridoxine, Pyridoxal and Pyridoxamine (Vitamin B_6)

Pyridoxine (14), pyridoxal (15) and pyridoxamine (16) are collectively designated vitamin B_6, pyridoxal phosphate (17) is a cofactor in many enzymes associated with amino acid metabolism, including decarboxylases and transaminases.

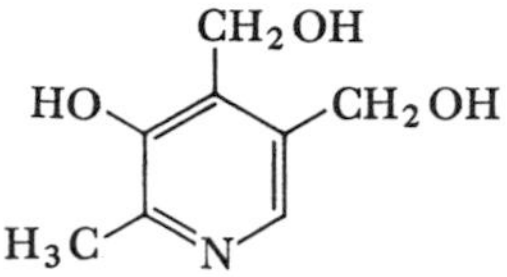

CH$_2$OH

HO CH$_2$OH

H$_3$C N

(14) Pyridoxine

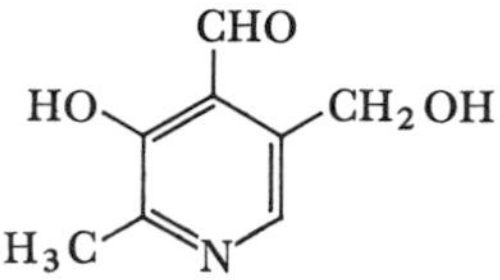

CHO

HO CH$_2$OH

H$_3$C N

(15) Pyridoxal

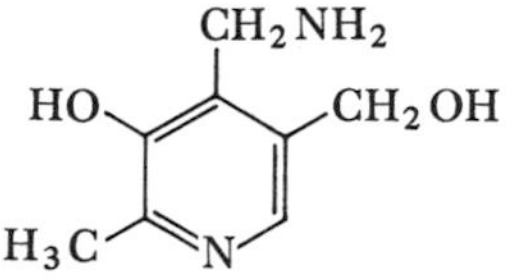

CH$_2$NH$_2$

HO CH$_2$OH

H$_3$C N

(16) Pyridoxamine

CHO

HO CH$_2$—O—P=O

H$_3$C N

OH

OH

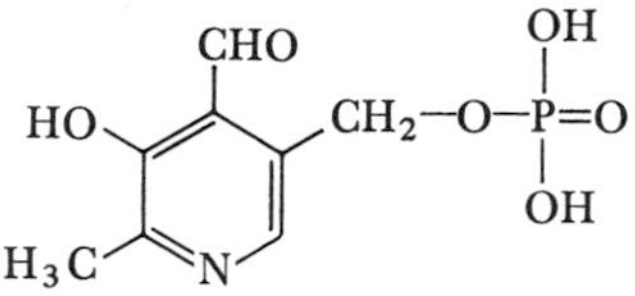

(17) Pyridoxal phosphate

With ruminants, rumen microorganisms synthesize this vitamin; in man, it appears that the intestinal microflora probably also synthesize vitamin B$_6$, providing partially for the host's requirements.

Concentrations of pyridoxine in pasture herbage have been little studied. Thomas and Walker (1949) report a range of 8-19 μg/g DM for herbage from 11 temperate pasture species.

E. Pantothenic Acid

Pantothenic acid (pantoyl-β-alanine) (18) is synthesized by green plants and most microorganisms but is an essential requirement for all animal species that have been investigated. It is a constituent of

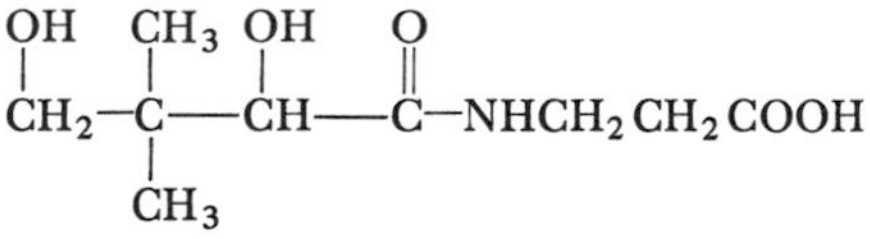

$$\underset{\underset{\displaystyle CH_3}{|}}{CH_2-\overset{\overset{\displaystyle CH_3}{|}}{C}-\overset{\overset{\displaystyle OH}{|}}{CH}-\overset{\overset{\displaystyle O}{||}}{C}-NHCH_2CH_2COOH}$$

OH   CH$_3$ OH   O

(18) Pantothenic acid

coenzyme A and acyl carrier protein. It is usually estimated microbiologically after enzymic digestion to release combined forms. Fresh and dried green feeds generally contain 10 to 25 μg pantothenic acid per g DM (Aitken and Hankin, 1970). Herbivores appear to receive adequate amounts from the intestinal microflora and the green-feed ingested.

F. Vitamin B$_{12}$

Vitamin B$_{12}$ (19) may be solely a product of bacterial synthesis; the amounts which have been reported in herbage may arise from microbiological contamination of plant surfaces, and small amounts

(19) Vitamin B_{12}

could also be taken up from the soil. The coenzyme form of vitamin B_{12}, cobamide coenzyme, participates in a series of apparently diverse reactions including dismutation of vicinal diols (e.g. propane-1,2-diol to propionaldehyde) and transfer reactions of the type glutamic acid to β-methylaspartic acid. The intestinal microflora of adult herbivora are able to synthesize adequate vitamin B_{12} for the animal's needs, provided that the diet contains adequate cobalt.

G. Folic Acid (Pteroylglutamic Acid)

Folic acid (20) is involved in several unique reactions in inter-mediary metabolism, including serving as a carrier for hydroxy-methyl and formyl groups and as an intermediate in *de novo*

(20) Folic acid (pteroylglutamic acid)

synthesis of the methyl group of methionine. Data on concentrations in herbage of folic acids are limited, but a range of 3-7·5 μg/g DM has been reported for dried lucerne (Aitken and Harkin, 1970). Concentrations of total pteroylglutamate derivatives in maize leaf and pea seedling extracts were 6-7 and 11·5 μg/g DM respectively (Cossins and Shah, 1972).

H. Biotin

Biotin (21) is the prosthetic group of several enzymes catalysing CO_2 fixation into organic linkage, including acetyl CoA carboxylase, propionyl CoA carboxylase, methyl-malonyl transcarboxylase and

(21) Biotin

pyruvate carboxylase. It is synthesized by plants; animals appear to receive their requirements from intestinal bacterial synthesis of this compound.

The concentrations of biotin in dried green feeds are reported to range from 16 to 40 μg/100 g DM.

REFERENCES

Aitken, F. C. and Harkin, R. G. (1970). "Vitamins in Feeds for Livestock." Tech. Comm. No. 25, Commonwealth Bureau of Animal Nutrition. C.A.B., Farnham Royal, Bucks., England.
Beaton, G. H. and McHenty, E. W., eds (1964). "Nutrition, A Comprehensive Treatise", Vol. 2. Academic Press, New York and London.

Chatterjee, I. B. (1970). *In* "Methods in Enzymology" (D. B. McCormick and L. D. Wright, eds) Vol. 18, Part A, pp. 29-34. Academic Press, New York and London.

Cossins, E. A. and Shah, S. P. (1972). *Phytochem.* 11, 587-593.

DeLuca, H. F. and Suttie, J. W., eds (1969). "The Fat-Soluble Vitamins." Univ. of Wisconsin Press, Madison, U.S.A.

Evans, W. C. (1964). *Vet. Rec.* 76, 365-372.

Forsyth, A. A. (1968). "British Poisonous Plants." Bulletin No. 161. Ministry of Agriculture, Fisheries and Food. H.M.S.O., London.

Goodwin, T. W. (1963). "The Biosynthesis of Vitamins and Related Compounds." Academic Press, London and New York.

Grant, A. B. and O'Hara, P. B. (1957). *N.Z. Jl Sci. Technol.* (A), 38, 548-576.

Green, J. and Bunyan, J. (1969). *Nutr. Abstr. Rev.* 39, 321-345.

Loewus, F. and Baig, M. M. (1970). *In* "Methods in Enzymology" (D. B. McCormick and L. D. Wright) Vol. 18, Part A, pp. 22-28. Academic Press, New York and London.

Loosli, J. K. (1949). *Ann. N.Y. Acad. Sci.* 52, 243-249.

McCormick, D. B. and Wright, L. D. (1970). *Methods in Enzymology,* Vol. 18. Part A. Academic Press, New York and London. (Parts B and C in preparation.)

McDonald, I. W. (1968). *Nutr. Abstr. Rev.* 38, 381-400.

Mitchell, H. H. (1964). "Comparative Nutrition of Man and Domestic Animals", Vol. 2. Academic Press, New York and London.

Thomas, B. and Walker, H. F. (1949). *Emp. J. exp. Agric.* 17, 170-178.

Thornton, P. A. and Moreng, R. E. (1958). *Poult. Sci.* 37, 1154-1159.

Weits, J. (1958). *J. Dairy Sci.,* 41, 1088-1093.

White, A., Handler, P. and Smith, E. L. (1968). "Principles of Biochemistry." Fourth Edition. McGraw-Hill Book Company.

CHAPTER 15

Water in Herbage

R. W. BAILEY

Applied Biochemistry Division,
Department of Scientific and Industrial Research,
Palmerston North, New Zealand

I. INTRODUCTION

Preceding chapters have been concerned with compounds which together nearly always compose less than half and usually only 15-20% of the herbage total fresh weight. The other constituent is water. While vital for all plant growth and of overriding interest to the plant physiologist, water is largely ignored by herbage chemists and biochemists and likewise of minor concern to many agronomists and animal nutritionists whose interest is with pasture "dry matter" production and digestion. Nevertheless some brief comments on the roles of water in herbage biochemistry, growth and utilization must be made here. The purpose of this chapter is to see that these aspects of water in herbage are not overlooked by those scientists more concerned with the other components. Detailed accounts of the various physiological aspects of herbage water are given by, for example, Slatyer (1967) and Kramer (1969) and the many references mentioned therein. One extreme of the herbage water situation, namely water removal to produce hay or dried grass, is discussed in detail in Chapter 27, Volume 3.

II. BIOCHEMICAL ASPECTS

A. Measurement and Distribution

1. Water Measurement

Water is usually measured by drying weighed plant tissue at 80-110°C in a forced draught oven for 18-24 or more hours to give the percent of "oven dry matter". Water is taken to equal the loss on drying. The soundest physiological basis for calculating the water is probably as percent of oven dry weight (Slatyer, 1967) (eq. 1). It is

$$(1) \quad \% \text{ Water} = \left(\frac{\text{Fresh weight} - \text{Dry weight}}{\text{Dry weight}} \right) \times 100$$

more commonly expressed as percent of fresh weight and obtained by simply taking $100 - \%$ of oven dry weight as this latter value is calculated as a percent of fresh weight (eq. 2). The results quoted in

$$(2) \quad \% \text{ Water} = 100 - \left(\frac{\text{Dry weight}}{\text{Fresh weight}} \times 100 \right)$$

this chapter are given as percent of fresh weight as they have been derived from published oven dry matter values. Measurement of water by loss on heating is only an approximation as this loss will include any other volatile constituents and any loss by heat decomposition of organic constituents. In addition the results are only meaningful if the green herbage is weighed immediately after cutting and is free from external soil (see Chapter 13, Vol. 1) and water. Freeze-drying does not always remove the last 5-7% of herbage moisture. For this reason it is recommended that, in addition to any other basis, analyses from freeze-dried material should be expressed on an oven-dry basis to permit comparisons with other oven-dried material.

In addition to the measurement of water levels *per se* other related measurements such as turgid water content or water potential when the tissue is in equilibrium with water or water vapour are largely relevant to plant physiology and will not be considered here (see Sullivan, 1971).

2. *Distribution of Water in Plant Tissues*

Apart from a small amount released in biochemical reactions nearly all herbage water is derived from a single external source, the soil, which in turn derives it from atmospheric precipitation of one kind or another. Smaller amounts are absorbed by the leaf tissues directly from incident water. Of the cell constituents the cell walls probably contain the lowest levels of water, particularly in maturing thickened walls. Cell protoplasm may contain up to 95% of water while the water content of the vacuole, up to 95%, is the highest in the plant cell. So far as the plant's vascular system is concerned the xylem sap represents the main water distribution system for the plant.

In herbage the main organized components in grasses are leaf blades, leaf sheaths and stems with in addition petioles in legumes and other broad leaved plants. A selection of water levels in these organs in various herbage species is given in Table I. The point of interest is that in grasses and legumes the leaf sheaths, petioles or stems tend to contain more water than the leaves at least until the stems mature and dry out. This difference no doubt reflects the higher rate of photosynthesis in the leaves and the transport role of the other organs.

B. *Functions of Water*

1. *Biochemical*

Water is essential for plant growth and is involved directly in the many biochemical reactions occurring in the plant, which need not be catalogued here, besides being the medium in which these reactions occur. With CO_2 it is one of the building blocks of all plant constituents. Presumably at the plant physiological pH values it may be regarded as the ultimate source of the H^+ and OH^- ions of many reactions.

The formation of polymer links by condensation reactions can involve the elemination of 1 mole of water per link formed. This is the case in the formation of protein peptide links but not for the repeating links in nucleic acids or the glycosidic links in sugars and polysaccharides when these latter involve synthesis from nucleotide sugars. In contrast the cleavage of all of these links by hydrolytic enzymes during remobilization always involves the uptake of 1 mole of water for each link cleaved.

TABLE I

Water content of herbage plant parts

Species	Water content (% of fresh weight at harvest)			References

1. Grasses

Species	Leaf blade	Leaf Sheath	Stem	References
Perennial ryegrass (*Lolium perenne*)	74·0	77·0	—	Bailey (unpublished)
Cocksfoot (*Dactylis glomerata*)	74·2	82·9	—	Bailey (unpublished)
Tall fescue (*Festuca arundinacea*)	71·5	75·4	—	Bailey (unpublished)
Prairie grass (*Bromus unioloides*)	74·9	74·3	75·4	Bailey (unpublished)
Digitaria smutsii	81·0	87·1	80·0	Bailey (unpublished)
Napier grass (*Pennisetum purpureum*)	89–75		88–83[a], 84–75[b]	Chatterjee et al. (1947)
Maize (*Zea mays*)	65·3[c] (71·3)	67·0	74·8	Bailey (unpublished)[c]

2. Legumes

Species	Leaflet	Petiole	Stem	References
Red clover (*Trifolium pratense*)	76–82	—	86–87	Bailey (1958)
White clover (*T. repens*)	80–82	89–91	—	Bailey (unpublished)
Lucerne (*Medicago sativa*)	75–74, 79–80[d]		69–74, 70–75[d]	Deny and Zaleski (1954) Vartha, N.Z. (unpublished)
Sainfoin (*Onobrychis viciifolia*)	71–75 80		74–82 85	Baker et al. (1952) Bailey (unpublished)

3. *Brassicas*

	Leaf	Petiole	Stem total	rind	Stem Top xylem	marrow	rind	Stem Bottom xylem	marrow	
Rape (*Brassica napus*)	78·7	86·0	81·0							Bailey (unpublished)
	90–89		80–84							Jones (1959)
Choumolier (*B. oleracea*)	80·0	84·2	81·4							Bailey (unpublished)
Marrow stem kale (*B. oleracea*)	89·1		89·0	85·0	85·8	90·5	85·1	78·2	90·5	Jones (1965)
Thousand head kale (*B. oleracea*)	89–84			79·0	74·6	85·7	77·0	69·2	85·6	Jones (1962)

[a] Soft stem.
[b] Coarse stem.
[c] Leaf blade minus midrib; in parentheses total leaf including midrib; material from fully matured plants.
[d] Early and late maturing varieties respectively, sampled at 20% flowering.

Probably of more interest in biochemical terms is the effect of water deficit or stress on the biochemical processes in plants (discussed in detail, Chapter 24, see also Slatyer, 1967, 1969; Laude, 1971). Virtually all aspects of plant growth are depressed as the water content falls to a stress level. Depression in DNA, RNA and protein synthesis leads ultimately to a breakdown of these polymers with a resultant increased transport of nitrogen and phosphorus compounds from the leaves to the stems. A rise in sucrose and fall in starch levels is also associated with water stress and is apparently the result of an increase in sucrose synthesis (Hiller and Greenway, 1968). This water stress lowers photosynthesis rates by its effects on the photosynthetic reactions and also by its inducement of stomatal closure with a consequent lowering of the CO_2 supply (see Chapter 17).

2. *Physiological*

The two major physiological roles of plant water may be defined as transport and cooling. As mentioned the xylem sap represents the main source of incoming water from the soil with the prime role of transporting the mineral nutrients absorbed from the soil. Phloem sap is largely responsible for transporting photosynthate and other metabolites. By far the largest portion of the water taken in through the plant roots is however transpired through the leaves. It seems that the main function of this water transpiration is that of cooling the plant to avoid thermal death of the leaves by heating from incoming radiation. The relationship of this water loss to photosynthesis is discussed in Chapters 16 and 17. This means of course that amounts of transpired water are directly related to the amounts of radiant and advective energy received by the leaves. According to Kerr (pers. commun.), in a high producing herbage growing under temperate spring-summer conditions some 500 kg of water pass through the herbage for every kg of dry matter produced.

III. AGRICULTURAL ASPECTS

A. *Agronomy of Water*

Water levels in herbage are of course dependent not only on the stage of growth of the herbage but also on the overall climatic conditions and water supply. In addition assessment of water levels from published dry matter figures is subject to the risks involved in the processing of the herbage samples as mentioned earlier. A selection of recorded water levels, based on dry matter figures, in tropical and temperate herbages either maintained as leafy grazing pasture or growing through the season to maturity is given in Table II. In spite of limitations the results illustrate several points. Firstly there is a consistent fall in the water level of total herbage as plants grow through to maturity reflecting in part the increase in mature stem tissue. Secondly the improved, cultivated temperate grasses appear to contain higher water levels than tropical grasses. Thirdly there can be very high levels of water indeed (up to 90%) in temperate pasture herbage maintained in a leafy state.

Water content of herbage is of course affected markedly by nutritional, environmental and even genetic conditions. Heavy nitrogen fertilizer treatments give herbage with a higher water content. For example an increase in nitrogen fertilizer from 44 kg to 300 kg/ha increased the water content of Italian ryegrass (*Lolium multiflorum*) from 78 to 85% (Harkness, 1966). Other authors have recorded similar effects with temperate grasses, e.g. Holmes (1951) and also with tropical grasses (Gomide *et al.*, 1969; Jung and Reid, 1966) although in this latter case the effect is not so marked or only evident in early stages of growth. According to Deinum (1966) increasing temperature with no humidity control lowers the water content of ryegrass, although growth cabinet studies with humidity control of Wilson and Ford (1971) suggest that this effect can be more marked and variable in tropical grasses. Lower light intensity also increases the water content (Deinum, 1966; Burton *et al.*, 1959) although this effect is not so marked under high nitrogen fertilization. It is also generally accepted (Harkness, 1966; Barclay and Vartha, 1966) that tetraploidy leads to a higher water content compared with the parent diploid plants.

So far as the standing herbage crop is concerned bulk water content figures are enlightening. Thus with lush herbage containing 80-85% of water the total herbage will contain 4-6 metric tons of water for every metric ton of dry matter. This water figure becomes important when it is necessary to cut and transport the herbage any distance to a grass drying or protein extraction plant. Kerr's (1973) figure of 500 kg of water passing through a pasture system for each kg of dry matter produced is however far greater than the 4-6 kg of water associated with the 1 kg of herbage dry matter at any point in time.

TABLE II

Fluctuations in herbage water during growth

Species	Water content (% of fresh weight)				Country	References
	leafy	→	mature	leafy during season		
1. Temperate (festucoid) grasses						
Perennial ryegrass (*Lolium perenne*)	85	→	64[a]		U.K.	Johnston and Waite (1965)
				84–86[a]	N.Z.	Ulyatt, MacRae and Grace (Unpublished)
Italian, tetraploid, ryegrass (*L. multiflorum*)				87–90	N.Z.	Ulyatt *et al.* (unpublished)
Cocksfoot (*Dactylis glomerata*)	90	→	75[b]		U.S.A.	Berg (1971)
	82	→	60		U.K.	Johnston and Waite (1965)
Timothy (*Phleum pratense*)	80·9	→	59·7		U.K.	Waite and Sastry (1949)
	73·6	→	59·4		Sweden	Kivimae (1959)
Tall fescue (*Festuca arundinacea*)	88	→	77[b]		U.S.A.	Berg (1971)
	75	→	68	80–85	N.Z.	Anderson (unpublished)
					U.K.	Fagan and Jones (1920–23)
2. Tropical (panicoid and eragrostoid) grasses						
Switchgrass (*Panicum virgatum*)	73	→	70[b]		U.S.A.	Berg (1971)
Guinea grass (*P. maximum*)	72	→	66[c]		Nigeria	Oyenuga (1960)
Pangola grass (*Digitaria decumbens*)	80·4	→	78·9 (28·2)[d]		Trinidad	Forster *et al.* (1960)
Paspalum dilatatum				66–77	Israel	Bondi and Meyer (1951)
Andropogon gayanus	72	→	62[e]		Nigeria	Haggar (1970)
	58	→	30[e]			

Species		Range	Country	Reference
Napier grass (Pennisetum purpureum)	86 → 82		India	Chatterjee et al. (1947)
	85·5 → 74·1[f]		Nigeria	Oyenuga (1959)
Pennisetum orientale		73–84	India	Nath and Das (1953)
Kikuyu grass (P. clandestinum)	76 → 62		U.S.A.	Gomide et al. (1969)
Sudan grass (Sorghum vulgare var. sudanensis)	84 → 63		U.S.A.	Jung and Reid (1966)
Spear grass (Heteropogon contortus)		59–76	India	Nath and Das (1953)
Maize (Zea mays)	87 → 74		India	Sharma and Mudgal (1966)
Bermuda grass	68 → 55		U.S.A.	Gomide et al. (1969)
(Cynodon dactylon)	81 → 70		Trinidad	Butterworth (1963)
3. Legumes				
White clover (Trifolium repens)		86–90	N.Z.	Ulyatt et al. (unpublished)
Berseem (T. alexandrinum)		82–89[g]	Israel	Bondi and Meyer (1951)
Subterranean clover (T. subterraneum)	86 → 78		W. Aust.	Hardwick (1954)
Red clover (T. pratense)	81 → 71[h]		Sweden	Kivimae (1959)
	82·6 → 74·6[i]		U.K.	Dent (1955)
Lucerne (Medicago sativa)	85 → 75		Israel	Bondo and Meyer (1951)
Sainfoin (Onobrychis viciifolia)	82 → 69		U.K.	Baker et al. (1952)
Phaseolus lathyroides	79 → 70		Queensland, Aust.	Milford (1967)
Desmodium uncinatum	78 → 70		Queensland, Aust.	Milford (1967)

TABLE II—continued

Species	Water content (% of fresh weight)		Country	References
	leafy → mature	leafy during season		
4. Brassicas				
Rapes (*Brassica napus*)		82–83[j]	U.K.	Jones (1959)
Marrow stem kale (*B. oleracea*)	84[j]		U.K.	Jones (1959)
Thousand head kale (*B. oleracea*)	86[j]		U.K.	Jones (1959)

[a] → indicates change during growth, — indicates range.
[b] No difference between 3 strains.
[c] Per cent stem changing from 7·4 → 38·3
[d] In parentheses, dried standing material.
[e] Wet and dry season respectively.
[f] Per cent stem changing from 7 → 51·6.
[g] Irrigated.
[h,i] Diploid and tetraploid respectively.
[j] Mature, whole plant.

B. Herbage Water and the Ruminant

Pasture water can represent a minor or major part of the ruminant's water and in some cases can be an over-supply. Under arid conditions or with hay or dried grass the herbage naturally only supplies a small portion of the animal's water. With green herbage containing 80-85% of water, however, the ruminant will consume 4-5·6 kg of water for every kg of dry matter eaten. Sullivan (1969) does not consider that this herbage water will be in excess of the animal's water needs unless the water content is for example above 78% at 4°C or 84% at 26·7°C. Ulyatt (pers. commun.) however calculated that for herbage with 87% of water, which is typical of New Zealand spring pasture, a sheep eating 2 kg per day of dry matter in this herbage consumes with it 13·4 kg of water. In contrast when the same sheep eats 2 kg of dried grass it only drinks 4 kg of water. Even allowing for the recirculation of endogenous water in the latter case it does seem that under these conditions pasture is supplying more water than the animal needs. In the case of a mature cattle beast in the same circumstances the daily consumption of 25 kg of dry matter in the herbage would entail the intake of 167 kg of water. Not only does this water probably represent an excess to the animal's requirement but considerable work is done both by the mere act of harvesting and by the processes of adsorption and circulation of the water.

Apart from supplying water *per se* herbage water can possibly affect pasture digestion in the rumen. Herbage of high water content is likely to be softer and broken down more quickly by chewing to give a rumen digesta of low dry matter, than herbage of low water content. This could mean a different digestion rate for the herbage of high water content or a more rapid flow of undigested (in the rumen) soluble constituents to the lower digestive tract. In this connection, however, Ulyatt and MacRae (1973) found little difference between rates of digesta flow in sheep fed low dry matter feed and published results for such flow in sheep fed high dry matter feed.

REFERENCES

Bailey, R. W. (1958). *J. Sci. Fd Agric.* **9**, 748-753.
Baker, C. J. L., Heimberg, M., Alderman, G. and Eden, A. (1952). *J. agric. Sci., Camb.* **42**, 382-394.
Barclay, P. C. and Vartha, E. W. (1966). *N.Z. Grassld Soc. Proc.* **28**, 184-194.

Berg, C. C. (1971). *Agron. J.* **63**, 785-786.
Bondi, A. and Meyer, H. (1951). *Bull. Res. Coun. Israel* 1(3), 26-36.
Burton, G. W., Jackson, J. E. and Knox, F. E. (1959). *Agron. J.* **51**, 537-542.
Butterworth, M. H. (1963). *J. agric. Sci., Camb.* **60**, 341-345.
Chatterjee, I. B., Hye, A. and Ali, S. (1947). *Ind. J. vet. Sci. anim. Husb.* **17**, 253-260.
Deinum, B. (1966). *X Intl Grassld Congr.* Helsinki, 415-418.
Dent, J. W. (1955). *J. Br. Grassld Soc.* **10**, 330-340.
Dent, J. W. and Zaleski, A. (1954). *J. Br. Grassld Soc.* **9**, 131-140.
Fagan, T. W. and Jones, H. T. (1920-23). *Reports of Welsh Plant Breeding Sta.* Series H, No. 3, 85-130.
Forster, R. H., Wilson, P. N. and Butterworth, M. H. (1960). *VII Int. Grassld Congr.*, Reading, 390-392.
Gomide, J. A., Noller, C. H., Mott, G. O., Conrad, J. H. and Hill, D. L. (1969). *Agron. J.*, **61**, 116-119.
Haggar, R. J. (1970). *J. agric. Sci., Camb.* **74**, 487-494.
Hardwick, N. E. (1954). *Aust. J. agric. Res.* **5**, 372-382.
Harkness, R. D. (1966). *X Intl Grassld Congr.*, Helsinki, 315-319.
Hiller, R. G. and Greenway, H. (1968). *Planta* **78**, 49-59.
Holmes, W. (1951). *J. agric. Sci., Camb.* **41**, 70-79.
Johnston, M. J. and Waite, R. (1965). *J. agric. Sci., Camb.* **64**, 211-219.
Jones, D. J. C. (1959). *J. agric. Sci., Camb.* **52**, 230-243.
Jones, D. J. C. (1962). *J. agric. Sci., Camb.* **58**, 265-275.
Jones, D. J. C. (1965). *J. agric. Sci., Camb.* **65**, 121-128.
Jung, G. A. and Reid, R. L. (1966). *W. Virginia Univ. agric. expt. Stn Bull.* 524T, 51 pp.
Kivimae, A. (1959). *Acta Agric. Scand.* Suppl. 5.
Kramer, P. J. (1969). "Plant and Soil Water Relationships", McGraw-Hill, New York and London.
Laude, H. M. (1971). *In* "Drought Injury and Resistance in Crops" (K. L. Larson and J. D. Eastin, eds), pp. 45-57. Crop Sci. Soc. Am., Madison, Wisconsin.
Milford, R. (1967). *Aust. J. expt. Agric. anim. Husb.* **7**, 540-545.
Nath, N. and Das, N. B. (1953). *Ind. J. vet. Sci. anim. Husb.* **23**, 185-204.
Oyenuga, V. A. (1959). *J. agric. Sci., Camb.* **53**, 25-33.
Oyenuga, V. A. (1960). *J. agric. Sci., Camb.* **55**, 339-350.
Sharma, C. B. and Mudgal, V. D. (1966). *Ind. J. dairy Sci.* **19**, 100-105.
Slatyer, R. O. (1967). "Plant Water Relationships", Academic Press, New York and London.
Slatyer, R. O. (1969). *In* "Physiological Aspects of Crop Yield" (J. D. Eastin, ed.), pp. 53-83. *Am. Soc. Agron.*, Madison, Wisconsin.
Sullivan, C. Y. (1971). *In* "Drought Injury and Resistance in Crops" (K. L. Larson and J. D. Eastin, eds), pp. 1-18. Crop Sci. Soc. Am., Madison, Wisconsin.
Sullivan, J. T. (1969). "Chemical Composition of Forages with Reference to the Needs of the Grazing Animal". U.S. Dept. Agric., ARS 34-107. Washington, U.S.A.
Ulyatt, M. J. and MacRae, J. C. (1973). *J. agric. Sci., Camb.* (in press).
Waite, R. and Sastry, K. N. S. (1949). *Emp. J. exp. Agric.* **17**, 179-189.
Wilson, J. R. and Ford, C. W. (1971). *Aust. J. agric. Res.* **22**, 563-571.

CHAPTER 16

Biochemistry of Photosynthesis

M. D. HATCH AND N. K. BOARDMAN

Division of Plant Industry, CSIRO, Canberra, Australia

I. INTRODUCTION

Although photosynthesis provides the carbon for all organic material of higher plants the potential for photosynthesis may not always be the prime limiting factor determining useful yield. For instance, the partitioning of photosynthetic products may be of overriding importance rather than gross dry matter production (Evans, 1971). However, amongst cultivated plants the useful yield from pastures and forage crops would appear likely to be directly reflected in the sum total of photosynthesis. An understanding of the photosynthetic process operative in these particular plants and the factors limiting and regulating this process is therefore of more than usual significance.

The primary purpose of this chapter is to provide an account of the photochemical and biochemical processes involved in photosynthetic conversion of CO_2 to organic cell constituents. Several aspects of the physiology of photosynthesis by plants and plant communities will be covered in Chapters 17 and 24. Brief comment on some aspects peripheral to the biochemistry of photosynthesis is, however, included here where it is considered that significant gaps would otherwise exist.

In the following account the light-dependent phase of photosynthesis that produces chemical energy, and the so-called dark reactions that utilize this energy to convert CO_2 to organic compounds will be considered separately in that sequence. The integration and continuity of this discussion will be assisted by the following prefacing remarks. Much of what is to follow will be considered in terms of the operation of two major biochemical variants for photosynthesis. In one of these, termed here the Calvin cycle, or C_3-pathway, CO_2 is initially fixed by carboxylation of RuDP[a] to yield 3-PGA. The alternative process also involves the operation of the Calvin cycle preceded in this case by a series of reactions commencing with the carboxylation of PEP to yield the C_4-dicarboxylic acids oxaloacetate, malate and aspartate. This process is termed the C_4-pathway. A unique feature of C_4-pathway species is that chloroplasts are distributed between two distinct cell types. These are radially arranged around the vascular bundles, the inner layer being termed bundle sheath cells and the outer layer the mesophyll cells. Broadly speaking, the reactions involved in PEP and C_4-dicarboxylic acid formation are located in the mesophyll cells while those of the Calvin cycle are in the bundle sheath cells. Depending upon the species, either malate or aspartate, or in some cases probably both these acids are transported to the bundle sheath chloroplasts and there decarboxylated to yield CO_2. This CO_2 is fixed via the Calvin cycle while the C_3 compound remaining after

[a] Abbreviations other than those that are commonly employed: CMU, 3-(p-chlorophenyl)-1,1-dimethylurea; DCMU, 3-(3,4-dichlorophenyl)-1,1-dimethylurea; PEP, phosphoenolpyruvate; 3-PGA, 3-phosphoglycerate; PMS, phenazine methosulphate; PS-I, photosystem I; PS-II, photosystem II; RuDP, ribulose-1,5-diphosphate; TNBT, tetranitro blue tetrazolium chloride.

decarboxylation serves as the precursor for the regeneration of PEP. For those species which transport malate to the bundle sheath chloroplasts malic enzyme is responsible for decarboxylation. As discussed below, these particular bundle sheath chloroplasts have some unusual structural and photochemical features.

There are also several physiological features which distinguish plants with the Calvin cycle from those with the C_4-pathway. The operation of the latter pathway is associated with reduced photorespiration, a greater capacity for photosynthesis, more economic use of water, and higher temperature and light optima for photosynthesis and growth.

For more specialized treatments of the topics in the following sections, readers are referred to reviews and other accounts of: photochemical reactions (Kok, 1965; Boardman, 1968; Hind and Olsen, 1968; Levine, 1969; Boardman, 1970); CO_2-fixation pathways (Bassham, 1964; Gibbs, 1967; Hatch and Slack, 1970b), including specifically the C_4-pathway (Hatch and Slack, 1970a; Karpilov, 1970; Wolfe, 1970; Hatch, 1971a); photorespiration and the glycollate pathway (Jackson and Volk, 1970; Tolbert, 1971); the physiology of photosynthesis and its relation to differing pathways (Hatch and Slack, 1970a; Wolfe, 1970; Loomis et al., 1971); as well as a recent general review (Walker and Crofts, 1970) and a symposium on these topics (Hatch et al., 1971). The use of additional citations in the following text will be limited.

II. LIGHT REACTIONS

Present knowledge of the energy conversion process of photosynthesis in which light energy is absorbed by the pigments of the chloroplast and converted into chemical free-energy in the form of NADPH and ATP has come mainly from studies on algae and green plants which fix CO_2 by the Calvin cycle. The light reactions of photosynthesis are performed on the internal membranes (termed lamellae or thylakoids) of the chloroplast and the utilization of this energy to fix CO_2 occurs in the stroma or soluble phase of the chloroplast. The stroma is bounded by the outer limiting membrane of the chloroplast and it penetrates throughout the interlamellar and intergrana spaces. The regions in which the internal membranes of the chloroplast are grouped together in parallel bundles are known as grana, and these are a characteristic feature of the mesophyll chloroplasts of higher plants. As mentioned in the introduction, plants with the C_4-pathway also contain chloroplasts in an inner cell layer, termed the bundle sheath, surrounding the vascular tissue. These bundle sheath chloroplasts show various degrees of grana development depending on the species.

A. Photosynthetic Unit

Because of the high concentration of chlorophyll in the grana of a mesophyll chloroplast each chlorophyll molecule receives a quantum of light energy only a few times a second even in full sunlight. Such a rate of absorption of quanta is several fold lower than the turnover of most metabolic reactions. However, the pigment molecules are so arranged that quanta absorbed by a large number of molecules are transferred to a special molecule of chlorophyll a, called the trap or reaction centre chlorophyll. Here, the primary conversion of light energy into chemical free-energy takes place. This is the concept of the photosynthetic unit which was first proposed to account for the observation of Emerson and Arnold (1932) that the maximum yield of CO_2 fixed, or O_2 evolved from *Chlorella* per single flash of intense light was one mole per 2500 moles of chlorophyll.

The light absorbing units of the plant chloroplast contain chlorophyll a, chlorophyll b and a number of carotenoids (see Chapter 10, Vol. 1). Energy is transferred between adjacent pigment molecules within a photosynthetic unit until it reaches the trap molecule. The absorption band of the trap chlorophyll a is at a longer wavelength than the absorption bands of the light harvesting pigments to ensure efficient trapping of the energy at the reaction centre.

The trap molecule is in close association with an acceptor molecule (A) and a donor molecule (D) in the membrane. The trap chlorophyll a, on excitation (Chl a*), is able to donate an electron to A. The acceptor molecule is reduced to A⁻ and this leaves the trap chlorophyll a deficient in an electron (eq. 1). The positively charged chlorophyll a then receives an electron from the donor D, and the chlorophyll molecule is restored to its ground state energy level (eq. 2). The net result is that the energy of a photon is used to transfer an electron from D to A, and thus the primary photochemical event of photosynthesis is an oxidation-reduction process.

$$\text{Chl } a^* + A \rightarrow \text{Chl } a^+ + A^- \tag{1}$$

$$\text{Chl } a^+ + D \rightarrow \text{Chl } a + D^+ \tag{2}$$

B. *Photosynthetic Electron Transport*

It is now established that chloroplasts contain two types of photosynthetic units (photosystem I, PS-I and photosystem II, PS-II). Reduction of NADP is driven by light absorbed by PS-I and PS-II, which act sequentially to transfer electrons from water to $NADP^+$ (Fig. 1) (Boardman, 1968; Hind and Olsen, 1968; Levine, 1969).

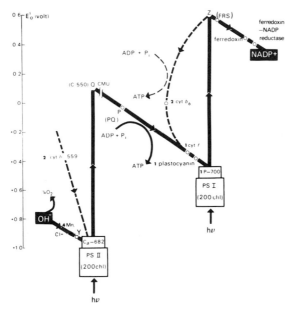

Fig. 1. Scheme for photosynthetic electron transport and photophosphorylation in chloroplasts (see text for explanation). The number beside a component indicates the number of molecules of that component per photosynthetic unit of 400 chlorophyll molecules. Q may be identical to C-550, and P identical to plastoquinone (PQ). Ferredoxin reducing substance (FRS) may mediate electron transfer from PS-I to ferredoxin. A scale of redox potentials is shown on the left.

Quanta of light absorbed by the pigments of PS-II are transferred to the trap pigment, which appears to be a form of chlorophyll absorbing at 682 nm and termed Chl a-682. Excitation of Chl a-682 catalyses the transfer of an electron from Y to Q, giving a strong oxidant, Y^+, and a reductant, Q^-. Electron flow from water to Y^+ requires manganese probably in the form of a manganese-protein complex, and chloride ion. Q and P are electron carriers which were first proposed to explain the fluorescence behaviour of isolated chloroplasts. The weak red fluorescence emitted by chloroplasts comes mainly from the chlorophyll molecules in the light harvesting centres of PS-II. The yield of fluorescence, however, is dependent on the redox state of Q. In dark-adapted chloroplasts, Q is oxidized and the fluorescence is low. On illumination of the chloroplasts Q becomes reduced and the magnitude of the fluorescence rises to a new steady state level. The kinetics of the rise in the fluorescence yield are biphasic and are interpreted in terms of two electron carriers (Q and P) on the reducing side of PS-II. P is present at a higher concentration than Q and it constitutes an electron carrier pool between the light reactions of PS-I and PS-II. Light induced absorbance changes of chloroplasts in the ultraviolet region have indicated that plastoquinone (PQ) is an electron carrier between the two light reactions, and it seems highly probable that P is identical to plastoquinone.

Recently, Knaff and Arnon (1969) discovered a light-induced absorbance change at 550 nm in isolated chloroplasts occurring both at room temperature and at liquid nitrogen temperature. They attributed the change to the reduction of a component in PS-II, termed C-550. More recent work has indicated that C-550 may be identical to the primary fluorescence quencher, Q (Boardman *et al.*, 1971; Erixon and Butler, 1971) but the chemical nature of C-550 is unknown.

The electron from Q^- is transferred through a series of electron carriers, which include plastoquinone (PQ), a *c*-type cytochrome (cytochrome *f*) and a Cu-containing protein (plastocyanin). At the present time, it is not known with certainty whether plastocyanin is closer to PS-I than cytochrome *f*. Neither plastoquinone nor cytochrome *f* can chemically reduce $NADP^+$.

Quanta absorbed by the light harvesting pigments of PS-I are transferred to P-700, a special form of chlorophyll absorbing at 700 nm. On excitation, P-700 donates an electron to an acceptor Z, giving oxidized P-700 and reduced Z. The oxidation of P-700 is accompanied by a negative absorption change at 700 nm. P-700 is restored to its reduced state by an electron from plastocyanin or cytochrome *f*. The strong reductant Z^-, which appears to have a redox potential in the vicinity of -0.5 to -0.6 volt, is able to reduce $NADP^+$. This reduction is mediated by the non-haem iron protein, ferredoxin, and the flavoprotein, ferredoxin-NADP reductase. A substance recently isolated by Yocum and San Pietro (1969) and Regitz *et al.* (1970) and named ferredoxin reducing substance (FRS) may be involved in transferring electrons from Z^- to ferredoxin.

Previously, it was considered that a b-type cytochrome (cytochrome b-559) was a component of the electron transport chain between the light reactions of PS-I and PS-II, but more recent studies suggest that cytochrome b-559 is on the oxidizing side of PS-II (Boardman et al., 1971). Because of its redox potential ($E_0' = 0.36$), cytochrome b-559 is placed on a side path from PS-II rather than on the pathway between water and PS-II (Boardman et al., 1971).

Artificial electron acceptors or Hill reaction oxidants accept electrons either from PS-II (e.g. trichlorophenolindophenol, ferricyanide) or from PS-I (e.g. benzyl viologen). The herbicides, 3(3,4-dichlorophenyl)1,1-dimethylurea (DCMU) and 3(p-chlorophenyl)-1,-1-dimethylurea (CMU) inhibit electron transport between Q and P. Photoreduction of $NADP^+$ can be restored in the inhibited chloroplasts by the addition of an artificial electron donor such as reduced 2,6-dichlorophenolindophenol. Electrons from the artificial donor enter the electron transport chain between the light reactions, and photoreduction of $NADP^+$ is then driven by PS-I.

PS-I is known as the far-red system because its absorption spectrum extends to longer wavelengths than that of PS-II. At wavelengths beyond 700 nm, PS-I receives a high fraction of the quanta absorbed by chloroplasts. Chloroplasts contain one molecule of cytochrome f and one molecule of P-700 per 430 chlorophyll molecules, from which it is concluded the photosynthetic unit contains about 400 chlorophyll molecules. As shown in Fig. 1 the chlorophyll molecules appear to be distributed about equally between the two photosystems.

C. Photosynthetic Phosphorylation

In non-cyclic phosphorylation, formation of ATP is coupled to electron flow from water to $NADP^+$, but there is still much uncertainty as to whether there are one or two sites of phosphorylation. Earlier measurements gave a P/e_2 ratio (moles of ATP per 2 electrons transported) of one, suggesting one site of phosphorylation, More recently, P/e_2 ratios in excess of one (between 1.2 and 1.8) have been reported for phosphorylation coupled to electron flow from water to $NADP^+$, or to ferricyanide. Interpretation of these higher ratios in terms of two sites of phosphorylation is complicated by the high rates of basal electron flow observed with chloroplasts under non-phosphorylating conditions i.e. in the absence of the cofactors required for ATP formation. If the basal rate of electron transport is subtracted from the rate of electron transport under phosphorylating conditions, then P/e_2 ratios of about 2 are obtained, but the validity of this subtraction is uncertain.

The Calvin cycle requires 3 moles of ATP and 2 moles of NADPH for each mole of CO_2 fixed, while the C_4-pathway would apparently require 5 moles of ATP and 2 moles of NADPH (Chen *et al.*, 1969; Hatch and Slack, 1970a). One site of phosphorylation would provide 2 moles of ATP per 2 moles of $NADP^+$ reduced, which is insufficient even for the Calvin cycle. Two sites of phosphorylation would provide more than enough ATP for the Calvin cycle, but not sufficient for the C_4-pathway.

Another type of phosphorylation has been observed with isolated chloroplasts. Known as cyclic phosphorylation, it requires an exogenous cofactor such as phenazine methosulphate, pyocyanin or ferredoxin. Unlike non-cyclic phosphorylation, cyclic phosphorylation is not accompanied by any net change in oxidation or reduction and it is not inhibited by DCMU. The process is driven by light absorbed by PS-I. Chloroplasts exhibit very low rates of cyclic phosphorylation in the absence of an exogenous electron carrier. It is, therefore, uncertain whether cyclic phosphorylation operates *in vivo*, but it seems likely that it does in view of the fact that algae show a light-dependent uptake of glucose or acetate in the presence of DCMU (Tanner and Kandler, 1969). Arnon (1967) considers that ferredoxin is the natural cofactor for cyclic phosphorylation. In Fig. 1, cytochrome b_6 is shown as a component of the cyclic pathway, but definitive evidence is lacking. On present evidence, it seems likely that the sites of cyclic phosphorylation and non-cyclic phosphorylation are different.

Recently, Knaff and Arnon (1969) have proposed a scheme of three light reactions in green plant photosynthesis. In their scheme, PS-II consists of two "short wavelength" light reactions (IIb and IIa) which operate in series and are connected by an electron chain containing C-550, plastoquinone, cytochrome b-559 and plastocyanin. Reduction of $NADP^+$ and non-cyclic phosphorylation is considered to be driven solely by PS-II. The "long wavelength" system (PS-I) is thought to be in parallel with photosystems IIb and IIa and its role confined to cyclic electron flow and cyclic phosphorylation. Cytochromes f and b_6 are components of cyclic electron transport, and according to Knaff and Arnon (1969) are not involved in non-cyclic electron flow from water to $NADP^+$. However, it is known that there is an antagonistic effect of far-red and red light on the oxidation and reduction of the electron carriers Q, plastoquinone, cytochrome f and P-700. This antagonism is consistent with the scheme shown in Fig. 1, but cannot be easily explained by the scheme of Knaff and Arnon (1969). At the present time, the scheme of Fig. 1 explains most of the experimental observations but it is certain that it will undergo modification and become more sophisticated as more experimental data become available.

D. Composition of the Photosystems

A partial fractionation of the photosystems was first obtained by incubating spinach chloroplasts with digitonin, a non-ionic detergent, and separating the resulting subchloroplast fragments by differential centrifugation (Boardman and Anderson, 1964). Separation of the photosystems has also been achieved by treatment of chloroplasts with Triton X-100 or sodium deoxycholate, while chlorophyll-protein complexes have been isolated from the photosystems after treating chloroplasts with sodium dodecylsulphate. More recently, physical methods (sonication or passage through a French pressure cell) have been used successfully to achieve some separation of the photosystems. Reviews of these studies have been published recently (Boardman, 1968, 1970) and the remarks here will be largely confined to a consideration of the composition of the photosystems.

Table I summarizes the composition of subchloroplast fragments obtained from spinach by the digitonin method. The small fragments (D-144) which sediment at 144,000 g have PS-I activity but little or no PS-II activity, and it has been estimated that these fragments are 95% PS-I. They contain a higher proportion of chlorophyll a relative to chlorophyll b than do chloroplasts and a lower ratio of xanthophyll to β-carotene. Their P-700 content (relative to chlorophyll) is approximately double that of chloroplasts, which supports the view that the photosynthetic unit size of PS-I is about one-half that of chloroplasts. Fractionation studies indicate that cytochromes f and b_6 are located in PS-I, and that cytochrome b-559 is in PS-II. Manganese, which is known from photochemical studies with Mn-deficient plants and algae to be required for electron flow from water to PS-II, is located in the D-10 fragments which are enriched in PS-II. These larger fragments, which sediment at 10,000 g are active in the Hill reaction with oxidants such as trichlorophenolindophenol or ferricyanide which can accept electrons from PS-II, but they show

TABLE I

Pigment composition of subchloroplast fragments from spinach

Ratio	Chloroplasts	Light Fragments (D-144) (PS-I)	Heavy Fragments (D-10) (Enriched in PS-II)
Chl[a]a/b	2·8	5·3	2·3
Xanthophyll/β-carotene	2·6	1·7	3·8
Chl/P-700	440	205	690
Chl/cyt f	430	363	730
Chl/Mn	73	250	52

[a] Chl = Chl a + Chl b; Chl = chlorophyll.

considerably less activity with NADP$^+$ which requires both PS-I and PS-II for reduction.

The larger fragments contain about 70% PS-II and 30% PS-I, (Boardman, 1968, 1970) but further enrichement in PS-II is obtained by treating the D-10 fragments with Triton X-100 (Huzisige *et al.*, 1969). From the observed chl *a*/chl *b* ratio of the D-10 fragments and their estimated content of PS-I and PS-II, it was calculated that PS-II would have a chl *a*/chl *b* ratio of about 1·7.

PS-I subchloroplast fragments are active in cyclic phosphorylation with PMS as cofactor, which supports the earlier conclusions that cyclic phosphorylation is driven by PS-I. D-10 fragments are active in non-cyclic phosphorylation, with a lower activity for cyclic phosphorylation.

E. Light Reactions in C_4-Plants

As mentioned earlier, chloroplasts of mesophyll cells of C_4-pathway species contain grana, but those of the bundle sheath show varying degrees of grana development, depending on the species. For example, the bundle sheath chloroplasts of *Sorghum bicolor, S: sudanense* and *Zea mays (Maize)* are agranal except for an occasional region in which two lamellae are appressed to form a rudimentary granum. On the other hand, C_4-species such as *Atriplex spongiosa* and *Panicum miliaceum* show good development of grana in both mesophyll and bundle sheath chloroplasts.

Downton *et al.* (1970) treated leaf sections of *S. sudanense* with the dye, TNBT, which acts as a Hill reaction oxidant. Photoreduction of the dye was observed in the mesophyll chloroplasts, but not in the bundle sheath chloroplasts. They concluded that the agranal bundle sheath chloroplasts lack non-cyclic electron flow from water. In leaf sections of *P. miliaceum*, TNBT was photoreduced in both mesophyll and bundle sheath chloroplasts. Black and Mayne (1970) reported that leaf extracts and isolated chloroplasts from a number of C_4-plants had a higher concentration of P-700, relative to chlorophyll and a higher ratio of chl *a*/chl *b* than several species of C_3-plants, suggesting that C_4-plants either have a more active PS-I or a smaller photosynthetic unit size.

The development of methods for separating mesophyll and bundle sheath chloroplasts (Woo *et al.*, 1970; Anderson *et al.*, 1971b) permitted a detailed study of the light reactions of isolated mesophyll and bundle sheath chloroplasts. The methods are based on the differential resistance of the bundle sheath and mesophyll cells to breakage. In the method of Woo *et al.*, (1970) the mesophyll chloroplasts were released first by grinding the tissue in a blender. A harsher treatment involving blending in a mill with glass beads was then used to obtain bundle sheath chloroplast fragments. A more gentle procedure was developed by Anderson *et al.* (1971a) for the isolation of intact bundle sheath chloroplasts from maize.

Mesophyll chloroplasts of C_4-plants reduce $NADP^+$ or artificial oxidants in the Hill reaction (Woo *et al.*, 1970). They also exhibit good rates of non-cyclic phosphorylation and of cyclic phosphorylation with PMS as a cofactor (Anderson *et al.*, 1971a). Thus mesophyll chloroplasts of C_4-plants show the same light reactions as chloroplasts of C_3-plants. Bundle sheath chloroplasts which have good development of grana (e.g. from *A. spongiosa*) are similar to mesophyll chloroplasts in their composition and photochemical activities (Woo *et al.*, 1970). In contrast, the agranal bundle sheath chloroplasts of *S. bicolor* and *Z. mays* are deficient in PS-II, as compared with mesophyll chloroplasts (Woo *et al.*, 1970, Anderson *et al.*, 1971a, b).

For instance, the Hill reaction activity of bundle sheath chloroplasts of *S. bicolor* using TCIP or ferricyanide as oxidant is only 6-13% of the corresponding activity of mesophyll chloroplasts (Anderson *et al.*, 1971b). Maize bundle sheath chloroplasts are 2-3 times more active than sorghum bundle sheath chloroplasts. Electron microscopy indicates that maize bundle sheath chloroplasts contain more regions of appressed lamellae than do sorghum bundle sheath chloroplasts, but it is not known whether there is any correlation between the level of PS-II activity in a bundle sheath chloroplast and the extent of appressed lamellae.

Maize mesophyll chloroplasts show rates of non-cyclic phosphorylation with ferricyanide as an electron acceptor, and cyclic phosphorylation with PMS as a cofactor which are comparable to the corresponding rates of spinach chloroplasts. The bundle sheath chloroplasts are very active in cyclic phosphorylation but rates of non-cyclic phosphorylation are low (Anderson *et al.*, 1971a).

Compositional and fluorescence measurements also demonstrate that the agranal bundle sheath chloroplasts of *S. bicolor* and *Z. mays* are deficient in PS-II. Cytochrome *b*-559 is barely detectable in sorghum bundle sheath chloroplasts and C-550 appears to be absent. As shown in Table II, sorghum bundle sheath chloroplasts have more P-700 and cytochrome *f* on a chlorophyll basis than mesophyll

TABLE II

Composition of mesophyll and bundle sheath chloroplasts

Ratio	Sorghum bicolor		Atriplex spongiosa	
	Mesophyll	Bundle Sheath	Mesophyll	Bundle Sheath
Chl^a a/b	3·1	5·7		
Xanthophyll/β-carotene	2·85	1·44		
Chl/P-700	496	254	387	414
Chl/cyt f	437	320		
Cyt b/cyt f	3·7	1·9		

^{a} Chl = chl a + chl b; cyt b = cyt b_6 + cyt b-559; cyt = cytochrome.

chloroplasts. They have a higher proportion of chlorophyll a and are richer in β-carotene. In fact, the composition of sorghum bundle sheath chloroplasts is strikingly similar to that of the PS-I subchloroplast fragments from spinach (Table I). The mesophyll chloroplasts of *S. bicolor, Z. mays* and *A. spongiosa* and the grana-containing bundle sheath chloroplasts of *Atriplex* resemble spinach (*Spinacea oleracea*) chloroplasts in composition.

Edwards and Black (1971) described a gentle procedure to separate mesophyll and bundle sheath cells from the C_4-plant, *Digitaria sanguinalis* (crabgrass). The mesophyll cells were twice as active as the bundle sheath cells in photochemical oxygen evolution. The bundle sheath cells had a higher chlorophyll a/b ratio, more P-700 and more ferredoxin-NADP reductase. Mayne *et al.*, (1971) concluded that the mesophyll cells of crabgrass are 2-4 times more active in non-cyclic electron flow, but the bundle sheath cells have 2-3 times greater capacity for cyclic electron flow. Grana development is very poor in the bundle sheath chloroplasts of crabgrass, although the amount of appressed lamellae appears to be greater than in maize bundle sheath chloroplasts.

To summarize, the photochemical systems in the mesophyll chloroplasts of C_4-plants or in the grana-containing bundle sheath chloroplasts do not appear to differ from those in C_3-plants. However, the bundle sheath chloroplasts which show poor development of grana are deficient in PS-II, the degree of deficiency depending on the species. Qualitatively, the species with the least amount of appressed lamellae in the bundle sheath chloroplasts (*S. bicolor*) shows the greatest deficiency in PS-II.

A deficiency of PS-II creates a problem for the reduction of $NADP^+$ in the agranal bundle sheath chloroplasts. It has been suggested that NADPH in these chloroplasts is provided by the oxidative decarboxylation of malate transported from the mesophyll chloroplasts (see following section). It thus seems a possibility that the prime photochemical function of the agranal bundle sheath chloroplast is the production of ATP by cyclic electron flow in PS-I.

III. PATHWAYS OF CO_2 FIXATION

A. Comparative Aspects

Carbon dioxide is incorporated into organic compounds in most living cells, and in all cases this occurs via carboxylation reactions. It is inherent that photosynthetic CO_2 fixation in autotrophic organisms must involve a cyclic reaction sequence since the compound initially carboxylated must ultimately be regenerated to sustain the process. As indicated in the introduction there are two major biochemical variants for photosynthetic CO_2 fixation in higher plants. One of these, the Calvin cycle, may be experimentally distinguished by the pattern of labelling of intermediate metabolites following the fixation of radioactive CO_2 (Bassham, 1965). The presence of several enzymes specific to the pathway is also a distinguishing feature. Although species with the C_4-pathway have markedly different labelling patterns and enzyme activities, the involvement of at least a large part of the Calvin cycle sequence in the terminal steps of the process has always been recognized (Hatch and Slack, 1970a, b). However, it is perhaps not surprising that in the earlier stages of investigation of the C_4-pathway the emphasis was upon differences rather than similarities. With the knowledge currently available, it is now possible to define clearly the basic unity existing between photosynthetic pathways and to discuss the pathways within this framework. For completeness, mention should be made of another group of plants whose overall photosynthetic assimilation includes a process known as Crassulacean acid metabolism (Beevers, *et al.*, 1966), although they are not amongst the important pasture species. Most of the CO_2 fixed by these plants is initially incorporated and accumulated into malic acid in the dark using carbohydrate as a source of PEP, the acceptor compound. During the light period closure of stomata restricts entry as well as exit of CO_2 and the Calvin cycle operates to form carbohydrates from CO_2 and a C_3 compound derived by the decarboxylation of malate. This process is primarily an adaptation for conservation of water. Its metabolic and functional similarities with the C_4-pathway will be apparent.

The common factor of these various pathways is the operation of the Calvin cycle as the sole mechanism for the net conversion of CO_2 to carbohydrate. Where they differ is in the processes involved in moving externally-derived CO_2 to the site of action of RuDP carboxylase, the first step in the Calvin cycle. Details of the mechanism of this cycle and the C_4-pathway are outlined below, together with some consideration of the significance of variations between these pathways.

B. *The Calvin Cycle*

The reactions of the Calvin cycle and the enzymes catalysing these reactions are depicted schematically in Fig. 2. All the reactions of the cycle itself and at least those leading to the formation of glycollate, starch and probably also sucrose and some amino acids, are located with chloroplasts. The net synthesis of carbohydrate from CO_2 involves a reaction sequence commencing with the carboxylation of the C_5 sugar RuDP giving rise to two molecules of 3-PGA, the

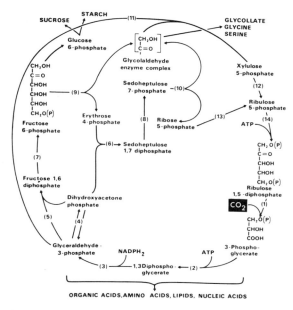

Fig. 2. The reactions and enzymes of the Calvin cycle. End-products are indicated in capital letters. Glycollate, glycine and serine are also intermediates of the glycollate pathway (see Fig. 4) which cycles carbon back to 3-PGA. The enzymes involved are: (1) RuDP carboxylase; (2) 3-PGA kinase; (3) NADP-glyceraldehyde phosphate dehydrogenase; (4) triose phosphate isomerase; (5) and (6) aldolases; (7) and (8) sugar diphosphatases; (9), (10) and (11) transketolases; (12) and (13) pentose phosphate isomerases; (14) ribulose 5-phosphate kinase.

conversion of 3-PGA to hexose phosphates by a well known series of reactions common to glycolysis, and then the synthesis of starch and sucrose via nucleosidediphosphate sugars. Quantitatively, sucrose and starch are the major end-products of photosynthesis. However, as indicated in Fig. 2, carbon may also leave the cycle as glycollate or phosphorylated C_3 compounds giving rise to several other products (Bassham, 1965; Bassham and Jensen, 1967). In order to maintain a carbon balance only one hexose molecule can be incorporated into sucrose or starch for each 6 CO_2 fixed. This would require six turns of the cycle and hence six molecules of RuDP. Thus, of the twelve molecules of 3-PGA so formed, ten molecules must be cycled through reactions 2 to 14 to form RuDP. The sites of ATP and NADPH utilization are also indicated, the theoretical requirement for each CO_2 fixed being 3ATP and 2NADPH.

Some remaining controversies surrounding the operation of the Calvin cycle, such as the site of sucrose synthesis, chloroplast membrane permeability to intermediates, and the general metabolic interrelations between chloroplasts and cytoplasm, are peripheral to its basic mechanism. These topics have been considered in recent reviews (Hatch and Slack, 1970b; Heber, 1969) and will not be discussed further here. The major deficiency in the understanding of the pathway itself relates to the capability of RuDP carboxylase to account for CO_2 fixation. As discussed elsewhere (Hatch and Slack 1970b) the activity of RuDP carboxylase in most leaves is only just adequate to account for photosynthesis provided the CO_2 concentration is near-saturating for the enzyme. CO_2 rather than bicarbonate is the primary substrate and the K_m for CO_2 is about 450 μM. It has always been tacitly assumed that CO_2 simply diffuses to the site of action of RuDP carboxylase. However, even at the concentration of CO_2 in solution in diffusion equilibrium with CO_2 in air (approximately 8 micromolar, μM), the activity of RuDP carboxylase would be only about 1% of its maximum value. In fact, since a CO_2 gradient must exist during steady-state photosynthesis the prevailing internal concentration of CO_2 due to diffusion must be less than 8 μM. Nevertheless, there is little doubt that RuDP carboxylase is responsible for the primary fixation of CO_2 in Calvin cycle species. This anomaly could be resolved either by the enzyme having vastly different kinetic characteristics *in situ* or by the operation of a mechanism to concentrate CO_2 in chloroplasts. With regard to the latter possibility, the evidence that the intervention of carbonic anhydrase is necessary to achieve maximum rates of photosynthesis in plants and algae (see Hatch and Slack 1970b) may be significant. Carbonic anhydrase catalyses the reversible hydration of CO_2 to bicarbonate and under certain conditions can accelerate movement of CO_2 across membranes to which it is bound. However, operating alone at least, it cannot provide the basis for a CO_2 concentrating mechanism. As discussed later, the unique reactions of the C_4-pathway, and possibly also Crassulacean acid metabolism, may serve as a mechanism for providing adequate concentrations of CO_2 for RuDP carboxylase.

C. The C_4-Pathway

There is now a large measure of agreement on the basic mechanism of operation of the C_4-pathway (see Hatch *et al.*, 1971). Figure 3 provides a scheme depicting this mechanism together with some more controversial interpretations that will be discussed below. Much

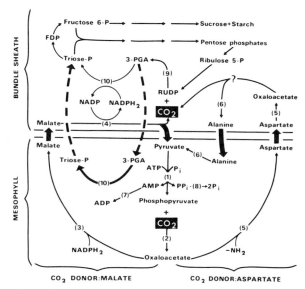

Fig. 3. Reactions, intercellular movement of metabolites (heavy arrows) and enzymes associated with the operation of the C_4-pathway. In different species either malate (left hand side) or aspartate (right hand side) is transported to the bundle sheath cells and decarboxylated. The movement of 3-PGA and triose phosphates between cells (dotted lines) is a proposed bypass, varying in quantitative importance in individual species. It probably only occurs in species which transport and decarboxylate malate (see text). The enzymes involved are: (1) pyruvate, P_i dikinase; (2) PEP carboxylase; (3) NADP-malate dehydrogenase; (4) malic enzyme; (5) aspartate aminotransferase; (6) alanine aminotransferase; (7) adenylate kinase; (8) pyrophosphatase; (9) RuDP carboxylase; (10) 3-PGA kinase and NADP-glyceraldehyde phosphate dehydrogenase. The remaining reactions are those operative in the Calvin cycle (see Fig. 2).

of the evidence for the various phases of the scheme, and for the partitioning of reactions between mesophyll and bundle sheath cells, has been considered in detail in reviews (Hatch and Slack 1970a, b) and a recent symposium (Hatch *et al.*, 1971). The leaf anatomy and chloroplast and cell ultrastructure of C_4-pathway plants has been reviewed most recently by Laetsch (1971).

The scheme proposes that the primary fixation of CO_2 occurs by carboxylation of PEP, catalysed by PEP carboxylase. For some C_4-pathway species there is direct evidence that essentially all the

CO_2 ultimately fixed into the C-1 of 3-PGA enters via this route (Hatch and Slack, 1970a); that is the Calvin cycle does not operate independently to a significant extent. Other evidence, including data suggesting that CO_2 movement to and from bundle sheath cells or chloroplasts is highly restricted, supports this view (Hatch, 1971a, 1971b). It is possible that the independent operation of RuDP carboxylases to fix directly externally-derived CO_2 may not be insignificant in all species or for all conditions.

Radiotracer studies using $^{14}CO_2$ indicate that the oxaloacetate, formed by carboxylation of PEP, is rapidly converted to malate and aspartate. During steady-state photosynthesis the pools of the latter acids are very much larger than the oxaloacetate pool. The enzymes that interconvert these C_4-acids are located in the mesophyll chloroplasts together with pyruvate, P_i dikinase, the enzyme responsible for PEP formation from pyruvate. The earlier suggestion that the C-4 of a C_4-acid may be transferred to the C-1 of 3-PGA by a transcarboxylation reaction (Hatch and Slack 1970a, b) was revised when it became apparent that RuDP carboxylase was sufficiently active in C_4-pathway species to account for this step. Thus, it now appears that one or both of C_4-acids are transported to the bundle sheath cells or chloroplasts where the C-4 is released as CO_2 and then refixed via RuDP carboxylase. The enzymes of the Calvin cycle, which are predominantly or solely located in the bundle sheath chloroplasts, then operate to form sucrose and starch and to regenerate RuDP (see Fig. 2).

As intimated earlier, malate is at least the predominant C_4-acid transported and decarboxylated in many species with the C_4-pathway. These species are characterized by having high levels of malic enzyme, located in the bundle sheath chloroplasts together with RuDP carboxylase. This enzyme would operate to release the C-4 of malate as CO_2 leaving the other three carbons as pyruvate. This compound can then be returned to the mesophyll chloroplasts to serve as a precursor of PEP. As indicated in Section IIE the species utilizing this particular route are those with bundle sheath chloroplasts deficient in grana, PS-II activity, and the capacity to photoreduce $NADP^+$. It is probably significant, therefore, that malic enzyme provides one mole of NADPH for each mole of CO_2 released. However, stoichiometrically, this is only half the NADPH necessary for the conversion of two moles of 3-PGA to triose phosphates. Since this deficiency is probably not made up by photoreduction of $NADP^+$, it would appear to be mandatory that part of the 3-PGA should be returned to the mesophyll chloroplasts for reduction to triose phosphates. Such a bypass is indicated in Fig. 3, and the degree to which it operates may vary with indicated species. An earlier observation consistent with this proposal is that the two enzymes responsible for the conversion of 3-PGA to triose phosphates, unlike other Calvin cycle enzymes, are distributed about equally between the mesophyll and bundle sheath chloroplasts (Hatch and Slack, 1970a).

Another group of C_4-pathway plants contain little malic enzyme activity (Hatch and Slack, 1970a; Downton, 1970; Andrews et al., 1971). The fact that this deficiency is compensated for by high asparatate and alanine aminotransferase activities (Andrews et al., 1971) supports the view that aspartate may replace malate as the CO_2 donor. It should be noted that in this group of plants the release of CO_2 would not be accompanied by NADPH formation. This correlates with the observation (see Section II) that their bundle sheath chloroplasts have a normal capacity to photoreduce $NADP^+$. Other observations suggest that a simple decarboxylation of aspartate is not involved. As indicated in Fig. 3, the best interpretation is that oxaloacetate, derived from aspartate in the bundle sheath cells, is the acid decarboxylated. This would conform with the observation that the aminotransferases are about equally distributed between the two cell types. Transport of alanine rather than pyruvate to the meso-phyll cells would be necessary if a balance of amino groups is to be maintained between the two types of cells.

Radiotracer studies have now provided additional support for the concept that different C_4-dicarboxylic acids act as CO_2 donors in different species. Chen et al., (1971) cited evidence correlating the rapid transfer of label from malate to 3-PGA in species with high malic enzyme activity, and rapid transfer from aspartate in species with low malic enzyme activity. These observations were extended in other studies showing that this label is transferred specifically from the C-4 of these acids to a large internal CO_2 pool and then to the C-1 of 3-PGA (Hatch, 1971b).

D. Function of Variant Pathways

In Calvin cycle species the movement of CO_2 to RuDP carboxylase has always been tacitly assumed to proceed by simple diffusion. The fact that the unique reactions of the C_4-pathway replace this phase of the process suggests that these reactions offer some advantage in this respect. Certainly, the advantage of the process represented in Fig. 3, where CO_2 is fixed only to be released and refixed again, is not immediately apparent. However, a good case can be made for the unique reactions of the C_4-pathway serving as a mechanism for concentrating CO_2 for utilization by RuDP carboxylase. The first clue to this interpretation was provided in the discussion on the operation of RuDP carboxylase in Calvin cycle plants (see Section IIIB). Here, it was pointed out that if RuDP carboxylase is to account for observed rates of photosynthesis it apparently must operate at a CO_2 concentration at least 50 times that arising by diffusion of CO_2 from air. The possible necessity of a mechanism to concentrate CO_2 above the ambient level was indicated. In C_4-pathway plants at least, the initial reactions of photosynthesis appear to provide the essential elements of such a mechanism. PEP carboxylase is present in sufficient amounts to support observed rates of photosynthesis even with CO_2 concentrations well below that formed at diffusion equilibrium with air (Hatch, 1971a). In species which transport malate to the bundle sheath chloroplasts malic enzyme is both sufficiently active and suitably located to release, and hence concentrate, CO_2 within these chloroplasts. A mechanism similar in principle could operate in those species which transport and decarboxylate aspartate instead of malate. Recent studies have provided direct evidence for the existence of a large intermediate pool of CO_2 in C_4-pathway leaves, formed from the C-4 of C_4-acids (Hatch, 1971b). This pool is much larger than could be accounted for by diffusion of CO_2 from air and, if exclusive to the bundle sheath cells or chloroplasts, would provide concentrations of CO_2 more than adequate to support the optimal operation of RuDP carboxylase.

There is evidence to support the view that the activity of RuDP carboxylase limits light-saturated photosynthesis in Calvin cycle plants (Björkman, 1971). Therefore, it would seem reasonable to suppose that the higher activity and efficiency of PEP carboxylase, the primary carboxylase in C_4-pathway plants, combined with the mechanism for concentrating CO_2 for RuDP carboxylase, would be a major contributing factor to the higher photosynthesis rates of these species. As is discussed later, a carboxylase with these characteristics would also allow higher photosynthesis rates per unit of stomatal opening. This could account for the reduced rates of water loss per unit of photosynthesis.

Crassulacean acid metabolism (briefly mentioned in Section IIIA) would appear to represent an even more extreme variant designed specifically to conserve water. In species with this process the Calvin cycle operates in the light while stomata are closed, fixing CO_2 provided by decarboxylation of malate. This malate is formed by fixation of externally-derived CO_2 during the previous dark period. It may be important that, with stomata closed, the CO_2 from malate could be concentrated in the leaf to a level adequate to saturate RuDP carboxylase. The concentration of CO_2 in this way may be an important additional factor in the economy of carbon assimilation in plants with Crassulacean acid metabolism.

E. Regulation of Photosynthesis

Regulatory processes may operate on photosynthetic metabolism for many purposes. In terms of the overall carbon economy of plants the supply of photosynthate must be balanced with demand. Accordingly, there must be feedback to regulate both the rate of CO_2 assimilation and the partitioning of assimilated carbon into the different end-products. Other processes operate to bring about relatively rapid changes in the level of certain key photosynthetic enzymes as part of the response to changes in the light and temperature at which plants are growing. Such effects could influence growth and hence yield of crops. Other regulatory processes must operate to integrate and balance the various metabolic phases of photosynthesis and to accommodate the profound changes in metabolism associated with both fluctuations in light intensity and light-dark transitions. Regulation of photosynthetic metabolism has been considered in recent reviews (Hatch and Slack, 1970a, b; Preiss and Kosuge, 1970). Some instances of this type of regulation are considered in more detail below.

It is perhaps surprising that well-defined regulatory processes operating directly on RuDP carboxylase in higher plants are not known. Of course, CO_2 fixation by this enzyme could be controlled indirectly by the supply of CO_2 or RuDP. For instance, ribulose-5-phosphate kinase in leaves is activated following illumination and inactivated when leaves are darkened (Latzko et al., 1970), providing the possibility of indirect control of CO_2 fixation via the supply of RuDP. Likewise, carbonic anhydrase activity responds to changes in light intensity (Everson, 1971) and this could influence the supply of CO_2 for RuDP carboxylase. According to a recent report by Ogren and Bowes (1971), O_2 directly inhibits RuDP carboxylase from soyabean by acting competitively with CO_2. They relate this effect to photorespiration (see Section IV) and O_2 inhibition of photosynthesis. Two other Calvin cycle enzymes glyceraldehyde-3-phosphate dehydrogenase and fructose diphosphatase are activated *in vivo* by light via NADPH and ferredoxin, respectively (Preiss and Kosuge, 1970). Control of these two enzymes could influence the partitioning of photosynthate between carbohydrate and other end-products including protein in particular.

In C_4-pathway plants both PEP carboxylase and pyruvate, P_i dikinase, enzymes concerned with the primary fixation of CO_2, are regulated by inhibition by products of the reactions they catalyse. In addition, pyruvate, P_i dikinase and NADP-malate dehydrogenase are rapidly and completely inactivated in darkened leaves and re-activated when leaves are illuminated. The activity of these enzymes also responds to fluctuations in light intensity within the normal daily range. With both enzymes this effect is ultimately mediated by the reversible oxidation of thiol groups, but with pyruvate, P_i dikinase at least the process is enzyme-catalysed and is influenced by orthophosphate, pyruvate, AMP and ADP.

Other effects of changing light intensity on enzyme activity are slower-acting and more adaptive in nature. For instance, changes in the light intensity at which plants are growing result in changes in the content of certain enzymes associated with the primary fixation of CO_2. Thus, with Calvin cycle plants grown at low light, RuDP carboxylase can be increased several fold in 5 or 6 days in response to increased light, and *vice versa*. The fact that these changes are associated with proportional changes in light-saturated photosynthesis rates supports the view that this enzyme is the prime limiting factor under these conditions (Bjorkman, 1971). Likewise, the content of PEP carboxylase and pyruvate, P_i dikinase in C_4-pathway leaves undergo changes of up to 8-fold in 5 days when the light intensity for growth was varied between full sunlight and about 1/10 of this intensity. It may be significant that the RuDP carboxylase level in C_4-pathway species does not alter in response to such changes in light intensity.

IV. PHOTORESPIRATION AND THE GLYCOLLATE PATHWAY

It is now clear that the process of photorespiration is intimately linked metabolically and functionally with photosynthesis. The biochemistry and physiology of photorespiration has been considered in detail in recent reviews (Jackson and Volk, 1970; Tolbert, 1971; also see Hatch *et al.*, 1971). This process apparently operates specifically in the light and is respiratory in the sense that O_2 is consumed and CO_2 evolved. The magnitude of photorespiration varies with light intensity, CO_2 and O_2 concentration, and temperature. However, under normal conditions it results in the return to the atmosphere of between 20% and 40% of the CO_2 just fixed by photosynthesis. When photorespiration is operative, plants photosynthesizing in a closed system reduce the CO_2 concentration to some finite and steady value. At this concentration, usually about 20% of the CO_2 level in air, fixation and evolution of CO_2 are balanced. This is termed the CO_2 compensation point. Enhancement of net photosynthesis when the external O_2 concentration is reduced to 1% (v/v) or less is believed to be due, in part at least, to inhibition of photorespiration. In spite of the profound effect of photorespiration on the carbon economy of plants there is still no convincing explanation for its basic function.

C_4-pathway plants do not show enhancement of photosynthesis when the O_2 concentration is reduced and have a CO_2 compensation point near zero. The extent to which this is due to a lack of photorespiration in these species, rather than efficient refixation of released CO_2, is not yet clear. However, there is evidence that many of the enzymes associated with photorespiration are markedly reduced in activity in C_4-pathway species (Osmond, 1971; Tolbert, 1971). The higher potential for photosynthesis characteristic of C_4-pathway plants, is very likely due, in part at least, to reduced photorespiration.

The metabolic process believed to be responsible for photorespiration is termed the glycollate pathway. As shown in Fig. 4, this reaction sequence can be considered as a cyclic bypass of the Calvin cycle. It utilizes glycollate probably derived from the C-1 and C-2 of a keto sugar phosphate. Glycollate is formed in the chloroplasts but all subsequent reactions up to glycerate formation, with the possible exception of those between glycine and serine, occur in organelles termed peroxisomes. Glycollate is converted to glycine by the combined operation of glycollate oxidase and a glycine aminotransferase and for every two molecules of glycollate oxidized there is a net consumption of one molecule of O_2. It is fairly certain that two molecules of glycine react to give serine and CO_2, the CO_2 being derived from a C-1 of glycine. However, the details of these transformations are not certain. Serine is then converted to glycerate which is phosphorylated in the chloroplasts. The 3-PGA so formed can then re-enter the Calvin cycle.

This process is costly. In terms of carbon some 20% to 40% of the total CO_2 assimilated by photosynthesis is lost via photorespiration. Energy cost in terms of total ATP and NADPH required per unit of

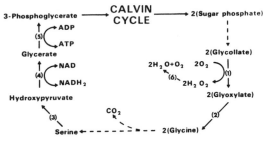

Fig. 4. Reactions of the glycollate pathway. The nature of the reactions indicated with dotted lines is not certainly established. The other enzymes are: (1) glycollate oxidase; (2) and (3) aminotransferases; (4) glycerate dehydrogenase; (5) glycerate kinase; (6) catalase.

net CO_2 assimilation would, on average, be almost doubled by the operation of photorespiration (Hatch, 1970). Resolution of the primary function of photorespiration must remain one of the outstanding problems of plant physiology.

V. CHARACTERISTICS VARYING WITH DIFFERING PATHWAYS

A. Ecology and Physiology

Geographically, Calvin cycle species occur naturally at all latitudes but predominate in the temperate and sub-temperate regions. On the other hand, C_4-pathway species are largely confined to the wet and arid tropical and subtropical regions while species with Crassulacean acid metabolism are restricted to the more arid parts of these regions. In terms of performance, the maximum photosynthesis and growth rates of C_4-pathway plants average about twice those for Calvin cycle plants. A large amount of data collated in recent reviews supports the view that this potential for photosynthesis can be manifested in the field performance of crop and pasture species (Cooper, 1970; Stewart, 1970; Loomis et al., 1971). Furthermore, the cost of this growth in terms of water-use is only about half of that for Calvin cycle species (Black, et al., 1969; Downes, 1969). The relationship between the differences in ecological distribution and field performance of these species and their particular physiological and biochemical features has been considered in recent reviews (Hatch and Slack, 1970a; Hatch et al., 1971).

Another feature of C_4-pathway plants is that CO_2 diffusion through stomata is usually the major limiting factor for photosynthesis at light saturation (see Hatch *et al.*, 1971). In contrast, it is the carboxylase which is the major limiting factor in Calvin cycle plants under these conditions. C_4-pathway plants also have higher light and temperature optima both for growth and for photosynthesis. This obviously correlates with the ecological distribution of these plants. Calvin cycle leaves generally attain optimal photosynthesis rates at about one third of full sunlight whereas C_4-pathway leaves are often not saturated at full sunlight. It is at high light intensity that the differences in photosynthesis rate between the two groups of plants is most evident. Likewise, the temperature optimum for photosynthesis by C_4-pathway leaves generally occur between 30°C and 45°C, about 10-15°C higher than for Calvin cycle plants. At the extremes, the optimum for *Caltha,* a Calvin cycle plant occurring in alpine areas, is about 10°C (Phillips and McWilliam, 1971) compared with 45-50°C for *Tidestromia,* a C_4 plant found in hot deserts (Bjorkman, 1971). For some plants, growth and photosynthesis responses to temperature have been linked with variations in the activity and temperature response (Q_{10}) of photosynthetic carboxylases (Phillips and McWilliam, 1971).

As already indicated (Section IIID), the primary carboxylase of C_4-pathway species, PEP carboxylase, has a higher activity and operates more efficiently with low CO_2 concentrations than RuDP carboxylase in Calvin cycle species, characteristics which could account for the higher potential for photosynthesis and the more economic use of water typical of these species. Other features of C_4-pathway plants can also be attributed to the characteristics of PEP carboxylase. For instance, a more prolonged response to increasing light, and therefore high light saturation values, would be a predictable consequence of increasing the capacity of the carboxylase reaction. Likewise, it is also predictable that the significance of CO_2 diffusion through stomata as an important factor limiting photosynthesis will increase as the capacity of the carboxylase increases. In general, the relative importance of stomatal resistance in limiting photosynthesis, as opposed to carboxylase capacity, is related to the existing gradient of CO_2 between the air and the leaf in the following way. When the difference between the internal and external CO_2 concentration is small, light-saturated photosynthesis will be primarily limited by the capacity of the carboxylation reaction. On the other hand, diffusion of CO_2 through stomata will be the major limiting factor when the gradient is large and the internal CO_2 concentration is small compared with the CO_2 in air.

Whether C_4-pathway plants use this excess potential for carboxylation to grow faster or to economize on use of water probably depends upon the prevailing conditions. In practice, the higher carboxylating capacity combined with a higher efficiency for utilizing CO_2 would allow greater photosynthesis for a given stomatal resistance. Equally, comparable photosynthesis could be achieved with higher stomatal resistance. In either case transpiration of water per unit of photosynthesis would be reduced relative to plants in which the capacity and efficiency of the carboxylation reaction is lower.

B. Taxonomic Distribution

Calvin cycle species are very likely represented in all higher plant families. C_4-pathway species are more restricted in distribution but are nevertheless now known to occur in at least ten families. These include the monocotyledon families, Gramineae, and Cyperaceae, and the dicotyledon families, Aizoaceae, Amaranthaceae, Chenopodiaceae, Compositae, Euphorbiaceae, Nyctaginaceae, Portulacaccac and Zygophyllaceae. Downton has recently compiled a list of these species (see Hatch *et al.*, 1971). Species with Crassulacean acid metabolism also occur in several dicotyledon families from various orders. When the occurrence of species with the C_4-pathway or Crassulacean acid metabolism is examined in an evolutionary arrangement of orders (see Evans, in Hatch *et al.*, 1971) several interesting generalizations become apparent. Thus, these species are broadly distributed in unrelated orders which are, however, amongst the more advanced orders. Furthermore, in spite of their wide distribution, there is an obvious taxonomic relationship between the C_4-pathway and Crassulacean acid metabolism.

To date, all families known to include C_4-pathway species also have representatives with the Calvin cycle. However, it would appear likely that most tribes or genera within these families are exclusive for one pathway or the other. Nevertheless, intra-generic variation has been shown within *Cyperus*, *Panicum*, *Atriplex*, *Kochia* and *Euphorbia*.

Amongst the grasses that are important pasture or forage species are representatives of both Calvin cycle and C_4-pathway types. Broadly speaking, those in the panicoid and eragrostoid sub-families have the C_4-pathway while those in the festucoid sub-family are

Calvin cycle species. It is possible to generalize further that most of the panicoid species transport CO_2 as malate, while the eragrostoid species belong to the group of C_4-pathway species which transport and decarboxylate aspartate (see Fig. 3). C_4-pathway species of agricultural importance occur within the genera *Chloris, Digitaria, Echinochloa, Eragrostis, Melinus, Panicum, Paspalum, Pennisetum Setaria, Sorghum* and *Zea*. Many C_4-pathway species in other genera make significant contributions to natural pastures. Legumes, the other important class of pasture-forage plants, are apparently all Calvin cycle species. None of the dicotyledonous C_4-pathway species are of major agricultural importance, although the Chenopods, *Atriplex* and *Kochia* make some contribution in natural pastures. Other species from this group of genera such as *Amaranthus,* species of *Cyperus* and many grasses amongst the C_4-pathway monocots, present serious weed problems in crops (Black *et al.,* 1969).

C. Evolution, Genetics and Hybridization

The C_4-pathway occurs in species from several advanced but widely separated orders. These species presumably evolved from Calvin cycle species in response to selection pressures associated with high temperature, high light and restricted water situations. Crassulacean acid metabolism could be regarded as an even more extreme adaptation to these conditions. Earlier evidence suggested that evolution of the C_4-pathway did not involve major genetic transformations. For instance, there was the evidence that the C_4-pathway almost certainly evolved separately in several families and that Calvin cycle and C_4-pathway species occur within the same genus. This view was confirmed when species of *Atriplex* with differing pathways were successfully hybridized (Bjorkman *et al.,* 1971). With the C_4-pathway species as the female parent the F_1 was intermediate anatomically, morphologically, and in terms of enzyme levels, but photosynthesis rates were much lower than for either parent. There was segregation of most characters in the F_2 but photosynthesis rates remained low in all individuals. Many geneticists and breeders remain optimistic that improved yields and environmental adaptability of pasture and crop species may result from hybridizations of this kind.

The nature of the qualitative genetic differences that exist between Calvin cycle and C_4-pathway species are not known. In terms of specific enzymes, only pyruvate, P_i dikinase appears likely to be unique to the C_4-pathway. Of course several other enzymes operative in this pathway occur at much higher levels than in Calvin cycle species. These include PEP carboxylase, malic enzyme, NADP-malate dehydrogenase, aminotransferases, adenylate kinase and pyrophosphatase (Hatch and Slack, 1970a; Andrews *et al.*, 1971). Some of these enzymes obviously differ from their counterparts in Calvin cycle species in respect to location and specific function. The possibility remains that some may be genetically unique iso-enzymes specific to the C_4-pathway.

REFERENCES

Anderson, J. M., Boardman, N. K. and Spencer, D. (1971a). *Biochim. biophys. Acta* **245**, 253-258.

Anderson, J. M., Woo, K. C. and Boardman, N. K. (1971b). *Biochim. biophys. Acta* **245**, 398-408.

Andrews, T. J., Johnson, H. S., Slack, C. R. and Hatch, M. D. (1971). *Phytochemistry* **10**, 2005-2013.

Arnon, D. I. (1967). *Physiol. Rev.* **47**, 317-358.

Bassham, J. A. (1964). *A. Rev. Pl. Physiol.* **15**, 101-120.

Bassham, J. A. (1965). *In* "Plant Biochemistry" (J. Bonner and J. E. Varner, eds), pp. 875-902. Academic Press, New York and London.

Bassham, J. A. and Jensen, R. G. (1967). *In* "Harvesting the Sun" (A. San Pietro, F. A. Greer, T. J. Army, eds), pp. 79-100. Academic Press, New York and London.

Beevers, H., Stiller, M. L. and Butt, V. S. (1966). *In* "Plant Physiology" (F. C. Steward, ed.), Vol. 4, pp. 119-242. Academic Press, New York and London.

Björkman, O. (1971). *In* "Photosynthesis and Photorespiration" (M. D. Hatch, C. B. Osmond and R. O. Slatyer, eds), pp. 18-32. Interscience, New York.

Björkman, O., Nobs, M., Pearcy, R., Boynton, J. and Berry, J. (1971). *In* "Photosynthesis and Photorespiration" (M. D. Hatch, C. B. Osmond, R. O. Slatyer, eds), pp. 105-119. Interscience, New York.

Black, C. C., Chen, T. M. and Brown, R. H. (1969). *Weed Sci.* **17**, 338-343.

Black, C. C. and Mayne, B. C. (1970). *Pl. Physiol., Lancaster*, **45**, 738-741.

Boardman, N. K. (1968). *Adv. Enzymol.* **30**, 1-79.

Boardman, N. K. (1970). *A. Rev. Pl. Physiol.* **21**, 115-140.

Boardman, N. K. and Anderson, J. M. (1964). *Nature, Lond.* **203**, 166-167.

Boardman, N. K., Anderson, J. M. and Hiller, R. G. (1971). *Biochim. biophys. Acta.* **234**, 126-136.

Chen, T M., Brown, R. H. and Black, C. C. (1969). *Pl. Physiol., Lancaster* **44**, 649-654.

Chen, T. M., Brown, R. H. and Black, C. C. (1971). *Pl. Physiol., Lancaster* **47**, 199-203.

Cooper, J. P. (1970). *Herb. Abstr.* **40**, 1-15.

Downes, R. W. (1969). *Planta.* **88**, 216-273.

Downton, W. J. S. (1970). *Can. J. Bot.* **48**, 1795-1800.

Downton, W. J. S., Berry, J. A. and Tregunna, E. B. (1970). *Z. Pflanzenphysiol.* **63**, 194-198.

Edwards, G. E. and Black, C. C. (1971). *Pl. Physiol., Lancaster* **47**, 149-156.

Emerson, R. and Arnold, W. (1932). *J. gen. Physiol.* **16**, 191-205.

Evans, L. T. (1971). *J. Indian bot. Soc.*, **50A**, 560-570.

Everson, R. G. (1971). *In* "Photosynthesis and Photorespiration" (M. D. Hatch, C. B. Osmond and R. O. Slatyer, eds), pp. 275-282. Interscience, New York.

Erixon, K. and Butler, W. L. (1971). *Biochim. biophys. Acta.* **234**, 381-389.

Gibbs, M. (1967). *A. Rev. Biochem.* **36**, 757-784.

Hatch, M. D. (1971a). *In* "Photosynthesis and Photorespiration" (M. D. Hatch, C. B. Osmond and R. O. Slatyer, eds), pp. 139-152. Interscience, New York.

Hatch, M. D. (1971b). *Biochem. J.*, **125**, 425-432.

Hatch, M. D., Osmond, C. B. and Slatyer, R. O., eds (1971). "Photosynthesis

Hatch, M. D. (1972). *Biochem. J.* (in press).

Hatch, M. D., Osmond, C. B., and Slatyer, R. O. eds (1971). "Photosynthesis and Photorespiration", Interscience, New York.

Hatch, M. D. and Slack, C. R. (1970a). *In* "Progress in Phytochemistry" (L. Reinhold and Y. Lowschitz, eds), Vol. 2, pp. 35-106. Interscience, London.

Hatch, M. D. and Slack, C. R. (1970b). *A. Rev. Pl. Physiol.* **21**, 141-162.

Heber, U. (1969). *In* "Proceedings of the International Symposium on Transport in Plants", Reinhordsbrunn, East Germany (in press).

Hind, C. and Olsen, J. M. (1968). *A. Rev. Pl. Physiol.* **19**, 249-282.

Huzisige, H., Usiyama, H., Kikuti, T. and Azi, T. (1969). *Pl. Cell Physiol., Tokyo* **10**, 441-455.

Jackson, W. A. and Volk, R. J (1970). *A. Rev. Pl. Physiol.*, **21**, 385-432.

Karpilov, Y. S. (1970). *In* "Proceedings of the Moldavian Institute for Research", Vol. 2, Pt. 3, pp. 1-67.

Knaff, D. B. and Arnon, D. I. (1969). *Proc. natn. Acad. Sci. U.S.A.* **64**, 715-722.

Kok, B. (1965). *In* "Plant Biochemistry" (J. Bonner and J. E. Varner, eds), pp. 903-960. Academic Press, New York and London.

Laetsch, W. M. (1971). *In* "Photosynthesis and Photorespiration" (M. D., Hatch, C. B. Osmond and R. O. Slatyer, eds), pp. 323-349. Interscience, New York.

Latzko, E., Garnier, R. and Gibbs, M. (1970). *Biochem. biophys. Res. Commun.* **39**, 1140-1144.

Levine, R. P. (1969). *A. Rev. Pl. Physiol.* **20**, 523-540.

Loomis, R. S., Williams, W. A. and Hall, A. E. (1971). *A. Rev. Pl. Physiol.* **22**, 431-468.

Mayne, B. C., Edwards, G. E. and Black, C. C. (1971). *In* "Photosynthesis and Photorespiration" (M. D. Hatch, C. B. Osmond, and R. O. Slatyer, eds), pp. 361-371. Interscience, New York.

Ogren, W. L. and Bowes, G. (1971). *Nature, Lond.* **230**, 159-160.

Osmond, C. B. (1971). *In* "Photosynthesis and Photorespiration" (M. D. Hatch, C. B. Osmond and R. O. Slatyer, eds), pp. 472-482. Interscience, New York.

Phillips, P. J. and McWilliam, J. R. (1971). *In* "Photosynthesis and Photorespiration" (M. D. Hatch, C. B. Osmond and R. O. Slatyer, eds), pp. 97-104. Interscience, New York.

Preiss, J. and Kosuge, T. (1970). *A. Rev. Pl. Physiol.* **21**, 433-466.

Regitz, C., Berzborn, R. and Trebst, A. (1970). *Planta* **91**, 8-17.

Stewart, G. A. (1970). *J. Aust. Inst. agric. Sci.* **36**, 85-101.

Tanner, W. and Kandler, O. (1969). *In* "Progress in Photosynthesis Research" (H. Metzner, ed.), Vol. III, pp. 1217-1223. Tübingen.

Tolbert, N. E. (1971). *A. Rev. Pl. Physiol.* **22**, 45-74.

Walker, D. A. and Crofts, A. R. (1970). *A. Rev. Biochem.*, 39, 389-428.

Wolfe F. T. (1970). *Adv. Frontiers of Plant Sciences* 26, 161-231.

Woo, K. C., Anderson, J. M., Boardman, N. K., Downton, W. J. S., Osmond, C. B. and Thorne, S. W. (1970). *Proc. natn. Acad. Sci. U.S.A.* 67, 18-25.

Yocum, C. F. and San Pietro, A. (1969). *Biochem. biophys. Res. Commun.* 36, 614-620.

CHAPTER 17

Physiology of Light Utilization by Swards

D. WILSON

Grasslands Division, D.S.I.R., Palmerston North,
New Zealand*

I. INTRODUCTION

The production of digestible energy and other nutrients in the form of herbage for consumption by ruminant animals is initially dependent upon the conversion of solar energy to chemical energy by photosynthesis. Light from solar radiation is of course also used by plants to regulate growth and development by phenomena such as phototropism, photomorphogenic induction and photomorphogenic reversal. This chapter is confined to a discussion of the utilization of photosynthetically active radiation for herbage growth.

* Present address: Welsh Plant Breeding Station, Aberystwyth, North Wales, U.K.

The most common species included in the term "herbage" are either grasses or legumes and most of our understanding of herbage productivity and constituents is based on studies of these two plant groups, particularly grasses. Morphological aspects of vegetative growth of grasses have been recently reviewed (Jewiss, 1965). Grass seedlings emerge as single shoots bearing leaves in two opposite ranks. Buds which may later grow into tillers form in the axil of each leaf, a process which can continue throughout the vegetative growth of the plant, giving rise to a hierarchy of tillers. The stem apex of a single grass tiller continually forms new leaves as long as it remains vegetative, and temperate grasses usually consist of a rosette of distichously arranged leaves arising from a series of nodes. Stem internodes may or may not elongate during vegetative growth, depending on the species. Each leaf comprises a long narrow distal lamina and a proximal leaf sheath which encloses the younger leaves and the stem apex. Higher order tillers which behave in the same way as primary tillers are produced exponentially until the plant meets some environmental limitation or enters the reproductive phase (Patel and Cooper, 1961). The grass leaf is short-lived compared with that of many dicotyledenous species and, in general, the rate at which leaves die is similar to the rate at which they appear, so that a fairly constant number of living leaves is carried on each tiller.

One important morphological feature differentiating pasture legumes from grasses is the usually horizontal orientation of the trifoliate leaf laminae which are supported on petioles of varying sizes arranged alternately on either erect or prostrate stems. There is a great variety of growth forms, ranging from the erect to the stoloniferous types but, in all, the leaf inclination confers rather different light trapping qualities than in the grasses, which in general orientate their leaves much more vertically than legumes when in a closed community.

The capture and photosynthetic utilization of light by herbage plant communities takes place at several different organizational levels. At the most basic level, photochemical and biochemical reactions absorb and transform electromagnetic energy into useful chemical energy for metabolic reactions. As shown in the previous chapter, a great deal of detailed information is available about these reactions which use light, water, carbon dioxide and minerals to produce organic matter and oxygen. In higher plants the fundamental apparatus seems similar for all species, although there are distinct ecological groups which differ in the details of the carbon pathway.

At the individual leaf level the availability of light and CO_2 at the reaction sites in the chloroplasts, and the distribution of assimilates from their sites of formation are both important. Light is scattered and absorbed within leaves and there are resistances to gaseous fluxes between the external air and the reaction sites. Light distribution within the leaf and leaf resistances can be affected by both leaf structure and external environment, although the effects of these are not yet fully understood. Furthermore, both leaf structure and the biochemical steps involved in photosynthesis can be modified genetically.

In the sward, interactions become even more complex. Within any grassland community, the patterns of reflection, absorption and transmission of light are strongly influenced by environment and by the architecture of the canopy. In addition, gradients of CO_2 and temperature within the sward can affect the rate of photosynthesis of individual leaves. Thus, at any moment in time a large proportion of the photosynthetic elements of the community may be suffering limitations of various environmental factors, which might effect the efficient operation of the photochemical and biochemical reactions within the leaf. It follows that light utilization in the established sward will be further complicated by changing growth patterns with time and by the management procedures adopted for sward maintenance.

In this chapter some of the factors influencing light utilization at these different levels of organization, and their effects on production of dry matter and digestible energy, are discussed.

II. LIGHT UTILIZATION BY THE INDIVIDUAL LEAF OR YOUNG SEEDLING

A. *Availability of Light Energy*

Solar radiation falls on the earth's surface with an intensity of about 2 cal/cm² per min at the perpendicular. It covers a range of wavelengths from ultravoilet (< 400 nm) through visible (medium-short, 400-700 nm) to infrared (> 700 nm). On average only about half of all the solar radiation falling on the outside of the atmosphere finally reaches the earth's surface. However, this proportion can reach as high as 70% in arid regions with little cloud, and as low as 40% in tropical or temperate rainy climates. The proportion also increases with altitude since the highest concentrations of dispersing and absorbing materials are found closest to ground level. The annual input of solar radiation over the surface of the world is shown in Fig. 1. Only about 45% of this total incident radiation is in the photosynthetically active region (400-700 nm) and in most higher plants, including grasses and legumes, the production of one gram of dry matter corresponds to the fixation of 4-5000 cal of chemically bound energy (Cooper, 1970).

1. Quantity

Annual energy input is greatest in sub-tropical regions, such as Central Australia, North Africa and the Near East where relatively low latitudes (20 to 30°C) coincide with little cloud cover. In contrast, inputs are very low in the temperate oceanic climates, although it is in these regions that grasslands are often most intensively farmed. Seasonal energy inputs in different climatic regions are associated with latitudinal effects on seasonal amplitude, persistence of cloud cover and angle of incidence of direct sunlight (Cooper, 1970). In temperate grassland climates the daily energy input in summer may be ten times greater (400-500 $cal/cm^2/day$) than that in winter (20-150 $cal/cm^2/day$). This contrasts with sub-tropical regions where there may only be four fold differences (180-290 cal to 550-660 $cal/cm^2/day$), and tropical climates where daily energy inputs of 400 to 500 cal/cm^2 can be expected throughout the year. Although the same total daily summer input may be received in high latitudes as in the tropics this is spread over daylengths of 18 hours or more in high latitudes whereas the tropical day is only of 12 to 13 h duration. Even with the same daily energy input these differences can have marked effects on the amount of carbon assimilated by leaves and swards during the day (de Wit, 1965).

The variation in energy input between different climatic regions is also associated with variations in temperature and potential evapotranspiration (Cooper and Tainton, 1968) which can limit the efficiency of light utilization under natural conditions. In temperate continental climates, for instance, winter temperatures can be less than 0°C for several weeks, while in the summer, photosynthesis can be limited by water deficits. In mediterranean climates utilization of light energy may be severely restricted for long periods in the summer by lack of rain. In sub-tropical regions also, seasonal water shortage, either in winter or summer, is an important limitation.

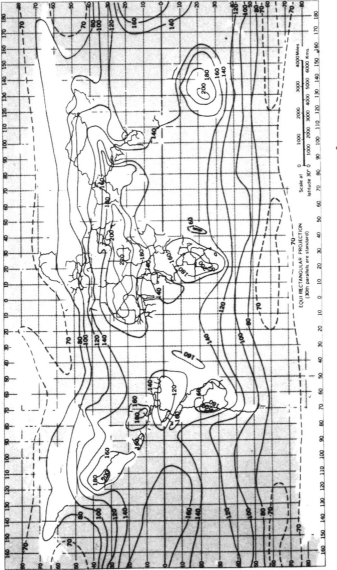

Fig. 1. Total solar radiation recorded at the earth's surface during the year (k cal/cm² per year) (reprinted from "An Introduction to Climate" by G. T. Trewartha, 1968, by permission of McGraw-Hill Book Company, © 1968 McGraw-Hill Inc.).

2. Quality

The proportion of the total incident radiation in the photo-synthetically active region varies with the light source, differing between direct sunlight, skylight (scattered or diffuse radiation) and cloudlight (light transmitted through a complete cloud cover). Direct sunlight and cloudlight have similar spectral distributions in the

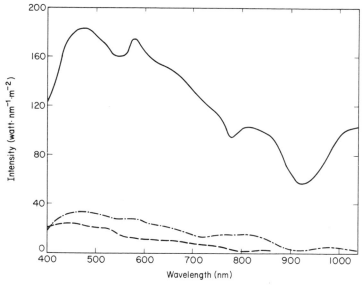

Fig. 2. Energy distribution of sunlight (——), skylight (- - -) and cloudlight (- · --) (after Scott et al., 1968).

visible regions, with their most marked absorption bands at about 480 nm, but sunlight has a higher proportion of infra-red (Fig. 2). Skylight, which can contribute 10 to 15% of the total on a sunny day, exhibits little radiation above 700 nm, absorbs maximally at about 480 nm but has a higher proportion in the ultraviolet (< 400 nm) which becomes even greater at high altitudes. Not only the total amount but also the spectral distribution of radiation reaching plants is thus affected by atmospheric conditions, latitude, altitude and time of day (Gates, 1965).

B. Photosynthesis and Light

Light is directly necessary only for the photochemical process of photosynthesis and the previous chapter has shown how these convert the light energy absorbed by pigments within the chloroplasts, (mainly chlorophyll a and b in higher plants) into the high energy phosphate bonds (ATP) and reducing power (NADPH) used subsequently for converting CO_2 to carbohydrates. The diffusion processes involved in the movement of gases between the chloroplasts and the external air, do not depend directly on light. However indirect effects may occur, for example, through the influence of light on stomatal movement and hence gas exchange of the leaf. The biochemical processes preceding and following reduction of CO_2 may or may not be influenced by light.

1. The Photosynthetic Unit

A great deal is now known about the fine structure and function of the chloroplast, the basic photosynthetic unit of higher plants, and I will only give a brief outline of some aspects relevant to this chapter (see also Chapter 24).

The site of the light absorbing chlorophyll is the aggregated granal stacks (thylakoids) contained in the lamellar system of the chloroplast. These are clearly shown in Fig. 3. Chlorophylls a and b have two major light absorption peaks. Chlorophyll a in ether absorbs light mainly at 429 and 660 nm in the blue and red respectively while chlorophyll b absorbs at 453 and 643 nm. The chloroplast lamellae contain the enzyme systems involved in photophosphorylation and reduction of TPN, whereas the stroma, the cytoplasm in which the lamellar system is embedded, is thought to contain the enzymes involved in CO_2 assimilation (Wettstein, 1967).

Chloroplast structure can vary between different ecological groups of plants and even between different sites within one leaf. As well as the biochemical differences between grasses with the C_3 and C_4 carbon pathways (discussed in Chapter 16) there is also a range of anatomical, physiological and ecological differences, including important differences in chloroplast distribution and ultrastructure (Laetsch, 1969) between these two classes. Leaves from C_4 plants are usually characterized by the presence of a well developed parenchyma bundle sheath which has larger chloroplasts, commonly containing much more starch than the chloroplasts found in the mesophyll tissues. In contrast, C_3 grasses have no well developed bundle sheath and any chloroplasts present in this region are reported to be similar to those of the mesophyll. A unique anatomical characteristic of chloroplasts in the C_4 group of plants is the peripheral reticulum (PR), "an anastomosing membrane system contiguous with the inner plastid membrane" (Laetsch, 1969). This is clearly shown in Fig. 3 which shows mesophyll chloroplasts of *Paspalum dilatatum,* a grass with the C_4-pathway, and *Lolium perenne,* a C_3 grass. This peripheral reticulum is a feature of both

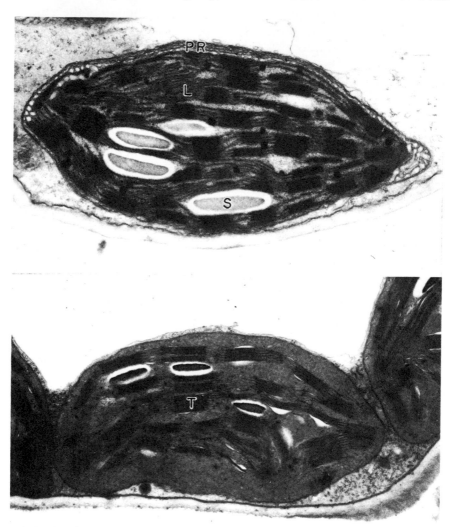

Fig. 3. Sections through chloroplasts of *Paspalum dilatatum* (upper) and *Lolium perenne* (lower) showing the lamellar system (L) with thylakoids (T), and the membrane peripheral reticulum (PR) of *P. dilatatum.* Some starch (S) grains are also present. (Glutaraldehyde-O_8O_4 fixation: stained with uranyl nitrate and lead citrate) x 18,000.

mesophyll and bundle sheath chloroplasts and while it is thought that it may be directly related to the C_4-pathway its function is not yet clear.

2. Photosynthetic Reactions to Light

The observed exchanges of CO_2 and O_2 between the leaf and the external air reflect the photochemical and biochemical reactions taking place within the chloroplasts. These exchanges are subject to modification by factors within the leaf, other parts of the plant and the external air. Under otherwise steady conditions however the

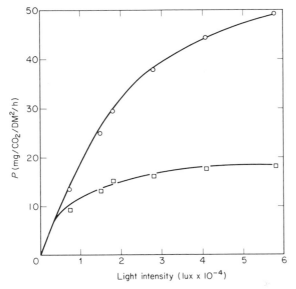

Fig. 4. Relationship between rate of apparent photosynthesis (P) and light intensity in *Cynodon dactylon* (o) and *Dactylis glomerata* (□) (after Chen *et al.*, 1969).

photosynthetic response of leaves to increasing light energy is well established (Gaastra, 1963). Figure 4 depicts the general response of leaves of two contrasting grasses, *Cynodon dactylon*, a C_4 species, and *Dactylis glomerata*, a C_3 species. At low light energies photosynthetic rate is proportional to light intensity, the degree of response is related to the quantum efficiency of photosynthesis of the leaf and should be characteristic for a given leaf. Under these conditions about 12-15% of the incoming light energy can be fixed (Cooper, 1970). This linear response to light will continue as long as variables other than light are not critical. Eventually other inputs, such as external CO_2 concentration, become limiting and the photosynthetic system becomes light saturated. The saturation light intensity and maximum rate of photosynthesis are therefore lower at low than at high concentrations of CO_2 and, similarly, the saturating CO_2 level is much lower at low light levels. At normal atmospheric carbon dioxide and oxygen concentrations light saturation of individual leaves of most species is reached at light intensities below that of full sunlight. In this situation either the level of CO_2 reaching the reaction sites in the chloroplasts or the activity of the carboxylating step will usually be limiting light utilization.

The light saturation level is about 20,000-30,000 lux in legumes and in grasses from temperate regions (C_3) but photosynthetic rates of sub-tropical grasses (C_4) increase in response to light intensities up to more than 60,000 lux (Fig. 4). In general, the photosynthetic rate is greater in C_4 than C_3 species, particularly at high temperatures. As a result maximum values of over 70 mg CO_2/dm^2 per h occur in C_4 grasses, representing an efficiency of energy conversion of 5-6%, whereas maximum photosynthetic rates of only 20-30 mg CO_2/dm^2 per h, corresponding to less than 2-3% conversion of light energy are the best achieved by C_3 species in high summer light intensities (Hesketh and Moss, 1963). This difference in reaction to light between these two groups of plants is however not apparent at temperatures less than about 15°C. However, even among species with similar responses to temperature appreciable variations in light saturation level and in maximum photosynthetic rate can occur (Cooper and Tainton, 1968). The response to light energy of net photosynthetic rate of young grass seedlings grown without mutual or self shading is similar to that of individual leaves, though the overall efficiency of conversion of light energy by the seedling tends to be lower than that of the individual leaf (Cooper, 1966).

Clearly, in different light environments, the limitations to photosynthetic rate may lie in the capacity of one or more of the processes outlined above. For example, at low light intensity the photochemical processes may be limiting but at high light intensity photosynthesis may be limited by the capacity of CO_2 diffusion or by biochemical processes.

C. Factors Affecting Rate of Photosynthesis

As we have seen (Chapter 16), two pigment systems (photosystems I and II) are responsible for the functioning of the photochemical process. These photosystems utilize preferentially light of different wavelengths and active functioning of both systems is required for efficient photosynthesis. Thus the rate at which these photochemical processes function is partly determined by both the quantity of light and its spectral distribution. On the other hand, the enzyme reactions preceding and following CO_2 reduction are strongly affected by both temperature and the concentration of CO_2 at the reaction sites. This contrasts with the process of CO_2 movement from the air outside the leaf to the chloroplasts which is only slightly influenced by temperature and depends mainly on the difference in CO_2 concentrations between the bulk air outside the leaf and the chloroplasts (Björkman, 1966). These effects of light, temperature and CO_2 interact with various anatomical, physiological and biochemical characteristics of the leaf and plant (Heath, 1969).

For convenience we will discuss these factors for the two main limiting situations (1) where light is the major limitation to photosynthesis and, (2) where other factors are limiting. However, between low, completely limiting, and high, completely saturating, light intensities a large transition range occurs where a number of environmental variables may limit photosynthesis simultaneously. In grasses with the C_4 carbon pathway, for example, photosynthesis may continue to respond to light energies approaching full sunlight yet CO_2 availability can be an important limitation over much of the light intensity range (Wilson and Ludlow, 1970). At least part of this transition can probably be explained by the unevenness of light reflection and transmission within leaves. Higher light intensities will be required to saturate the chloroplasts throughout the full depth of the leaf than those nearest the radiation source. Incomplete penetration of CO_2 into the leaf would also influence the responses to light.

1. Light Limitation

Under conditions of light limitation the amounts of light absorbing pigments in the leaf might be expected to affect the rate of photosynthesis. However extensive studies by Gabrielsen (1948) suggested that this occurs only in very weak light and at relatively low chlorophyll levels, i.e., at a light intensity of about 1400 lux and as little as 4 to 5 mg of chlorophyll per dm^2 of leaf. In old leaves the onset of chlorosis with ageing can affect light limited photosynthesis (Saeki, 1959) and chlorotic effects resulting from exposure to temperature extremes (Cooper and Tainton, 1968) may have similar results. Reported associations between leaf chlorophyll content and the assimilatory capacity of grass leaves of plants at higher light intensities may reflect a relationship between chlorophyll amount and numbers of photosynthesizing cells, since the amount of chlorophyll per cell may be much less variable than cell size (Wilson and Cooper, 1969a).

(a) *Genetic Effects.* Recent investigations indicate that C_3 and C_4 species have distinctly different contents of chlorophylls *a* and *b* and of P700, the pigments currently considered the reaction centre for Photosystem I. Black and Mayne (1970) found that both the ratio of P700 to total chlorophyll and of chlorophyll *a* to *b* was greater among C_4 than C_3 species. They proposed that this might indicate a more active Photosystem I or different size photosynthetic unit in C_4 compared with C_3 plants and that the greater energy (ATP) requirements of the former might be supplied by more active photophosphorylation in this way.

There are other differences in rates of photochemical processes or in photosynthetic activity in weak light between or within species. For example, differences in the initial slope of the response curve of photosynthesis to light, in the rate of light limited photosynthesis and in leaf quantum efficiencies have been found between different species, including two grasses, *Zea mays* and *Triticum vulgare* (Brown, 1969). Even within *L. perenne* there are differences in the net rate of photosynthesis at low light intensities among plants from contrasting climatic regions (Wilson and Cooper, 1969b). The reason for such differences is not always clear. In *L. perenne* these differences in light limited photosynthesis were unrelated to independent variations in the slopes of response curves of photophosphorylation to light (Treharne, pers. comm.). On the other hand, the differences in the saturation rate of the Hill reaction between mutant and normal chloroplasts of tobacco is associated with a corresponding difference in photosynthetic activity (Homann and Schmid, 1967). However, at the point where the light response curve deviates from linearity, differences between plants in rate of light limited photosynthesis could well be mediated by CO_2 availability.

The possible existence of genetic variation in photosynthetic rate in weak light has important implications in herbage communities since a large proportion of potentially active leaves in a sward might suffer light limitation at certain times. Heritable variation in isolated chloroplast activity of photophosphorylation reactions has been suggested in *Z. mays* (Miflin and Hageman, 1966). However only about 17% of the observed variation in light limited photosynthesis among diverse *L. perenne* genotypes was genetic and additive (Wilson and Cooper, 1969c).

(a) *Environmental pretreatment.* Leaves of herbage plants often develop in quite different environmental conditions to those in which they will later be photosynthetically active so that preconditioning can be an important factor to a leaf's potential. Leaves of tropical grasses can suffer damage to their photophosphorylation potential by short term exposure to relatively low temperatures. For example, exposure of *Digitaria decumbens* to one cold night (10°C) resulted in a marked drop in the Hill reaction potential, apparently associated with physical impairment of chloroplast integrity (West, 1970). In contrast, light preconditioning of white clover (*Trifolium repens*) appeared to be ineffective in altering the initial slope of the light response curve (McCree and Troughton, 1966a). However, there is evidence from other species that response of light limited photosynthesis to light preconditioning can be determined by genetic adaptation. Thus, plants of *Solanum dulcamara,* genetically adapted to shade, respond to prolonged exposure to high light intensity by subsequently exhibiting much slower rates of both light limited and light saturated photosynthesis (Gauhl, 1970). Similarly, Tieszen and Helgager (1968) have reported adaptation of the Hill reaction in alpine and arctic species, as have Treharne and Eagles (1970) in *Dactylis glomerata.*

2. *Complete or Partial Light Saturation*

(*a*) *Carbon dioxide transport into the leaf.* In air of normal CO_2 and O_2 content, once there is significant departure from linearity in the light response curve, factors affecting rate of photosynthesis through CO_2 transport or utilization become important. These may be environmental or genetic in origin and can act directly or indirectly through physiological or anatomical mechanisms (Heath, 1969).

Movement of CO_2 through the leaf is considered to be a diffusion process which has been analysed in terms of Fick's law of diffusion (Gaastra, 1959). Such analysis predicts that, under steady state conditions, photosynthesis (*P*) will be proportional to the difference between the CO_2 concentration outside the leaf (C_a) and at the reaction sites in the chloroplasts (C_c), and inversely proportional to the total "resistance" (*R*) to CO_2 transfer:

$$P = \frac{C_a - C_c}{R}$$

R comprises a series of consecutive "resistances" through which CO_2 must pass, firstly in the gaseous phase and then in the liquid phase. The term "resistance", by analogy with an electrical resistance, has been used as a convenient expression, though the analogy is not strictly accurate when discussing the biochemical reactions involved in photosynthesis. Figure 5 illustrates the physical pathway which CO_2 must traverse through stomata, intercellular spaces and cell wall before entering the chloroplasts.

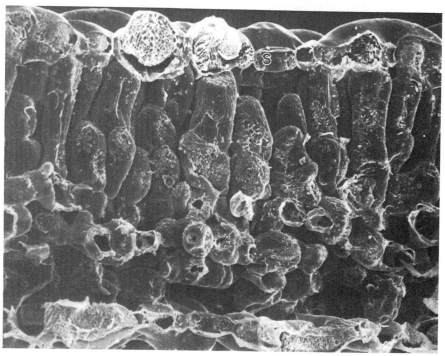

Fig. 5. Transverse view of a section of *Medicago sativa* leaf broken across a stomata (S). Scanning electron micrograph. x 570.

Carbon dioxide from the bulk air first diffuses across a boundary layer to the entry points in the leaf. The size of the resistance at the boundary layer (r_a) depends mainly on wind-speed (Slatyer and Bierhuizen, 1964), acting either directly or indirectly through leaf flutter. The geometry of the leaf may also contribute to r_a. Most CO_2 enters the leaf through the stomata, although some may be absorbed directly through the cuticle. CO_2 then passes through the intercellular spaces, still in the gaseous phase, to the mesophyll cell walls where it goes into solution. It then passes in the aqueous phase through the cell wall, cytoplasm and chloroplast membrane to the thylakoids. Free diffusion of CO_2 in the leaf is restricted in the gaseous phase by stomatal (r_s) and cuticular (r_c) resistances and in the aqueous phase by the mesophyll resistance (r_m) (Gaastra, 1959).

The stomatal resistance is largely dependent on stomatal geometry and density, and the amount of CO_2 absorbed directly through the cuticle is usually only a small proportion of total uptake. Cuticular absorption can, however, vary with environmental conditions and with genotype and may be considerable in some species (Freeland, 1948). r_c is often included in r_s for the purposes of analysis but it is less confusing to refer to r_c and r_s as the two components of the leaf resistance, r_l (Slatyer and Bierhuizen, 1964).

The mesophyll resistance, r_m, in the widely used sense of Gaastra involves both a resistance to CO_2 transfer and a component associated with the primary carboxylation event in photosynthesis.

It is now considered that the physical mechanisms of both diffusion and convective transfer of CO_2 or HCO_3^-, (convective by cytoplasmic streaming), are involved in liquid phase carbon movement. At least part of the transport process, however, may be mediated by one or more biochemical mechanisms which have been shown to influence the rate of transfer. For example, the presence of bicarbonate ions and, more particularly, of the enzyme carbonic anhydrase have been shown to enhance CO_2 transport (Enns, 1967). Carbonic anhydrase is thought to act by influencing the rate of the reaction between CO_2 and water (producing HCO_3^- and H^+) at the cell wall, and differs in level and location between plant species with the C_3 and C_4 carbon pathway. In C_3 species there appears to be much more carbonic anhydrase than in C_4 species. In the former it is located mainly in the chloroplasts, whereas the cytoplasm is the main site in C_4 species. The significance of this variation is not yet clear. Thus there may well be a biochemical component in the liquid phase of carbon movement in the leaf, although for analysis it is usually treated as entirely diffusive.

The carboxylation component of Gaastra's r_m has been treated as a separate "carboxylation resistance", r_x, by others (cf. Monteith, 1963; Osmond et al., 1969; Chartier et al., 1970), although, strictly speaking, such an electrical analogy cannot be used to describe a biochemical reaction. The resistance r_x is related only to the biochemical processes and has been taken as inversely proportional to the velocity of carboxylation. The total intracellular resistance, r_i

(Osmond *et al.*, 1969), which is equivalent to the mesophyll resistance (r_m) of Gaastra (1959), can thus be regarded as made up of two main components, r_m and r_x. Thus photosynthesis can be described by the equation:

$$P = \frac{C_a - C_c}{r_a + r_l + r_i}$$

The rate of photosynthesis can therefore be influenced through r_a, r_l or r_i. Factors affecting the gaseous phase leaf resistance (r_l) can act through the stomatal (r_s) or cuticular (r_c) resistance, and those affecting the liquid phase intracellular resistance (r_i) may influence the resistance to transport of carbon to the reaction sites (r_m) or the primary carboxylation event (r_x).

(*b*) *Environmental Effects.* The common environmental variables affecting rate of photosynthesis can act through one or more of the resistances to CO_2 transport and hence individual resistances may be relatively more or less important depending on the environment. For example, the contribution of the boundary layer resistance, r_a, to the total may decrease rapidly as windspeed increases (Bierhuizen and Slatyer, 1964). However a number of general environmental effects on net photosynthesis of the individual leaf or on net assimilation rate of young seedlings are now well known (Cooper and Tainton, 1968).

(i) *Temperature.* Photosynthetic activity of herbage species usually exhibits an increasing response to temperature, reaches an optimum which extends over several degrees and eventually declines. For temperate (C_3) grasses such as *L. perenne* and *D. glomerata*, this optimum is about 20°C and rates are markedly less at temperatures below 10° and above 30°C. The optimum for tropical and sub-tropical (C_4) grasses is usually about 35°C, and although rates of photosynthesis at this temperature are faster than many temperate species at their optimum, the C_4 grasses are at a relative disadvantage at temperatures below about 15°C (Cooper, 1970) (Fig. 6), possibly because of differences in the temperature optima of their carboxylating processes (Treharne and Cooper, 1969). However high temperature optima for photosynthesis also occur in tropical legumes which have the C_3 carbon pathway (Wilson and Ludlow, 1970).

The response in net photosynthesis of C_3 plants to temperature is, of course, affected by the rate of respiration. Species with the C_4 carbon pathway do not appear to respire in the light. Although dark respiration usually continues to increase steadily with temperature after net photosynthesis declines, respiration in the light (photo-respiration) generally shows an optimum about 10°C lower than that of dark respiration. It also appears as if photorespiration may have temperature optima similar to or slightly higher than net photo-synthesis (Hofstra and Hesketh, 1969). In C_3 plants the actual rates

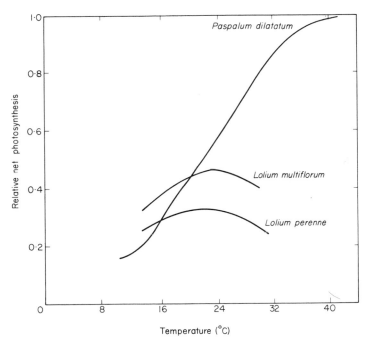

Fig. 6. Response of net photosynthesis to temperature in *Paspalum dilatatum, Lolium multiflorum* and *L. perenne* (after De Jager, 1968).

of respiration in light and dark are very different at most tempera-tures in high light intensities. In the range 15°-25°C, for example, apparent light respiration at a light intensity of 200 W/m² (400-700 nm) of *L. perenne* leaves may be up to twice as rapid as dark respiration (Wilson, 1972b).

The reasons for the marked temperature optima of net photo-synthesis are not clear. It is possible that enzyme denaturation is responsible for the decline in rate of photosynthesis of C_4 plants at high temperatures. In species with lower temperature optima, this is less likely to be the main determinant, and a greater Q_{10} for respiration than for photosynthesis may be responsible. In C_4 species, provided there is no water stress, changes in r_s are unlikely to be responsible for the decline of photosynthetic rate at high temperature since stomatal conductance for CO_2 may increase well beyond the temperature at which net photosynthesis declines (Raschke, 1970). However, over a wide temperature range (15-35°C) net photosynthesis and stomatal conductance of *Z. mays* were found by Raschke (1970) to be proportional to one another possibly because within this temperature range stomata regulated intracellular CO_2 concentrations.

Nevertheless, some temperature effects on photosynthesis may act through r_s, since temperature affects humidity, which can in turn control stomatal movement.

(ii) Light. The overall response of photosynthesis to light is partly caused by indirect effects of light acting on components of CO_2 transfer and assimilation. One important mechanism for the control of photosynthesis is the effect of light on stomatal opening, and thus on r_s (Meidner and Mansfield, 1968). However, many other factors also influence stomatal movement. Stomata open in response to an increase in turgor of guard cells relative to the adjacent epidermal cells and the degree of stomatal opening is determined by the difference between the osmotic value of guard cell and epidermal cell. In most species increased light results in increased osmotic value and stomatal opening. This may be because of photosynthetic production of organic solutes by chloroplasts of guard cells or because of transport of ions or other solutes into the guard cells. The spectral distribution of light may also have some effect, since experiments have shown that stomata may open widest when exposed to the wavelengths maximally absorbed by chlorophyll (Zelitch, 1969). Nevertheless, high levels of light energy do not necessarily mean maximal opening. Even in the light, other inhibitory factors such as high concentrations of CO_2 or water stress may cause complete stomatal closure (Meidner and Mansfield, 1968).

(iii) Water stress. Slatyer (1969) has fully discussed the internal water relations of the plant in relation to the physiology of growth. Net photosynthesis is usually reduced progressively by water stress (Fig. 7) and current evidence suggests that the most usual cause of this reduction is stomatal closure. Direct effects on the biochemical processes of photosynthesis can occur, though perhaps only under severe stress. In cotton leaves, for example, Troughton (1969) was able to detect an increase in the intracellular resistance (r_i) only at relative water contents approaching the wilting point. The decline in photosynthetic rate to this stage was related only to increasing r_s.

Similarly, among a number of tropical grasses and legumes a declining rate of photosynthesis, when vapour pressure deficits were increased from 15 mm Hg, was associated solely with increasing r_s (Wilson and Ludlow, 1970).

Among plants with the C_3 pathway, one complication to the study of the effects of water stress on net photosynthesis is the

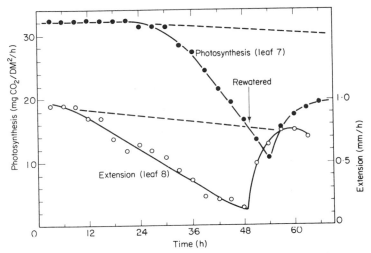

Fig. 7. Rate of apparent photosynthesis and rate of leaf extension of *Lolium temulentum* leaves, in response to increasing water stress, and recovery following rewatering. Results based on continuous recordings on one plant. Broken lines represent expected response without stress (after Wardlaw, 1969).

possibility of a different effect on photorespiration; this is difficult to assess because photosynthesis and photorespiration are proceeding at the same time. Slatyer (1969) has also drawn attention to the possible indirect effect of stomatal closure on net photosynthesis. When stomata close, the leaf temperature usually rises, causing an increase in respiration rate (particularly photorespiration) and a consequent reduction in net photosynthesis which is possibly greater than that resulting from direct dehydration. Because of their lack of photorespiration this type of effect would not necessarily be experienced by C_4 species.

Once leaf photosynthesis has been affected by water stress for several hours, recovery is often incomplete. Figure 7 depicts the response of a leaf of *Lolium temulentum,* an annual grass, to increasing water stress and shows the poor recovery of photosynthesis following rewatering (Wardlaw, 1969). Some growth components, in this case rate of leaf elongation, were reduced by water stress some time before any effect on photosynthesis. In addition while extension growth had almost ceased at a relative leaf turgidity of 75%, net photosynthesis was still about one third of its maximum rate. Other characteristics, such as carbohydrate deposition in seeds, may however take longer to show symptoms of water stress than photosynthesis (Wardlaw, 1967). Severe stress causes complete dislocation of the growth system so that it is not surprising that, even after recovery of turgor, normal metabolism, including cell division and photosynthesis, takes some time to re-establish and may not fully recover. In many cases the increased rate of senescence found in previously active leaves after a period of water stress, may be associated with partial loss of stomatal function (Slatyer and Bierhuizen, 1964).

(iv) Mineral nitrogen. Photosynthetic rates appear to be sensitive to deficiencies in soil nitrogen. Ryle and Hesketh (1969) found that, in *Z. mays,* a reduction in net photosynthesis over several days, associated with increasing nitrogen deficiency, was accompanied by a steady increase in the intracellular resistance (r_i). In contrast, the combined stomatal and cuticular resistance (r_l) did not decline consistently with time, although by the end of the experiment plants with sufficient nitrogen had lower r_l values than deficient plants.

(v) Oxygen levels. The rates of photosynthesis of C_3 plants can increase by 20-60% when atmospheric oxygen concentrations are reduced to less than 1%. This stimulation is presently attributed to an inhibition of photorespiration in these plants (Downes and Hesketh, 1968). These authors found such enhancement in several tribes of temperate grasses, but not in tropical C_4 grasses. Among tropical legumes, which appear to have only the C_3 pathway, net photosynthesis can be increased by 44% as a result of such reduced O_2 concentrations (Ludlow, 1970).

(vi) Environmental pretreatment. The potential photosynthetic activity of a leaf or young seedling can be greatly modified by the light and temperature conditions under which it develops. The degree of physiological plasticity depends on the genotype, and the range of adjustment of photosynthesis to growth conditions may differ considerably among populations genetically adapted to different environments (Björkman, 1966).

When leaves of herbage species develop in weak, compared with strong light, their potential rate of photosynthesis is usually reduced (Cooper, 1970). The reason for this reduction may lie in either the intracellular (r_i) or combined stomatal and cuticular resistances (r_l).

In *D. glomerata*, Eagles and Treharne (1969) found that the greater light saturated photosynthesis among leaves grown in strong compared with weak light was accompanied by greater carboxydismutase activity. However over the lower range of light intensities, the photochemical efficiency of the chloroplast was related to the respective photosynthetic rates of the populations. On the other hand, data from other species suggests that physical factors associated with both r_l and r_i may be involved (Holmgren, 1968) and it seems that in *L. perenne*, a change in stomatal geometry which could increase r_s, occurs when leaves are grown in weak light (Wilson and Cooper, 1969d). In the tropical *Panicum maximum* a progressive decline from 72·2 to 22·2 mg/dm^2 per h in maximum photosynthetic rate, as a result of a reduction from 100 to 11% relative light intensity during growth, was accompanied by increase in both r_s and r_i (Wilson and Ludlow, 1970). A similar situation was also found in the legume *Phaseolus atropurpureus*. Just which particular component of r_i changes in such cases is hard to say, but in *Z. mays* Hatch *et al.*, (1969) found large increases in the activity of a number of enzymes associated with the C$_4$ pathway (PEP carboxylase and P$_i$ dikinase) when plants were grown in strong compared with weak light. These increases in enzyme activity, adapting to new light conditions within 6 days, were associated with increases in maximum photosynthetic rates and with changes in the light saturation characteristics of leaves.

Plants may also adapt to the temperatures in which they are grown. Thus, Mooney and West (1964) found that among five contrasting species plants acclimated to a cold environment were more efficient, in terms of percentage of maximum photosynthetic activity, at lower temperatures than were plants grown in warmer conditions. A similar effect has also been found in a number of grasses (Treharne *et al.*, 1968; Charles-Edwards *et al.*, 1971).

The apparent response of photosynthesis to growth temperature may depend however on the units in which photosynthesis is measured, (Woledge and Jewiss, 1969; Wilson, D., 1970). Thus, leaves of *Festuca arundinacea* grown in day/night temperatures of 20°C/15°C had faster rates of photosynthesis per unit dry weight, but not per unit leaf area, than leaves from 10°C/5°C (Woledge and Jewiss, 1969). Furthermore, in this case leaves grown at the higher temperature also had a higher optimum temperature for photosynthesis, a shorter life, and a lower dark respiration rate at any one temperature than leaves from the cold regime. As with light preconditioning, there is evidence that such changes in potential photo-synthesis with growth temperature may be associated with both extra, and intracellular, resistances to CO_2 transport and utilization.

(c) *Variation in photosynthetic response.*

(i) Age and position of leaves. A progressive decline in net photosynthesis with leaf age after maturity has been recorded in many species and, in old leaves, this is accompanied by visible chlorosis. The onset of this decline is often assumed to start at, or soon after, complete leaf expansion. In some conditions, however, the rate of photosynthesis of grass leaves may not start to decline until 2 or 3 weeks after the leaf is fully expanded (Jewiss and Woledge, 1967; Treharne et al., 1968; Wilson and Cooper, 1969a). If this is a common situation then many leaves within a sward may have a potentially rapid rate of light saturated photosynthesis, even when they are shaded by younger fully expanded leaves on the same tiller, since temperate grass leaves may appear at the rate of one every 5 days in good conditions. This decline in net photosynthesis with age has been associated with increases in both r_s and r_i in the tropical grass *Sorghum almum* and in r_i only in the tropical legume *Glycine javanica* (Wilson and Ludlow, 1970).

The position of the leaf on the tiller can also apparently affect the photosynthetic rate of grass leaves. In *L. perenne* for example, later formed leaves are usually larger with fewer but bigger mesophyll cells, resulting in an apparent position effect when photosynthesis is expressed on the basis of unit mesophyll volume (Wilson and Cooper, 1969a). Even on a leaf area basis however the position of grass or legume leaves on the tiller or runner can influence photosynthesis, operating largely through some component of r_i (Wilson and Ludlow, 1970).

(ii) Size of metabolic "sink". Another possible influence is the size of the "sink" to which assimilates are transported. The sink hypothesis, reviewed by King et al. (1967) and Neales and Incoll (1968), considers that if assimilate is not transported to suitable sinks the rate of photosynthesis is depressed, while if new sinks are provided the rate is increased. The hypothesis is not altogether proven but in wheat, for example, the developing ear does act as an effective "sink", since its removal can result in a 50% reduction in photosynthetic rate of the "source" leaf (King et al., 1967).

(iii) Genetic variation. Genetic variation in the level of light saturated photosynthesis is considerable, and the existence, both between and within herbage species, of variation, in the response of apparent photosynthesis to light, CO_2 and temperature is well established. Although there are large differences between species with the C_3 and the C_4 carbon pathway respectively even within these groups marked variation exists both at the inter- and intra-specific level. For example, Böhning and Burnside (1956) found that a group of species genetically adapted to "sun" conditions exhibited light saturation values of 20,000-25,000 lux whereas another group adapted to shade had values of only 4,000-10,000 lux. Genetically adapted "sun" and "shade" races also occur within species (Björkman, 1968a, b). Within C_3 species a relationship between optimum temperatures of photosynthesis of different populations and the temperature conditions of their natural environment has also been demonstrated. In *L. perenne,* for example, Eagles (1964) found that at lower temperatures (5-10°C) maximum apparent photosynthesis was greater in a Danish than in an Algerian population, while at 20-30°C the converse was true. This was the net effect of changes in both true photosynthesis and respiration. Even in the absence of such ecological adaptation, genetic variation in photosynthetic rate can be high within herbage species (Cooper and Wilson, 1970).

Although there are large differences in photosynthetic responses both between and within species, the reason for many of these differences is not always clear. Differences exist in both extra- and intracellular resistances (Gaastra, 1959; Holmgren *et al.,* 1965), but the relative importance of these appears to vary with genotype and environment. To clarify this point, I have extracted from the literature the available data on the maximum rates of photosynthesis recorded for various levels of r_l and r_i (Figs 8 and 9), in a wide range of different C_3 and C_4 species. These include many grasses and legumes, often grown in different conditions, so that both genetic and environmental variation is involved. In addition theoretical curves, based on equation 11 of Monteith (1963), are plotted showing the influence of r_l and r_i on the rate of light saturated photosynthesis of a leaf at atmospheric CO_2 concentrations. From these figures, it appears that species with the C_4 pathway usually have low intracellular resistances (0·3-2·0 sec/cm) compared with C_3 plants (1·5-14·0 sec/cm), and it seems that this is the general basis for the faster rates of photosynthesis often recorded among the former species. There is also a good general relationship among many of the C_3 species between photosynthetic rate and the intracellular resistance (r_i) over a wide range of conditions. However, within C_4

species, in conditions giving rates of CO_2 uptake in excess of 50 mg/dm^2/h, quite large differences in photosynthesis seem to occur with the same r_i values. In contrast the relationship between rate of photosynthesis and the leaf resistance (r_l) differs for different r_i values, so that the C_4 species, with their low intracellular resistances,

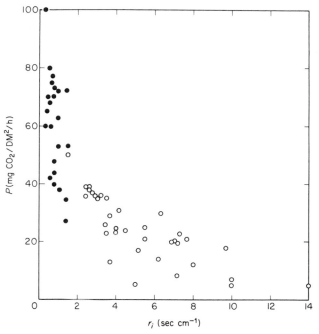

Fig. 8. Relationship between rate of apparent light saturated photosynthesis (P) at normal atmospheric CO_2 concentrations and the intracellular resistance to CO_2 transfer (r_i) among species with the C_4 (•) and C_3 (○) carbon pathways (data from Bull, 1969; Boyer, 1970; El-Sharkawy and Hesketh, 1965; Holmgren, 1968; Holmgren et al., 1965; Ludlow, 1970; McPherson, 1970; Osmond et al., 1969; Troughton, pers. comm.; Troughton and Slatyer, 1969; Wilson and Ludlow, 1970).

generally have much faster rates at the same r_l than do C_3 species. This can be seen from the data and from the theoretical curves. Note the good general agreement between the form of these curves and data taken from so many different sources. The general fit is better among the C_4 species since only a limited range of r_i values is involved. As predicted from theory, at low r_i values small changes in r_l can have very pronounced effects on the rate of photosynthesis, particularly in species and conditions where the intracellular resistance is low. Thus, an increase of only 2 sec/cm in stomatal resistance at low r_l values could bring rates of photosynthesis of C_4 plants within the range achieved by C_3 species.

It is often not certain which plant characteristics contribute most to the separate resistances, particularly to the intracellular resistance (r_i). The leaf resistance (r_l) is entirely diffusive and is mostly determined by the resistance of the stomata which changes in

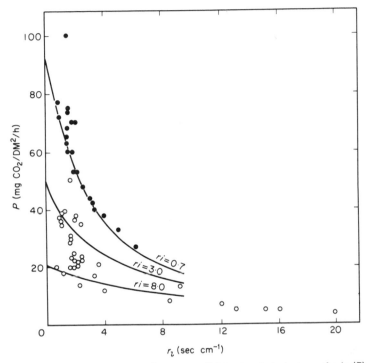

Fig. 9. Relationship between rate of apparent light saturated photosynthesis (P) at normal atmostpheric CO_2 concentrations and the leaf resistance (r_l) (stomatal + cuticular) to CO_2 transfer, among species with the C_4 (●) and C_3 (○) carbon pathways. Theoretical curves, supplied by J. H. Troughton, for three values of r_i are shown. These were calculated from equation 11 of Monteith (1963) assuming a visible-light intensity of 420 w/m^2 and 530/µg/1 CO_2 (data from same sources as that in Fig. 9).

response to various environmental factors. Differences in r_s between plants are affected by differences in stomatal geometry and density, and by differences in response of stomatal aperture to the environment (Woods and Turner, 1971). In some grass species the rate and extent of stomatal opening are known to be genetically controlled, although no biochemical explanation for these differences are yet known (Zelitch, 1965). Even within one grass species genetic variation in stomatal geometry and density and in r_s have been found (Wilson, 1971, 1972a).

Variation in the intracellular resistance may be physical (r_m) and/or biochemical (r_x). Thus, the faster rates of photosynthesis found among tropical grasses may be associated with a greater affinity for CO_2 by enzymes of the C_4 pathway compared with those of the C_3, although the site of these enzymes may also be involved because it has been suggested that PEP carboxylase may be located in the chloroplast envelope (Hatch and Slack, 1970). The lack of apparent photorespiration in C_4 plants, which is accompanied by a very low value (< 10 ppm) for the CO_2 compensation point, is thought to be the result of either (a) the substrate of CO_2 production being used in the C_4 pathway instead of being carboxylated (Zelitch, 1967), or (b) photorespired CO_2 being immediately reassimilated via PEP carboxylase in the mesophyll chloroplasts (Downton and Tregunna, 1968). The function of the photosynthetically active parenchymatous bundle sheath in C_4 plants might be to aid assimilate translocation out of the leaf (Downton and Tregunna, 1968). On the other hand, the bundle sheath, together with the large exposed cell surface to volume ratio of C_4 plants, might facilitate rapid gas exchange and thus a strong reassimilation of CO_2 released in photorespiration (Stoy, 1969).

Photorespiratory losses in C_3 plants may reach 40-60% of net photosynthesis in some conditions and it is thought that this may account for a large proportion, if not all, of the difference between C_3 and C_4 species. However, other limitations associated with differences in chloroplast or leaf anatomy or in the sites or rates of carbonic anhydrase activity, which would be part of the diffusive term of the intracellular resistance, may be involved (Hesketh, 1967). Although C_3 species exhibit marked photorespiration and usually have CO_2 compensation points greater than 40 ppm, there may be genetic variations in these characteristics even within agronomically valuable C_3 herbage species (Wilson, 1972b).

Among C_3 species themselves, variation in net photosynthesis associated with differences in r_i exists and relationships between rates of light saturated photosynthesis and various physical and biochemical characteristics have been observed. For example, *in vitro* activity of the major carboxylating enzyme ribulase-1,5-diphosphate carboxylase (RuDP) has exhibited close correlations with *in vivo* rate of light saturated photosynthesis; (a) among genetically adapted sun and shade ecotypes of *Solidago virgaurea* (Björkman, 1968a, b), (b) in response to partial defoliation of bean, maize and willow (Wareing *et al.*, 1968) and to application of gibberellin in red clover (Treharne and Stoddart, 1968), and (c) in *D. glomerata* grown in different light and temperature regimes (Eagles and Treharne, 1969; Treharne and

Eagles, 1970). Activity of the major C_4 carboxylating enzyme PEP carboxylase has also been correlated with rate of photosynthesis of individual leaves among populations of *Cenchrus ciliaris* (Treharne *et al.*, 1971). It has therefore been suggested that at saturating light intensities and normal CO_2 concentrations, photosynthetic rate may be limited by the level of carboxylating enzymes (r_x) and not only by diffusion resistances (r_m) to CO_2 transport in the leaf (Treharne, 1971). On the other hand, Holmgren (1968) found that differences in r_m, apparently associated with number and size of mesophyll cells, were correlated with differences in light saturated photosynthesis of *S. virgaurea*, and kinetic evidence indicated that any variation in enzyme activity (r_x) would only have a small effect. Similar relationships between photosynthesis and mesophyll structure have also been observed among other non-herbage (El-Sharkawy and Hesketh, 1965) and herbage species (Wilson and Cooper, 1969b). More recently, Chartier *et al.* (1970) have assessed the relative significance of the diffusion and carboxylation terms of the intracellular resistance of *Phaseolus vulgaris* as rate-determining processes in photosynthesis. When plants were growing in near-optimal conditions, the diffusive portion (r_m) placed a strong limitation on leaf photosynthetic rate, and r_x was comparatively unimportant. The relative importance of r_x and r_m may, however, differ with plant material and with growth conditions. It is also possible that an increase in content of carboxylating enzymes could be accompanied by an increase in the number of sites of carboxylation, and a consequent decrease in the diffusive resistance (Chartier *et al.*, 1970).

D. Distribution and use of Assimilates

The useful shoot yield of a grass or clover plant depends not only on the photosynthetic efficiency of unit leaf tissue but also on the ability of the plant to translocate and distribute assimilates. These are transferred from the photosynthesizing cells into the conducting tissues of the phloem which distributes them to other parts of the plant, where they may be stored as reserves or used for plant growth.

1. Transport

The process of movement to the phloem may require energy in the form of ATP, may take only a few minutes and appears to be selective (Biddulph, 1969). In addition to exporting assimilate actively photosynthesizing cells may, at the same time, store some within the chloroplasts in the form of starch which can be later degraded and exported during dark periods. The amount of photosynthate temporarily stored in this way varies between and within species and with environmental conditions. For example, Wilson and Bailey (1971) have found leaf starch levels of up to 50% of the total hot water soluble carbohydrate in *D. glomerata* after 9 h light, but only 16% in *L. perenne* grown in the same conditions. C_4 species can store large amounts of starch in their bundle-sheath chloroplasts which, it is thought, might act as sinks for end-products of photosynthesis, allowing photosynthesis to continue rapidly at times when translocation may be relatively slow (Downton and Tregunna, 1968). The proximity of the bundle sheath to the vascular tissue may also facilitate rapid transport of assimilates to the phloem. The importance of the bundle sheath to the translocation patterns of these plants is emphasized by Moss and Rasmussen (1969). They found that isotopic activity in maize (C_4) leaves was localized in the bundle sheath parenchyma only three minutes after initial exposure to $^{14}CO_2$, whereas activity in sugar beet (C_3) leaves after this length of time was generally distributed in the mesophyll cells.

The principal carbohydrate translocated by the phloem appears to be sucrose, together with smaller quantities of other sugars, sugar phosphates and various metabolites. The rate of movement of sucrose in the phloem commonly ranges from 50-300 cm/h. Such rates are, of course, much greater than would be expected from a simple diffusion process although the exact mechanism of translocation is still not fully understood. Rate of translocation in the whole plant is strongly affected by environmental variables such as temperature, particularly root temperature, and plant water status. Thus warmer roots relative to shoots increase translocation and of course the movement of solutes implies movement of water (Wardlaw, 1968). Although the level of complex carbohydrates such as starch increases in leaves during the light, the movement of sucrose out of leaves takes place mainly in light. For example, Hodgkinson and Veale (1966) found that the decline in ethanol-soluble ^{14}C photosynthate (simple sugars) in lucerne (*Medicago sativa*) leaves was more rapid in light than in the dark, whereas the converse was the case with the ethanol-insoluble portion. However, transport and distribution of assimilates seem to be mainly governed by source-sink relations and most environmental influences are thought to act by modifying the size and activity of sinks.

2. *Distribution*

Two main types of sink for photosynthate can be distinguished. These are (a) the accumulating sinks where parenchymatous cells become repositories for assimilates which may later be mobilized for use in growth and development, and (b) metabolic sinks where active growth is proceeding and sucrose is used in the underlying biosyntheses. Major accumulating sinks in herbage plants are found in basal stems, roots and seeds, and the form of the main reserve polysaccharide in these regions may be starch, as in grasses with the C_4 carbon pathway, or fructosan in C_3 grasses (Chapter 3, Vol. 1). The main metabolic sinks are meristematic regions such as are found in the main stem, tillers and roots, and young expanding leaves. In general, assimilates seem to move freely within grass plants and the direction of movement is governed mainly by the site and size of the sink (Chapter 25).

Individual fully expanded leaves can retain a substantial proportion of assimilated $^{14}CO_2$. Although 47-64% of incorporated ^{14}C may be exported within 24 h from *L. multiflorum* leaves much may still remain in the leaf even after 2-3 weeks (Marshall and Sagar, 1968; Sagar and Marshall, 1966). However, the rate of export seems to be much greater in C_4 species. Thus Hofstra and Nelson (1969) found that young tropical grass leaves exported 70% or more of their assimilated ^{14}C within 6 h, whereas a number of C_3 dicotyledons exported only 45-50%.

Among tillers, small young ones tend to import assimilates from larger ones, although once they can assimilate carbon for themselves there appears to be some interchange of materials between tillers of different age (Sagar and Marshall, 1966). The movement of carbon between tillers of such grasses seems to be largely through the root system, although some may find its way through the vascular system of the stem. In rhizomatous grasses, such as *Agropyron repens*, ^{14}C may be translocated from large to small tufts via the rhizome, but the reverse has not been found among plants in light (Forde, 1966).

The degree of interdependence of tillers depends on the size and age of the plant. In young vegetative grasses of the *L. multiflorum* and *P. pratense* type the whole shoot seems to be an integrated unit with mutual exchange of metabolites occurring among all tillers. As the plant gets larger, the principal tillers become more independent of one another, although young, daughter, tillers on each main unit are still obtaining material from it. Large tillers may eventually reach a stage of complete independence which only continues as long as the whole plant is intact (Sagar and Marshall, 1966). Thus in *L. multiflorum,* defoliation of all tillers except the main one immediately re-establishes a flow of assimilates from the main shoot to the cut tillers (Chapter 23) which may continue until growth of the defoliated tillers is well advanced. This surge in export from the main shoot appears to be achieved without reduction in assimilate incorporation into the roots and therefore seems to be a real increase in efficiency (Marshall and Sagar, 1968). The onset of the reproductive phase provides more active meristems as sinks. While ears themselves assimilate when still green they also import large quantities of assimilate from the flag leaf during development (Williams, 1964; Rawson and Hofstra, 1969).

Movements of assimilates to similar metabolic sinks also take place in legumes (Chapter 25). For example Hoshino *et al.* (1966) exposed leaves of *Trifolium repens* to $^{14}CO_2$ and found quite extensive, although not intensive, distribution of ^{14}C after only 2 h, mostly in roots and growing points. After one day, ^{14}C occurred all over the plant but especially at the growing points, lateral buds, roots and root nodules. After five days, ^{14}C had accumulated even in the newly opened leaves on the main stem. and the leaves on the lateral buds.

Partitioning of assimilates between different meristems is dependent on plant type, stage of growth and environment. For example, perennial ryegrasses have been shown to accumulate more ^{14}C in the roots than do more annual types (Ryle, 1968), a difference which also appears to be reflected in shoot/root dry weight ratios (Wilson and Cooper, 1969c) and seems to be a logical association with their respective longevity. The annual *L. temulentum* partitions a larger proportion of assimilates in tillers and roots at the seedling stage than later. High light intensities and cool temperatures may also promote the export of assimilate to roots and tillers and, in *D. glomerata,* short days have the same effect (Ryle, 1968). These effects may differ between populations from contrasting climatic regions. For example, Eagles and Østgård (1971) found that diversion of assimilates to roots at low temperatures (5°C) and short days was more marked in Norwegian than in Portuguese populations of *D. glomerata.* In contrast, at 25°C the Portuguese populations diverted more to the roots. Similarly, the change from the vegetative to reproductive state also had marked effects on the activity and relative requirements of the various meristems. The proportion of shoot relative to root can obviously have a marked effect on the plant's capacity for trapping light. It is highly sensitive to the external environment (Brouwer, 1963; Davidson, 1969a, b) and in some temperate herbage species the shoot/root ratio shows a definite temperature optima which is greatest at temperatures only slightly above those optimal for shoot growth. In subtropical species the proportion and yield of shoot may still be increasing at 35°C (Davidson, 1969a).

III. LIGHT UTILIZATION BY THE WHOLE SWARD

A. Light Interception and Distribution

The amount of photosynthetically useful light energy available to individual leaves of a grassland community is affected not only by geographical location, climate and weather but also by the ability of the community to reflect, transmit and absorb light. The radiation distribution within crop stands is significantly affected by the spatial orientation of leaves and stems. In a sward the total solar radiation received by shaded leaves will be less in quantity and of different spectral composition to the light falling on unshaded leaves. The principal change in the quality of radiation is caused by preferential absorption by leaves of light with a wavelength smaller than 750 nm, and within the visible range in particular in the blue and red regions of the spectrum (Moss and Loomis, 1952). Consequently, the light falling on shaded leaves is of poorer quality for photosynthesis. Figure 10 depicts transmission patterns of photosynthetically useful and infra-red radiation to ground level through two graminaceous and two leguminous canopies. Scott et al. (1968) found that this selective absorption at different wavelengths was concentrated in the top layer of the canopy, and that there were differences between species in their light transmission patterns.

Absorption, transmission and reflection of light from leaves depends on leaf structure and composition. Such features as the humid air in the intercellular spaces, the structural material of the cell wall, the solution within the cells, and the pigments within the chloroplasts are all important in determining light and radiation distribution (Gates et al., 1965). In a sward, losses of total radiation by transmission and reflection from leaves may occur more at the top of the canopy. Within the canopy, receipt of reflected and transmitted radiation will more nearly balance the corresponding losses.

For efficient utilization of light by the whole community complete interception and effective distribution of light within the sward are both important. Crop growth and its relationship with light

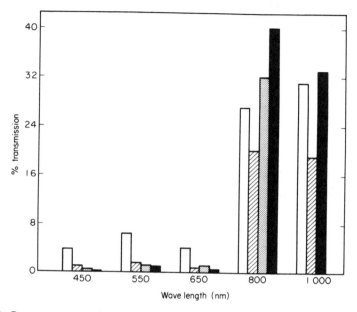

Fig. 10. Percentage transmission, at five wavelengths, of incident radiation reaching ground level through communities of lucerne (□), oats (▨), tall fescue (▨) and white clover (■) (data from Scott *et al.*, 1968).

interception and utilization has been extensively examined (see for example, Brown and Blaser, 1968). Seedling growth rates are initially limited by the small leaf surface area exposed to light and net assimilation rate (NAR) per unit leaf area is high. However, as the canopy develops, self and mutual shading of leaves cause a reduction in mean NAR, although crop growth rate (CGR) continues to increase in response to the increasing quantity of light trapped. The rate of increase in CGR gradually declines with increasing leaf area until most of the available light is intercepted. Subsequent events are explained by two alternative models.

In the first model, CGR reaches a maximum at an optimum leaf area per unit of soil surface area (leaf area index = LAI), when most available light is intercepted and the photosynthesis/respiration ratio is maximal. With further increase in LAI, CGR declines because heavy shading then causes respiration of basal parts of the sward to exceed photosynthesis; the rate of decline depends on how much foliage is below the light compensation point. According to this model the optimum LAI (LAI_{opt}) is less with reduced incident light (Black, 1963).

In the second model there appears to be no optimum LAI. Once the critical LAI (LAI_{crit}) has been achieved, at which most of the available light is intercepted, CGR appears to be independent of LAI and remains relatively constant even at high LAI values (Brougham, 1956). Net photosynthesis per unit area of ground (P_{NG}) has also been shown to follow this pattern. For example, McCree and Troughton (1966b) found that all the available light was intercepted by white clover plants, when LAI was 3. Thereafter, the area of green leaf increased until the rate of leaf death equalled the rate of leaf production, with LAI values of 7 to 11. Even at this stage, P_{NG} and rate of leaf production remained close to their maximum.

These two apparently conflicting models are difficult to reconcile but may be special cases of a general relationship. It is possible that in some experiments the rate of increase of the respiratory load at the base of the sward was greatly exaggerated or that material dying between harvests was not taken into account. Differences in methods used to estimate LAI values can also lead to apparently contrasting results (Brown and Blaser, 1968). Fulwood and Puckridge (1970) thought that the absence of an optimum LAI in their studies on *Lolium rigidum* might be the result of slow respiration rates, and that higher temperatures might well change the relationship between P_{NG} and LAI by increasing respiration rate.

The critical or optimal leaf area index (LAI_{crit} or LAI_{opt}) is clearly an important concept, since at this point CGR and P_{NG} are at or close to maximum for the conditions in question. Whether or not CGR declines at greater LAI a high LAI_{crit} or LAI_{opt} is of obvious value since, in general, increased LAI results in more efficient light interception. The foliage area required to intercept all or most incoming radiation is also affected by the angle of incident light and Brougham (1958) has suggested that in New Zealand, where there is marked seasonal amplitude in the sun's elevation, the amount of foliage required to intercept 95% of the light in midwinter is approximately half that required in summer.

1. Light Penetration

The penetration of light down the canopy has been shown to follow approximately Beer's Law for extinction in a homogeneous medium,

$$\frac{I_L}{I_0} = e^{-KL}$$

where I_0 = incident light, I_L is the light penetrating a leaf area index of L, and K is the light extinction coefficient. Light intensity is thus reduced logarithmically with distance or with accumulated leaf area from the top of the canopy. The value of K depends on the angle of incident light and on the geometrical arrangement and light transmission characteristics of leaves within the community. Although pasture canopies are not completely homogeneous this treatment serves as a useful approximation, although it neglects effects of sun-flecking, of the different spectral composition of light transmitted through leaves and of the quality of the incident light (Verhagen and Wilson, 1969).

2. Foliage Angle

One important character affecting K is the foliage angle, and grasses, because of their more vertically inclined leaves, have lower K values than clovers and therefore require a greater LAI to intercept the incoming light. For example, *Trifolium repens,* which has horizontal leaf laminae and K approaching 1·0, intercepts 95% of the light at an LAI of about 3·5, whereas LAI values for many grasses range from about 5 to 15. Some grasses with more horizontal leaves, such as *Cynodon dactylon* have LAI_{crit} values as low as those of clovers (Brown and Blaser, 1968). In *Lolium,* maximum crop growth rates are therefore higher in the more erect varieties and even within *L. perenne,* prostrate types have been shown to reach complete light interception at lower LAI and have lower CGR than more erect plants (Rhodes, 1971). In general, erect leaves may be expected to give greater crop production at large values of LAI, but when LAI is small horizontal leaves are advantageous (Alberda and Sibma, 1968; Cooper *et al.,* 1971).

3. Defoliation

It is not surprising that the morphological characters associated with the most rapid pasture growth differ with the intensity and frequency of defoliation. When defoliation is so frequent that most herbage is produced prior to complete light interception, canopies of short-leaved prostrate ryegrass plants are more productive than those with more erect, longer-leaved types. The former have higher K and lower LAI_{crit} values, and greater concentration of leaf area below the defoliation level. In contrast, the long-leaved more erect canopies are more productive under infrequent defoliation (Cooper et al., 1971). The proportion of LAI at the base of the sward may also change with time, decreasing during regrowth after defoliation (Rhodes, 1971).

4. Leaf Distribution

K is also strongly affected by the horizontal and vertical distribution of leaves within the canopy. Lower values of K are achieved in communities with well separated and regularly distributed leaves than in those where the distribution is more random (Loomis and Williams, 1969). The relationship between light penetration and LAI in mixed grass-clover pastures will depend on the relative proportions and dispersal of leaves of the two species and is generally intermediate between the pure grass and pure clover situation (Brougham, 1958). In such mixtures the leaves tend to be dispersed in a regular manner which promotes the more efficient interception of light than in similar groups of completely randomized leaves (Warren-Wilson, 1959).

The geometry of the sward may change during plant development. Thus, the marked stem extension at flowering in grasses is accompanied by a strong decrease in K and increase in LAI as the leaves become distributed over a greater vertical distance on the stems. Even during vegetative growth, the proportion of horizontal leaves may change considerably (de Wit, 1965).

The relationship between canopy architecture, light distribution and sward production clearly changes with time of day, weather, season, stage of crop development and individual plant components. In addition, herbage production may also be affected by the influence of leaf arrangement on air circulation, which affects CO_2, water vapour and heat transfer within the canopy.

B. Relative Importance of the Size and Efficiency of the Assimilatory System

The growth rate of plants or communities is dependent on both the total amount of photosynthetically active light harvested and the efficiency with which it is used by the individual leaves. Thus, the rate of dry matter accumulation (Relative Growth Rate = RGR) of a single plant can be analysed in terms of the contribution of net assimilation rate (NAR) and of the leaf area ratio (LAR, the proportion of the plant which is actively photosynthetic), such that RGR = NAR x LAR. Similarly, the rate of dry matter increase of swards (CGR) can be related to NAR and the leaf area index (LAI), so that CGR = NAR x LAI (Watson, 1952). Many of the factors contributing to NAR and LAI and their effectiveness have already been discussed. In terms of total crop yield, the potential importance to the size of the photosynthetic surface of the rate of leaf area development and leaf longevity (leaf area duration) must be emphasised (Watson, 1952). These features are likely to be particularly important to the annual grass crop and it might be expected that yields may be more affected by them in short growing seasons.

1. The Isolated Plant

Both NAR and LAR of the isolated plant can be affected independently by different environmental factors and differences between plants in either or both these parameters may occur. The effect on growth rate of changes in either NAR or LAR will depend on the direction and extent of concurrent changes in the other. For example, Wilson and Cooper (1969e) found that differences in RGR between seedling populations within *L. perenne* and within *L. multiflorum* were associated more with differences in NAR than LAR, but a large difference in LAR *between* the two species resulted in the *L. multiflorum* group having the more rapid growth rate. In general, differences in dry matter yield between species with the same carbon pathway seem to be more closely associated with leaf area than with NAR, and even where differences in NAR are positively associated with RGR they are often accompanied by relatively greater variation in LAR (Watson, 1952; Thorne *et al.,* 1967). Many of the differences in growth responses which occur between C_3 and C_4 herbage grasses are however based on differences in their photosynthetic responses, rather than on the relative development of the photosynthetic surface (Cooper and Tainton, 1968).

Differences between plants in all the foregoing parameters may be modified by the growing environment. Although in general, temperature changes affect NAR less than LAR the effect of temperature on

NAR and its influence on growth may depend on whether the prevailing temperature lies above or below the optimum for the species under consideration. In C_4 species, for example, the lack of growth below 15°C is associated with a greatly reduced NAR. Climatic ecotypes within species also show interactions of NAR with temperature. Thus, Eagles (1967) detected differences in RGR between N. European and Mediterranean ecotypes of *D. glomerata*, at low (5-10°C) and at high (30°C) temperatures. These differences were associated with variation in NAR which was greater in the N. European ecotype than the Mediterranean at 30°C but less at 5°C. However, changes in both NAR and LAR were responsible for the overall temperature effects on growth rate. Even so, climatic ecotypes from these two regions usually show distinct differences in LAR at extremely high or low temperatures. Mediterranean plants grown in a northern winter usually have more rapid leaf growth and larger leaves than the northern types (MacColl and Cooper, 1967; Robson and Jewiss, 1968).

In general, LAR tends to be greater at higher temperatures and low incoming radiation; seasonal trends show that LAR increases from spring to autumn. Consequently effects of small temperature changes on growth rates can often be the result of changes in LAR (Thorne *et al.*, 1967). In contrast, seasonal changes in incoming radiation appear to influence plant growth through NAR rather than LAR. In spaced seedling studies of *L. perenne, D. glomerata* and *F. arundinacea,* MacColl and Cooper (1967) found that variation in RGR throughout the year closely followed changes in NAR and incoming radiation. However there is an optimum light intensity for growth which can vary among species. The increase in NAR with increasing light intensity is accompanied by a fall in LAR and the net result is that RGR rises from the compensation point (where NAR is zero) to a peak at an intermediate light intensity then falls as light is increased further (Blackman and Wilson, 1951). Irrespective of light intensity, the length of the photoperiod may affect RGR and Eagles (1971) has shown that differences in LAR among *D. glomerata* populations may play a larger part than NAR in determining differential population responses in RGR to photoperiod. In a constant environment both NAR and LAR of spaced plants may fall with time. Thus Thorne (1960) found that the rate of decline in NAR with age varied among species but that LAR and RGR decreased with time at similar rates for all species they studied.

2. The Sward

During canopy development, or in the established sward, the relative importance of NAR and of the light trapping ability of the canopy will depend on the response of each plant to water stress and temperature variation. NAR can clearly be limited by factors such as water shortage or extreme temperatures which influence the photosynthetic process. Photosynthetic activity of plants with the C_4 carbon pathway may often, however, be less affected by moisture stress than that of C_3 plants, and at temperatures above 30°C the more rapid rates of net photosynthesis of C_4 plants may confer a distinct advantage. However, apart from such broad comparisons, current evidence indicates that, in the absence of marked stress or temperature extremes, NAR may be a less general limitation to growth in the vegetative phase than the ability of the canopy to efficiently intercept and distribute photosynthetically active radiation. Correlations between CGR and NAR have been recorded only among C_3 herbage grasses under primary canopy conditions where there was no appreciable senescence of older leaves and where 80% of the incoming light was intercepted by all material (Hunt and Cooper, 1967). On the other hand, Cooper et al. (1971) found that differences in crop growth rates of from 23 to 43 g/m^2/day, among primary canopies of six forage grasses, were based on differences in canopy structure rather than on differences in photosynthetic rates of individual leaves (Fig. 11). Brougham (1960) found that differences in maximum crop growth rate, at critical LAI, between a number of clovers and grasses, was significantly correlated with LAI_{crit} over periods of up to 4 weeks. Puckridge and Ratkowsky (1971) examined the relationship between LAI and CO_2 uptake per unit of ground of an established wheat community at different plant densities and light intensities, and with two different cultivars. They found that at each light intensity all measurements of CO_2 uptake when plotted against LAI fell on a single curve with no indication of an optimum LAI. Although CO_2 uptake per unit LAI showed some variation during the growth season, variation in LAI appeared to be the dominant factor.

Nevertheless, it is extremely difficult to distinguish cause and effect in the relationships between canopy architecture and herbage yield, particularly since environmental limitations which can affect the photosynthetic process are often present in the field. For example, in closed corn crops, the increase of CO_2 uptake per unit

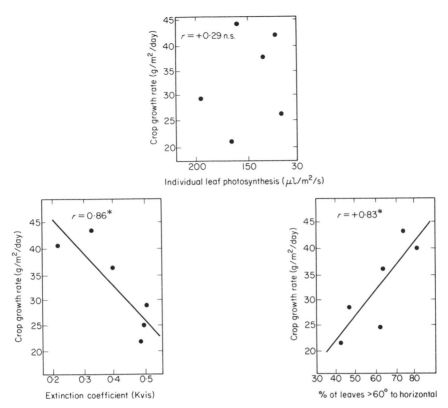

Fig. 11. Crop growth rate and canopy characteristics in six forage grasses (after Cooper *et al.*, 1971).

of ground area with increasing light intensity can be affected by the degree of light interception (Baker and Musgrave, 1964) but may also be modified by reactions of photosynthesis to drought and to variations in temperature and CO_2 (Moss *et al.*, 1961).

C Potential Production of Dry Matter and Digestible Energy from Established Swards

Cooper (1970) has recently reviewed this topic and I will only outline some aspects relevant to the present discussion. The efficient production of herbage from pasture on a continuing basis must take into account the short- and long-term requirements for maintenance and growth of the sward. While it is desirable to have sufficient foliage to intercept all or most incoming light, such a situation often cannot be maintained for long periods without harmful effects on potential production. In this condition the load of aged and senescing leaves at the base of the sward would increase, and initiation and development of new tillers and leaves would be retarded. As Brown and Blaser (1968) have pointed out, the appropriate cutting or grazing height for achieving highest annual yields and maintaining a vigorous sward may differ at different seasons and may not even be closely related to concurrent growth rates. While continued interception of all the incoming light may be best in swards cut for hay or conservation, grazing or cutting to low LAI values may be necessary at times to promote tiller development. In some cases, the morphology of the species may dictate the most appropriate management procedure. Thus, lucerne apparently yields most if LAI is reduced from very high levels at the bloom stage to almost zero thereby encouraging basal-bud development. On the other hand, yield of the stoloniferous *Cynodon dactylon* may be similar under either lax or intensive defoliation (Brown and Blaser, 1968). Therefore, as Cooper (1970) has pointed out, estimates of annual production based on potential growth rates or photosynthesis of young closed swards, in which all leaves are active and plants are all in the vegetative state, may be misleading.

Potential annual production in conditions of ample soil moisture and nutrients and optimum management has been determined on small experimental plots grown throughout the year at different localities. In temperate climates, maximum herbage yields recorded in this way from highly productive varieties of *L. perenne* range from approximately 17 to 25 metric tons/ha/year in the United Kingdom and up to about 27 metric tons/ha in New Zealand where energy inputs, particularly in winter, are much greater. These production levels represent annual efficiencies of light energy conversion of from 2 to 3%. In tropical regions, species with the C_4 carbon pathway are able to take advantage of the greater annual energy input with their high rates of maximum photosynthesis and high light saturation values. In these climates production of up to 85 metric tons/ha/year (*Pennisetum purpureum*), with an efficiency of energy conversion of about 5%, has been recorded. However, in many subtropical areas, seasonal temperatures below about 15°C may severely limit pasture production of C_4 species, even with adequate soil moisture and high nutrient inputs. In S.E. Queensland, for example, species such as *Paspalum plicatulum* and *Digitaria decumbens* may have daily growth rates of up to 200 kg/ha in summer, declining to only 5 kg/ha in winter, even with irrigation and high light energy and fertilizer inputs. In these conditions, annual production of such species rarely exceeds 25-30 metric tons/ha, which, because of the high light energy inputs, represents energy conversions of only $1\frac{1}{2}$ to 2% (Cooper, 1970).

Such yields are usually much greater than those obtained in farming practice. Herbage growth is often reduced by high or low temperatures, inadequate nutrient input or water stress. The low temperature effects on potential yield which are more severe in subtropical regions where they coincide with relatively high light, may be modified to some extent by breeding varieties more tolerant to low temperature or, in terms of total farm production, by growing temperate species as winter crops. Water deficits are most severe where seasonal water shortages coincide with high potential evapo-transpiration (P_e). In subtropical regions P_e may reach over 1500 mm/year and in Georgia, U.S.A., annual yields of C. dactylon may vary from 14 to 30 metric tons/ha between dry and wet years (Prine and Burton, 1956). Even when P_e is relatively low water deficits may be sufficient to affect grass growth. In S.E. England, for example, with P_e of only 400-500 mm yield responses to irrigation of 29% in grasses and 56% in clover have been achieved (Tayler, 1965). However, one of the most widespread limitations to grassland production is the soil nutrient status, particularly nitrogen. Tropical species may continue to respond to up to 1800 kg/ha of nitrogen and even in temperate Britain, grass production usually responds linearly to nitrogen inputs up to at least 500 kg/ha (Cooper, 1970).

In practice, the requirement for high dry matter yields may have to be modified by the energy and nutritional requirements of the ruminant animal. For example, the proportion of easily digested organic matter is much lower in tropical than temperate grasses, although the extent to which this is the result of the relative climates in which they are grown or of factors inherent in the species, is not clear (Minson and McLeod, 1970). Even among temperate species of similar digestibility there may be large differences in intake of carbohydrate components which can markedly affect rate of animal live-weight gain. For example, Ulyatt (1970) found that lambs fed *ad lib* on white clover gained weight at a rate of 331 g/day whereas animals similarly fed on perennial ryegrass only increased their weight by 227 g/day. The difference was attributable to the low structural carbohydrate content and higher ratio of readily fermentable to structural carbohydrate in the clover. Since the growth rate of ryegrass is usually more rapid than that of white clover (Brougham, 1960), the farmer concerned with animal weight gain must achieve a balance between the high dry matter production of one and the higher quality of the other. Furthermore, management procedures aimed at achieving high yields might also have to be modified in the light of relationships which exist between growth stage of the sward and its nutritive value to the animal.

ACKNOWLEDGMENTS

I would like to thank Dr J. P. Cooper, Welsh Plant Breeding Station, Aberystwyth for his careful criticism of the manuscript and for supplying data from publications then in preparation. Dr J. H. Troughton, D.S.I.R., Lower Hutt, kindly provided figure 5.

REFERENCES

Alberda, T. and Sibma, L. (1968). *J. Br. Grassld Soc.* 23, 206-215.
Baker, D. N. and Musgrave, R. B. (1964). *Crop Sci.* 4, 127-131.
Biddulph, O. (1969). *In* "Harvesting the Sun: Photosynthesis in Plant Life" (A. San Pietro, F. A. Greer and T. J. Army, eds), pp. 143-168. Academic Press, New York and London.
Bierhuizen, J. F. and Slatyer, R. O. (1964). *Aust. J. biol. Sci.* 17, 348-389.
Björkman, O. (1966). *Brittonia* 18, 214-224.
Björkman, O. (1968a). *Physiologia Pl.* 21, 1-10.
Björkman, O. (1968b). *Physiologia Pl.* 21, 84-99.
Black, C. C. and Mayne, B. C. (1970). *Pl. Physiol., Lancaster,* 45, 738-741.
Black, J. N. (1963). *Aust. J. agric. Res.* 14, 20-38.
Blackman, G. E. and Wilson, G. L. (1951). *Ann. Bot.* 15, 373-408.
Böhning, R. H. and Burnside, C. A. (1956). *Am. J. Bot.* 43, 557-561.
Boyer, J. S. (1970). *Pl. Physiol. Lancaster* 46, 236-239.
Brougham, R. W. (1956). *Aust. J. agric. Res.* 7, 377-387.
Brougham, R. W. (1958). *Aust. J. agric. Res.* 9, 39-52.
Brougham, R. W. (1960). *Ann. Bot.* 24, 463-474.
Brouwer, R. (1963). *Jaarb. Inst. biol. schelk. Onderz. LandbGewass.* 31-39.
Brown, K. W. (1969). *Physiologia Pl.* 22, 620-637.
Brown, R. H. and Blaser, R. E. (1968). *Herb. Abstr.* 38, 1-9.
Bull, T. A. (1969). *Crop Sci.* 9, 726-729.
Charles-Edwards, D. A., Charles-Edwards, J. and Cooper, J. P. (1971). *J. exp. Bot.* 22, in press.
Chartier, P., Chartier, M. and Catsky, J. (1970). *Photosynthetica* 4, 48-57.
Chen, T. M., Brown, R. H. and Black, C. C. (1969). *Pl. Physiol., Lancaster* 44, 649-654.
Cooper, J. P. (1966). *Proc. X Int. Grassld Congr., Helsinki* 715-720.
Cooper, J. P. (1970). *Herb. Abstr.* 40, 1-15.
Cooper, J. P. and Tainton, N. M. (1968). *Herb. Abstr.* 38, 167-176.
Cooper, J. P. and Wilson, D. (1970). *Proc. XI Int. Grassld Congr.,* 522-527.

Cooper, J. P., Rhodes, I. and Sheehy, J. E. (1971). *Ann. Rep. Welsh Pl. Breed. Stn*, 1970, 57-69.

Davidson, R. L. (1969a). *Ann. Bot.* **33**, 561-569.

Davidson, R. L. (1969b). *Ann. Bot.* **33**, 571-577.

De Jager, J. M. (1968). Unpublished Ph.D. thesis, Univ. Wales.

Downes, R. W. and Hesketh, J. D. (1968). *Planta* **78**, 79-84.

Downton, W. J. S. and Tregunna, E. B. (1968). *Can. J. Bot.* **46**, 207-215.

Eagles, C. F. (1964). *Nature, Lond.* **215**, 100-101.

Eagles, C. F. (1967). *Ann. Bot.* **31**, 31-39.

Eagles, C. F. (1971). *Ann. Bot.* **35**, 75-86.

Eagles, C. F. and Østgård, O. (1971). *J. appl. Ecol.*, in press.

Eagles, C. F. and Treharne, K. J. (1969). *Photosynthetica* **3**, 29-38.

El-Sharkawy, M. and Hesketh, J. D. (1965). *Crop Sci.* **4**, 514-518.

Enns, T. (1967). *Science, N.Y.* **155**, 44-47.

Forde, B. J. (1966). *N.Z. Jl. Bot*, **4**, 496-514.

Freeland, R. O. (1948). *Pl. Physiol., Lancaster* **23**, 595-600.

Fulwood, P. G. and Puckridge, D. W. (1970). *Proc. XI Int. Grassld Congr.* p. 530-534.

Gabrielsen, E. K. (1948). *Physiologia Pl.* **1**, 5-37.

Gaastra, P. (1959). *Meded. LandbHogesch. Wageningen* **59**, 1-68.

Gaastra, P. (1963). *In* "Environmental Control of Plant Growth" (L. T. Evans, ed.), pp. 113-140. Academic Press Inc., New York and London.

Gates, D. M. (1965). *Ecology* **46**, 1-13.

Gates, D. M., Keegan, H. J., Schleter, J. C. and Weidner, V. R. (1965). *Appl. Optics* **4**, 11-20.

Gauhl, E. (1970). *Corn. Inst. Yearbook.* **68**, 633-636.

Hatch, M. D. and Slack, C. R. (1970). *A. Rev. Pl. Physiol.* **21**, 141-162.

Hatch, M. D., Slack, C. R. and Bull, T. A. (1969). *Phytochem. Newsl.* **8**, 697-708.

Heath, O. U. S. (1969). "The Physiological Aspects of Photosynthesis", p. 309, Heinemann Educational Books Ltd., London.

Hesketh, J. D. (1967). *Planta* **4**, 371-374.

Hesketh, J. D. and Moss, D. N. (1963). *Crop Sci.* **3**, 107-110.

Hodgkinson, K. C. and Veale, J. A. (1966). *Aust. J. biol. Sci.* **19**, 15-21.

Hofstra, G. and Hesketh, J. D. (1969). *Planta* **85**, 228-237.

Hofstra, G. and Nelson, C. D. (1969) *Planta* **88**, 103-112.

Holmgren, P. (1968). *Physiologia Pl.* **21**, 676-698.

Holmgren, P., Jarvis, P. G. and Jarvis, M. A. (1965). *Physiologia Pl.* **18**, 557-573.

Homann, P. H. and Schmid, G. H. (1967). *Pl. Physiol., Lancaster*, **42**, 1619-1632.

Hoshino, M., Mishimura, S. and Okubo, T. (1966). *Proc. Crop Sci. Soc. Japan* **35**, 137-141.

Hunt, L. A. and Cooper, J. P. (1967). *J. appl. Ecol.* **4**, 437-458.

Jewiss, O. R. (1965). *In* "The Growth of Cereals and Grasses". 12th Easter Sch. agric. Sci. Univ. Nott. 39-54.

Jewiss, O. R. and Woledge, J. (1967). *Ann. Bot.* **31**, 661-671.

King, R. W., Wardlaw, I. F. and Evans, L. T. (1967). *Planta* **77**, 261-276.

Laetsch, W. M. (1969). *Sci. Prog., Oxf.* **57**, 323-351.

Loomis, R. S. and Williams, W. A. (1969). *In* "Physiological Aspects of Crop Yields" (J. D. Eastin, F. A. Haskins, C. Y. Sullivan and C. H. M. Van Bavel, eds), pp. 27-52. Publ. Amer. Soc. Agron. and Crop Sci. Soc. Amer. Madison, Wis., U.S.A.

Ludlow, M. M. (1970). *Planta* **91**, 285-290.

MacColl, D. and Cooper, J. P. (1967). *J. appl. Ecol.* 4, 113-127.

McCree, K. J. and Troughton, J. H. (1966). *Pl. Physiol., Lancaster,* 41, 559-566.

McCree, K. J. and Troughton, J. H. (1966). *Pl. Physiol., Lancaster,* 41, 1615-22.

McPherson, H. (1970). Unpublished Ph.D. thesis, Australian National University, Canberra.

Marshall, C. and Sagar, G. R. (1968). *J. exp. Bot.* 19, 786-794.

Meidner, H. and Mansfield, T. A. (1968). *Physiology of Stomata,* pp. 179. McGraw-Hill, London.

Miflin, B. J. and Hageman, R. H (1966). *Crop Sci.* 6, 185-187.

Minson, D. J. and McLeod, M. N. (1970). Proc. XI Int. Grassld Congr. pp. 719-723.

Monteith, J. L. (1963). *In* "Environmental Control of Plant Growth" (L. T. Evans, ed.), p. 95. Academic Press Inc., New York and London.

Mooney, H. A. and West, M. (1964). *Am. J. Bot.* 51, 825-827.

Moss, R. A. and Loomis, W. E. (1952). *Pl. Physiol., Lancaster,* 27, 370-391.

Moss, D. N., Musgrave, R. B. and Lemon, E. R. (1961). *Crop Sci.* 1, 83-87.

Moss, D. N. and Rasmussen, H. P. (1969). *Pl. Physiol., Lancaster* 44, 1063-1068.

Neales, T. F. and Incoll, L. D. (1968). *Bot. Rev.* 34, 107-125.

Osmond, C. B., Troughton, J. H. and Goodchild, D. J. (1969). *Z. Pflzenphysiol.* 61, 218-237.

Patel, A. S. and Cooper, J. P. (1961). *J. Br. Grassld Soc.* 16, 299-308.

Prine, G. M. and Burton, G. W. (1956). *Agron. J.* 48, 296-301.

Puckridge, D. W. and Ratkowsky, D. A. (1971). *Aust. J. agric. Res.* 22, 11-20.

Raschke, K. (1970). *Planta* 91, 336-363.

Rawson, H. M. and Hofstra, G. (1969). *Aust. J. biol. Sci.* 22, 321-331.

Rhodes, I. (1971). *J. Br. Grassld Soc.* 26, 9-15.

Robson, M. J. and Jewiss, O. R. (1968). *J. appl. Ecol.* 5, 179-190.

Ryle, G. J. A. (1968). *A. Rep. Grassld Res. Inst. Hurley,* pp. 125-134.

Ryle, G. J. A. and Hesketh, J. D. (1969). *Crop Sci.* 9, 451-454.

Saeki, T. (1959). *Bot. Mag., Tokyo* 72, 494-498.

Sagar, G. R. and Marshall, C. (1966). *Proc. IX Int. Grassld Congr.* 493-497.

Scott, D., Menalda, P. H. and Brougham, R. W. (1968). *N.Z. Jl. Bot.* 6, 427-449.

Slatyer, R. O. (1969). *In* "Physiological Aspects of Crop Yield" (J. D. Eastin, F. A. Haskins, C. Y. Sullivan and C. H. M. Van Bavel, eds), pp. 53-88. Publ. Amer. Soc. Agron. and Crop. Sci. Soc. Amer. Madison, Wis., U.S.A.

Slatyer, R. O. and Bierhuizen, J. F. (1964). *Aust. J. biol. Sci.* 17, 115-130.

Stoy, V. (1969). *In* "Physiological Aspects of Crop Yield" (J. D. Eastin, F. A. Haskins, C. Y. Sullivan and C. H. M. Van Bavel, eds), pp. 185-206. Publ. Amer. Soc. Agron. and Crop Sci. Soc. Amer. Madison, Wis., U.S.A.

Tayler, R. S. (1965). *Outl. Agric.* 4, 235-243.

Thorne, G. N. (1960). *Ann. Bot.* 24, 356-371.

Thorne, G. N., Ford, M. A. and Watson, D. J. (1967). *Ann. Bot.* 31, 71-101.

Tieszen, L. J. and Helgager, J. A. (1968). *Nature, Lond.* 219, 1066-1067.

Treharne, K. J. (1971). *In* "Crop Processes in Controlled Environments". (A. R. Rees, K. Cockshull, D. W. Hand and R. G. Hurd, eds), pp. 285-302. Academic Press, London and New York.

Treharne, K. J. and Cooper, J. P. (1969). *J. exp. Bot.* 20, 170-175.

Treharne, K. J. and Eagles, C. F. (1970). *Photosynthetica* 4, 107-117.

Treharne, K. J. and Stoddart, J. C. (1968). *Nature, Lond.* 220, 457-458.

Treharne, K. J., Cooper, J. P. and Tayler, T. H. (1968). *Crop Sci.* 8, 441-445.

Treharne, K. J., Pritchard, A. J. and Cooper, J. P. (1971). *J. exp. Bot.* 22, in press.

Troughton, J. H. (1969). *Aust. J. biol. Sci.* 22, 289-302.

Troughton, J. H. and Slatyer, R. O. (1969). *Aust. J. biol. Sci.* 22, 815-827.

Ulyatt, M. J. (1970). *Proc. XI Int. Grassld Congr.* pp. 709-713.

Verhagen, A. M. W. and Wilson, J. H. (1969). *Ann. Bot.* 33, 711-727.

Wardlaw, I. F. (1967). *Aust. J. biol. Sci.* 20, 25-39.

Wardlaw, I. F. (1968). *Bot. Rev.* 34, 79-105.

Wardlaw, I. F. (1969). *Aust. J. biol. Sci.* 22, 1-16.

Wareing, P. F., Khalifa, M. M. and Treharne, K. J. (1968). *Nature, Lond.* 220, 453-457.

Warren Wilson, J. (1959). *New Phytol.* 58, 92-101.

Watson, D. J. (1952). *Adv. Agron.* 4, 101-144.

West, S. H. (1970). *Proc. XI Int. Grassld Congr.* p. 514-517.

Wettstein, D. von (1967). *In* "Harvesting the Sun: Photosynthesis in Plant Life" (A. San Pietro, F. A. Greer and T. J. Army, eds), pp. 153-190. Academic Press, New York and London.

Williams, R. D. (1964). *Ann. Bot.* 28, 419-428.

Wilson, D. (1970). *Planta* 91, 274-278.

Wilson, D. (1971). *N.Z. Jl agric. Res.* 14, 761-771.

Wilson, D. (1972a). *New Phytol.* 71, 811-817.

Wilson, D. (1972b). *J. Exp. Bot.* 23, 517-524.

Wilson, D. and Bailey, R. W. (1971). *J. Sci. Fd Agric.* 22, 335-337.

Wilson, D. and Cooper, J. P. (1969a). *New Phytol.* 68, 645-655.

Wilson, D. and Cooper, J. P. (1969b). *New Phytol.* 68, 627-644.

Wilson, D. and Cooper, J. P. (1969c). *Heredity, Lond.* 24, 633-649.

Wilson, D. and Cooper, J. P. (1969d). *New Phytol.* 68, 1125-1135.

Wilson, D. and Cooper, J. P. (1969e). *Ann. Bot.* 33, 951-965.

Wilson, G. L. and Ludlow, M. M. (1970). *Proc. XI Int. Grassld Congr.* pp. 534-538.

de Wit, C. T. (1965). *Verslagen Landbouwkundige Onderz.* 663, 57p. Centre Landbouwpubl., Wageningen.

Woledge, J. and Jewiss, O. R. (1969). *Ann. Bot.* 33, 897-913.

Woods, D. B. and Turner, N. C. (1971). *New Phytol.* 70, 77-84.

Zelitch, I. (1965). *Biol. Rev.* 40, 463-482.

Zelitch, I. (1967). *In* "Harvesting the Sun: Photosynthesis in Plant Life" (A. San Pietro, F. A. Greer and T. J. Army, eds), pp. 231-248. Academic Press, New York and London.

Zelitch, I. (1969). *A. Rev. Pl. Physiol.* 20, 329-350.

CHAPTER 18

Mineral Absorption and its Relation to the Mineral Composition of Herbage

J. F. LONERAGAN

School of Biology,
Murdoch University,
Perth, Western Australia

I. INTRODUCTION

The mineral composition of herbage is determined by 2 sets of rate processes—those governing the rate of change in its mineral content and those governing the rate of change in the amount of herbage, i.e. its rate of growth. The rate of change in the mineral content of herbage depends primarily on the processes governing the absorption of minerals into roots, but it is also modified by transport of minerals to herbage and the re-export and leaching of minerals from herbage. This essay briefly outlines the nature of these

103

processes and the way they interact with each other and with growth in determining mineral composition of herbage.

II. ABSORPTION

Plants obtain mineral elements from soils by absorption from solution. The rate of mineral absorption from soils depends upon the physiological processes governing mineral absorption from solution, the physico-chemical and biological processes governing mineral supply from soil to solution at the root surface, and upon the interactions between plants and soils on these processes.

A. *Mineral Absorption from Solution*

Absorption of minerals may be defined as their entry into any cell, tissue, or organ by any mechanism. In the present context, the absorption process of prime interest is the entry of ions from the medium into the xylem sap of plant roots. It involves a whole series of events whose mechanisms and interrelationships are not understood: even the physical location of some events is unknown. Moreover, the mechanisms may vary with mineral species and may even vary from a single mineral species with changing mineral concentration. As a consequence there is often too little basic information to satisfactorily interpret most of the large volume of descriptive data which has been published on the relationships between various environmental factors and the absorption by plant roots of minerals from solution.

1. *Absorption into Cells*

Much of the basic information which is available has come from studies of ion absorption by large, single, algal cells and by slices of tissues containing aggregates of uniform cells. The findings from these studies will be reviewed briefly.

Plant cells have walls consisting of interweaving microfibrils of cellulose with relatively large spaces (100 Å) between them and impregnated with pectic substances, with proteins and occasionally also with waxy substances, the cutins and suberins. In multicellular tissues the cellulose of each cell remains discrete but the pectic materials, which coat all cells, coalesce at the junction of adjacent cells forming the "middle lamella" (Fig. 1). The pectic materials thus form a continuous system throughout the whole plant body while

the cellulose forms a series of discrete boxes. The cellulose microfibrils are composed of polyglucose units which are electrostatically
neutral. The pectins are gels of polygalacturonic acid residues
containing a variety of other uronic acids and saccharides and which
are stabilized in some way by calcium ions. They form a relatively
open structure but are strongly charged due to the dissociation of the
carboxylic acid residues whose average pK is about 3: at pH values
above 4 they have a charge concentration of the order 0·5 to 1
equivalent per litre.

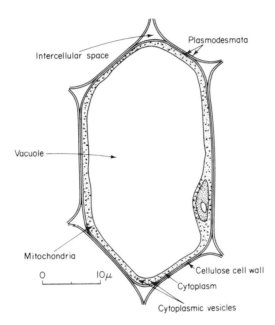

Fig. 1. A vacuolated cell in parenchymous plant tissue. Reproduced with permission from
Sutcliffe 1962.

Each living cell is lined with a thin layer of cytoplasm surrounding
a central vacuole. The cytoplasm moves continuously at an astonish-
ing pace, carrying mitochondria and other inclusions rapidly around
them. Between all cells, thin cytoplasmic strands, the plasmodesmata,
pass through adjacent walls and connect the cellular cytoplasms into
a continuous system, the symplasm. Whether cytoplasmic streaming
continues in the plasmodesmata between cells is a matter of dispute.
    At its outer surface, the symplasm forms a membrane, the
plasmolemma, which can be clearly recognized in electron micrographs and which is believed to be composed of a mixture of lipids

and proteins. Another membrane, the tonoplast, forms at the junction of cytoplasm and vacuole.

Earlier workers thought of biological membranes as consisting of proteins and lipids arranged in continuous layers, like bread and meat in a sandwich, but more recent workers picture membranes as an assemblage of sub-units, possibly cubes, in each of which proteins and lipids occur in some pattern (Briggs *et al.*, 1961; Esau, 1965; Clowes and Juniper, 1968; Rees, 1969; Kramer, 1969; Cook, 1971).

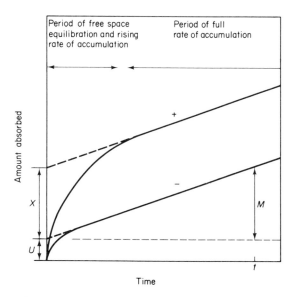

Fig. 2. Diagrammatic representation of absorption of anions and cations, plotted against time. U = amount of ion pairs in the free space, X = amount of cation exchange, M = amount absorbed within cells in time t. Reproduced with permission from Briggs *et al.*, 1961.

When transferred from water to mineral salt solutions, single cells or thin slices of plant tissue generally absorb cations and anions rapidly. Two distinct phases are evident—an initial, rapid phase which reaches equilibrium within 20 or 30 minutes and a slower phase which continues at a constant rate for a longer period (Fig. 2). The first phase is insensitive to temperature, to anaerobiosis, and to metabolic poisons. It is also rapidly reversed either by washing with water or by salt solutions. It is thought to represent movement into spaces within and between the cell walls (the water free space) and adsorption on to negative and positive charges within the cell wall and on the cytoplasmic surface (the Donnan free space). The first phase of cation absorption from single salt solutions is usually very large as the result of adsorption by the negatively charged groups of the pectic substances. Anions may also be adsorbed but to a much smaller extent: positively charged groups on proteins are probably responsible for anion adsorption.

The second, slower phase of mineral absorption by plant cells is sensitive to temperature, to anaerobiosis, and to metabolic poisons. This phase cannot be easily washed out and is believed to represent the absorption of mineral elements into living cells by metabolic processes. Under some conditions the rate of metabolic absorption of minerals into the cell proper is affected by the rate of movement of minerals through the spaces in the cell walls. In this event, interaction may be encountered among the absorbing mineral, other charged ions, and the charged surfaces of the cell walls. But there is no evidence to suggest that there is any functional relationship between metabolic absorption into the cell and adsorption phenomena in cell walls. It is therefore important to delineate movement of substances into plant cells by absorption into cytoplasm from their movement into free space. It is especially important in short-term experiments where free space phenomena may make relatively large contributions to the total ion movement into tissues (Fig. 2).

For absorption to proceed from the medium into the cytoplasm or vacuole, influx of minerals across the plasmolemma must exceed efflux. Efflux from cells into the medium does occur but is usually only important after the cells have been exposed to very high concentrations from ions. Influx sometimes proceeds along and sometimes against a concentration gradient of mineral ions. Positive concentration gradients arise from removal of the absorbed mineral ion from the cytoplasm by incorporation into various compounds required by the growing cell or by electrostatic adsorption on to various cell constituents. But influx can also proceed against a concentration gradient resulting in accumulation of mineral species in the vacuole to concentrations higher than those present in the surrounding medium. In such accumulation, cell membranes must play a key role.

The nature of the cytoplasmic membranes and of the basic mechanisms which establish concentration differences across them is a subject of lively research and debate as already indicated. Passage of minerals across membranes has been suggested to proceed in a variety of ways—either through pores, by reaction with groups in the membrane, or by pinocytosis of the membrane. Much of the early work was concerned with the permeability characteristics of membranes but there is currently greater emphasis on their metabolic properties and the presence within them of enzymic proteins. Whatever the mechanisms, the membrane must be efficient at discriminating between unlike minerals and in keeping like minerals separated except at the point of transport or of exchange diffusion.

Differences in concentrations of mineral ions across membranes can result from either passive or active movement of the ions. Ions can accumulate in response to electrochemical potentials arising from differential permeability of the membrane or of active absorption of other ions. Ions can also accumulate from their active movement across membranes against electrochemical potentials.

Active accumulation of mineral ions involves expenditure of metabolic energy but the link between energy and movement is not clear. Energy supply for accumulation of anions has been related to electron transfer processes while that for accumulation of cations has been related to mechanisms involving adenosine-tri-phosphate (ATP) in reactions with specific, cation-activated, ATPase enzymes within the membrane. Alternative proposals envisage that electron transport through the membrane is linked to a separation of charge which establishes a H^+ ion gradient across the membrane. The resulting electrochemical potential provides the driving force for the movement of an anion or a cation which counter-balances the H^+ formation. In such a scheme the movement of ions could still involve exchange with anions or cations bound to specific sites within the membrane. Several proteins with capacities to bind specific ions have now been identified in membranes from animal, microbial, and plant cells (Robertson, 1958, 1968; Briggs *et al.*, 1961; Sutcliffe, 1962; Dainty, 1969; Macrobbie, 1970, 1971; Packer *et al.*, 1970).

The mechanisms by which mineral ions move in the plant cytoplasm are also not clear. At first sight protoplasmic streaming would seem to provide a suitable vehicle for rapid movement of minerals. Yet the rapid movement of minerals over long distances through the symplasm proceeds uninterrupted even when protoplasmic streaming stops (Helder, 1967; Arisz, 1969).

2. *Absorption into Roots*

From the viewpoint of absorption for translocation to plant herbage, 3 main regions may be recognized in the root—a region at the root tip where xylem elements have not differentiated, an intermediate region where xylem elements are mature and are not enclosed by impermeable outer layers, and a region towards the base of the root where xylem elements are mature but where the cells of the outer layers of the root have become impermeable (Fig. 3).

The region at the root tip which has no mature xylem elements comprises the root cap, the meristematic cells, and the region of cell elongation. The cells, especially those of the meristem, have an unusually high capacity to absorb ions. They pass few of the absorbed ions to the xylem since they have a very high growth rate

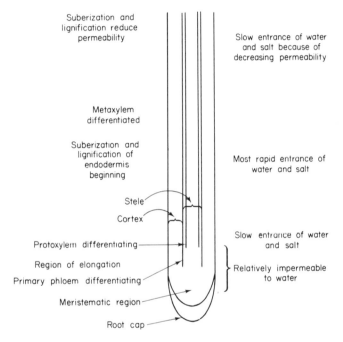

Fig. 3. Diagram of a young root, showing relation between anatomy and absorbing regions for water and salt. Reproduced with permission from Kramer, 1969.

and since they can only transport ions to the xylem over relatively long distances in the symplasm.

Towards the base of the root, cells in the endodermis, pericycle, or the cortex, may become heavily thickened and suberized (see Chapter 5, Vol. 1); in advanced stages, they form a thick periderm of suberized cells which appear impermeable to water and minerals.

Most of the ions which reach plant herbage are absorbed and transported to the xylem in the region of the root lying between its heavily suberized base and its immature tip. In this region a central stele is clearly delineated (Figs 3, 4). It contains fully differentiated, dead xylem cells which provide a direct and uninterrupted connection with xylem vessels of the herbage. The xylem cells are surrounded by densely-packed but thin-walled parenchyma whose outer layer constitutes a continuous ring—the pericycle. The stele is surrounded by a cortex consisting of a continuous cylinder of cells, the endodermis, and several layers of loosely arranged, thin-walled parenchyma. The cells of the endodermis have a characteristic band of material, the casparian strip, running around their anticlinal walls (Fig. 5). The casparian strip is initiated in the middle lamella and impregnates and encrusts the primary wall of the cell. It is thought to consist of lignin or suberin or both, thus forming a chemical

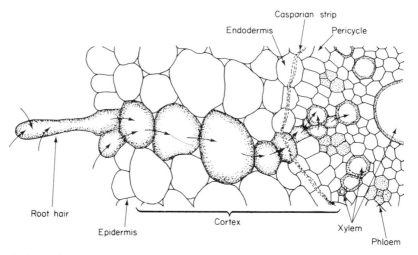

Fig. 4. Part of transection of wheat root in region of mineral absorption and transport to xylem. Reproduced with permission from Esau, 1965.

discontinuity within the continuous pectic framework of the root and within the capillary spaces of the cell walls. The casparian strip also attaches firmly to itself the cytoplasm of the endodermal cells (Fig. 5).

The endodermis thus appears to present a cylindrical barrier which, except at the root tip and at occasional passage cells, compels water and dissolved minerals to pass through plant cytoplasm before reaching the xylem. Outside the cortex an epidermis forms a single compact layer of cells. Many of the epidermal cells develop long, fine root hairs which protrude for several mm into the medium. A thin skin of some mineral, possibly pectin, covers the outside of the hairs and the epidermal cells.

Concentrations of free ions in the xylem vessels frequently exceed concentrations of the external medium manyfold. For example, Russell and Shorrocks (see Russell and Barber, 1960) reported 70 and 150 fold increases in the concentrations of rubidium and phosphate ions. Clearly a permeability barrier must separate xylem sap from external solution. Since the dead cells of the xylem cannot provide such a barrier it must be located in surrounding tissues. Some evidence suggests that the epidermis forms a barrier to the entry of some minerals which can only pass through it following absorption into the cytoplasm of the root hairs or of the epidermal cells.

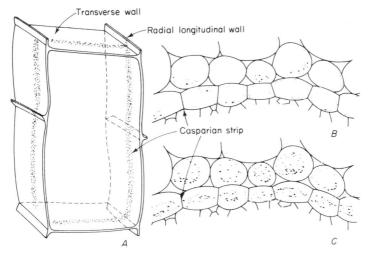

Fig. 5. Endodermal cells. *A*, entire cell showing location of casparian strip. *B, C,* effect of treatment with alcohol on cells of endodermis and of ordinary parenchyma. *B*, cells before treatment, *C*, after. Casparian strip is seen only in sectional views in *B, C.* Reproduced with permission from Esau, 1965.

However, most mineral ions appear to move freely through the cell walls and intercellular spaces of the epidermis and of the cortex stopping at the endodermis. Here the casparian strips are believed to form an impermeable barrier forcing water and all minerals which reach the xylem to pass through the cytoplasm. If this is so, the plant cytoplasm could absorb minerals from the root free space at any point from the root hair to the endodermis. Where absorption actually occurred would depend largely upon the rate at which a mineral moved inwards in the free space in relation to the rate at which the outermost cells depleted it from solution.

Wherever it may be located, the permeability barrier probably achieves its control of mineral absorption by forcing minerals to pass through cytoplasm. Absorption of minerals into the xylem would thus involve three essential steps—movement of minerals across an outer plasmalemma into the cytoplasm, movement through a series of cells, probably through the symplasm, and movement across an inner plasmalemma into the xylem vessels (Briggs *et al.*, 1961; Brouwer, 1965; Fried and Broeshart, 1967).

Several authors have suggested that this movement of ions across the root into the xylem may be analogous to the movement of ions across the cytoplasm into cell vacuoles. Movement of minerals across the outer plasmalemma and through the cytoplasm is likely to be similar in both systems. The mechanisms involved have already been discussed. However, movement across the tonoplast into the vacuole could be very different from movement across the inner plasmalemma into the xylem. Indeed recent research indicates that this is so (Dunlop and Bowling, 1971).

Any theory of mineral accumulation in the xylem sap must accord properties or activities to the plasmalemma bounding the medium different from those of the plasmalemma bounding the xylem. An early theory of mineral accumulation in xylem sap suggested that a lower oxygen tension at the xylem rendered the plasmalemma there more permeable than that bounding the medium (Crafts and Broyer, 1938). This latter hypothesis now appears untenable in the light of the demonstration that the electrochemical potential for potassium ions does not change across the root from epidermal cells to the innermost cells of the pericycle (Dunlop and Bowling, 1971). The behaviour of phosphate may provide a more relevant model. Loughman (1969) has clearly demonstrated that inorganic phosphate is rapidly phosphorylated to sugar compounds during or shortly after passage across the plasmalemma bounding the medium and is released from these compounds at the plasmalemma bounding the xylem sap. But the reactions involve different enzyme systems at each boundary—phosphorylation near the medium and dephosphorylation near the xylem sap. As a result, environmental factors and metabolic inhibitors may have differential effects on the absorption of phosphate into the cytoplasm and its secretion into the xylem sap. The behaviour of phosphate is often regarded as atypical because of its rapid metabolism into organic compounds. It may yet prove to be a useful model for the behaviour of other minerals which form less stable and hence less easily defined organic compounds. Such carrier molecules or "ionophores" have long been postulated.

While the movement of most minerals into xylem sap fits a theory involving absorption into and movements through the symplasm, calcium and strontium ions seem to behave in an anomalous way. When roots are immersed in solution of calcium salts very little calcium moves into the cytoplasm and large amounts appear quickly in the xylem sap. Moreover, in a region where the pericycle had developed tertiary thickening and suberization, barley roots absorbed and transported phosphate and rubidium ions rapidly to the xylem provided that the pericycle still retained symplastic connections to the endodermis. In the same region, barley roots did not transport any strontium to the xylem (Clarkson and Sanderson, 1971). It therefore seems possible that the alkaline earth cations may reach the

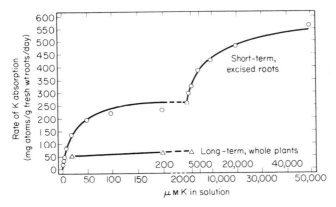

Fig. 6. Effect of potassium concentration in solution on the rates of potassium absorption in (a) short-term study with excised roots of barley (data of Epstein et al., 1963) and (b) long-term study with whole plants of barley. Reproduced with permission from Johansen et al., 1968.

xylem sap by a different route from that used by minerals which have been studied in more detail.

Absorption of many minerals by whole roots behaves as if different mechanisms operate at different mineral concentrations (Epstein, 1966; Laties, 1969). For example, with increasing concentrations of potassium ions up to 0·1 mM, potassium absorption increases asymptotically, reaching a maximal value which remains unchanged to 1 mM. With increasing concentrations above 1 mM, potassium absorption again increases asymptotically (Fig. 6). The low concentration mechanism of potassium absorption (frequently referred to as System 1) is characterized by its high affinity for potassium and its insensitivity to sodium ions and to the nature of the anions in solution. By contrast, the low affinity mechanism or mechanisms for potassium absorption (System 2) is inhibited by sodium and calcium ions and is strongly affected by the nature of the anion, being enhanced by chloride and depressed by sulphate. The characteristics of System 1 are compatible with the operation of a highly selective potassium absorbing system at the plasmolemma while those of System 2 may represent passive accumulation in vacuoles following accumulation there of anions.

Short-term studies on the absorption of chloride and phosphate support such an interpretation. Thus the isotherms of chloride and phosphate absorption in whole roots are very similar to those of potassium absorption. However, isotherms of chloride absorption into non-vacuolated root cells indicate that only System 1 operates whereas vacuolated cells show characteristics of both systems. In keeping with this finding the characteristics of chloride and phosphate absorption into the xylem are those of System 1 and increasing concentrations of chloride or phosphate ions above those which saturate System 1 results in linear rather than hyperbolic increases in chloride and phosphate absorption and transport to tops. Laties (1969) has interpreted these results to indicate that at high mineral concentrations, diffusive movement of minerals through the plasmolemma and into the xylem governs mineral absorption into the xylem. Epstein (1966) disputes this interpretation but both authors agree that, at concentrations of minerals where only System 1 operates, absorption into the xylem is governed by active mechanisms located at the outer plasmolemma.

The high selectivity of System 1 is compatible with the known high selectivity of plants for accumulation of minerals in their herbage. However plant roots are not completely selective and cannot discriminate between closely related ions. For example, roots do not distinguish between potassium and rubidium ions. Nor do they discriminate in absorption between chloride and bromide, between calcium and strontium, or between copper and zinc ions. Each ion of these pairs mutually and competitively inhibits the absorption of the other. Many examples of non-competitive inhibition of ions on the absorption of minerals are also known (Epstein, 1966).

Changes in mineral form have large effects on ion absorption as evidenced by effects of chelating agents and of pH on ion absorption. For example chelating agents markedly depress zinc absorption from solutions (Guinn and Joham, 1962). Increasing pH values above 6 also markedly depresses phosphate absorption probably by decreasing the concentration of univalent dihydrogen phosphate ions

(Hagen and Hopkins, 1955). However, pH also has large effects on the ionization of many cellular constituents and its action is not limited to effects on the nature of the mineral ion: for example, decreasing pH strongly inhibits absorption of many cations including potassium, calcium, manganese, and zinc.

Many ions have synergistic effects on the absorption of other minerals. The effect of chloride ions in enhancing potassium absorption from high concentrations of potassium has already been mentioned. Calcium also enhances the absorption of many ions. Omission of calcium ions from solution may seriously inhibit absorption of both cations and anions possibly as a result of membrane breakdown. But even at concentrations adequate for plant growth additional calcium may enhance the absorption of anions including phosphate: thus increasing calcium from 0·25 to 2·5 mM trebled phosphate absorption by *Medicago truncatula* (Robson *et al.*, 1970). The effect was thought to result from the action of calcium ions on electronegative charges near the sites of phosphate absorption.

Several mechanisms in the process of mineral absorption into the xylem sap are metabolically controlled and hence sensitive to the gaseous and temperature environment. Most work fails to differentiate accumulation in the xylem from incorporation into cytoplasm but the results suggest that mineral accumulation ceases under anaerobic conditions and responds to increasing oxygen tensions to about 10%. Temperature coefficients vary from 2 at temperatures above 10°C to 6 or more at lower temperatures. Climatic factors influencing transpiration sometimes also influence both the rate of movement of minerals to the root surface and the rate of mineral accumulation in the xylem as discussed in the next sections (Sutcliffe, 1962).

In conclusion it should be emphasized that plant species and cultivars differ greatly in their mineral absorption (Epstein and Jefferies, 1964). Collander has provided an excellent and much quoted example in his comparison of sodium absorption with potassium and rubidium absorption of a wide range of plant species (see Sutcliffe, 1962).

3. Problems of Relevance

Studies of the mechanisms of ion absorption have, of necessity, involved short-term experiments with selected tissues prepared under artificial conditions. When the findings from such studies have been applied to problems of mineral absorption by plants growing for long periods they have sometimes seemed completely irrelevant. Recent studies have shown that many of these problems arise from failure to appreciate the limitations of the tissues, their pre-treatment, and their treatment on the applicability of the findings. When sufficient congnisance of all factors is taken, short-term experiments can sometimes provide valuable help in the resolution of problems of long-term mineral absorption.

Difficulties have sometimes arisen when mineral absorption by the root has been assumed to indicate accumulation in the xylem sap. But, as we have seen, mineral absorption is only one of the processes involved in the absorption of minerals by roots. It competes for minerals with adsorption on cell walls (Fig. 2), with incorporation into constituents of the cytoplasm, and with absorption into cell vacuoles as has been clearly shown in exuding onion roots. In onion roots, previous treatment with chloride had no effect on the rate of chloride absorption by roots but resulted in most of the chloride going directly to the xylem exudate, whereas in low-chloride roots it accumulated in the vacuoles of cortical cells (Hodges and Vaadia, 1964).

Accumulation of potassium in vacuoles of root cells may also partly explain the huge discrepancy between rates of potassium absorption by low-salt roots and by whole plants (Fig. 6). But other factors also contribute since, unlike chloride in onion roots, potassium absorption by low-salt wheat roots rapidly slows down as their potassium content increases (Fig. 7). As a result, plants grown with a continuous supply of potassium showed little effect of increasing potassium concentrations in enhancing potassium absorption which so characterized potassium absorption by low-salt barley roots (Fig. 6). Present evidence indicates that the presence of potassium depressed the operation of both Systems 1 and 2 of potassium absorption (Johansen, 1971). At the same time the characteristics of System 2 change with potassium status of roots. Pitman et al. (1968) have shown that the low selectivity of system 2 for potassium changes quickly to a high selectivity as roots accumulate potassium or sodium.

Like potassium, the absorption of nitrogen and of phosphorus are also strongly influenced by pre-treatment (Humphries, 1951). Pre-starved tissues show strongly enhanced absorption of these minerals probably as a result of their accelerated incorporation into cellular compounds. With all these minerals, experimental results from pre-starved tissues will have only restricted relevance to long-term absorption by pasture plants.

The use of excised roots for measurement of absorption has also sometimes raised serious problems. For example, calcium passes rapidly into the xylem sap following absorption and accumulation of calcium in the root. Measurement of calcium absorption by excised roots thus gives a completely false picture of absorption into the xylem (Moore *et al.*, 1965). On the other hand direct measurement of calcium contents in the xylem sap from excised roots could prove

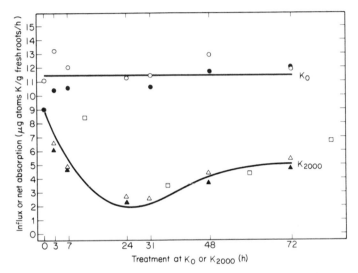

Fig. 7. Change with time of ^{86}Rb influx into intact ($\circ$,$\triangle$) and excised ($\bullet$,$\blacktriangle$) roots of low-salt barley seedlings after transfer to solutions with (K_{2000}) and without (K_0) potassium. Net potassium absorption rates ($\square$) were also estimated for low-salt, intact barley seedling transferred to K_{2000}. Reproduced with permission from Johansen *et al.*, 1970.

useful. Microbial activity may also affect experimental results producing serious anomolies in short term experiments when microbial absorption competes with the plant for a limited supply of minerals (Barber and Loughman, 1967; Barber, 1968).

Despite these difficulties, short term experiments of mineral absorption have been successfully related to long term mineral accretion in plants. Where this has been done it has proved essential to recognize the importance of pre-treatment and of experimental conditions to the particular minerals and plants under study. In the case of potassium, pre-treatment conditions with potassium were paramount as already discussed (Fig. 7). In the case of zinc, pre-treatment conditions had little effect but the temperature and composition of the absorbing solution were extremely important (Chaudhry and Loneragan, 1972). In both cases, and in contrast to calcium, mineral absorption by excised roots closely followed mineral absorption by whole plants. When the appropriate conditions were adjusted, short-term absorption of both of these minerals could be related to long-term absorption by the whole plant from culture solutions.

B. *Mineral Absorption from Soils*

The composition of the soil solution at the absorbing surface of the root can be expected to have a dominating influence on mineral absorption. In addition, soil organisms influence mineral absorption by modifying the structure and absorbing properties of the root system.

1. *The Root-Soil Interface*

In soils, the surface of young roots appear to have a coating of mucilaginous material, possibly pectin, extending for some distance into the soil (Scott *et al.,* 1958; Jenny and Grossenbacher, 1963). This and other plant secretions modify the environment at and near the root surface. As a consequence, the bacterial and fungal populations in this "rhizosphere" region usually differ greatly from those characteristic of the bulk soil. Many plants also form intimate associations with fungi which not only modify conditions at the soil-root interface but may also modify the form of roots and may act as if they are an intimate physiological part of the root absorbing surface (Barber, 1968; Bowen and Rovira, 1969). The modified root together with its fungus is known as a "mycorrhiza". The form and nutrient absorbing activities of ectotrophic mycorrhiza of forest trees has been known and studied for some time (Harley, 1969). Less well known are the form and function of the endotrophic mycorrhiza which occur widely on crop and pasture plants (Mosse, 1963).

The fungal hyphae of the ectotrophic mycorrhiza form a thick layer on the unsuberized root surface of forest trees and send hyphae outwards into the soil and inwards into the intercellular spaces but not the cells of the root cortex. Most of the minerals absorbed by the plant pass first through the fungus. The ectotrophic mycorrhiza increase the area of the absorbing surface of the root system by increasing its diameter, promoting its branching, prolonging the life of the short roots, and by providing extensions of fungal hyphae into the soil.

The fungal hyphae of the endotrophic mycorrhiza ramify through the soil and the root cortex where they enter cells and form outgrowths of various types. They do not form a fungal sheath outside the root and do not usually cause characteristic changes in root morphology. Recent evidence (Hayman and Mosse, 1971; Sanders and Tinker, 1971) suggests that the fungal hyphae may be highly effective in promoting mineral absorption by roots. They appear to do so by acting as an extension of the root absorbing surface.

2. Mineral Concentrations at the Root-Soil Interface

The mineral composition of the bulk soil solution has been studied balance between the rates of arrival of minerals at the absorbing surface and the rates at which they are absorbed by the plant.

The mineral composition of the bulk soil solution has been studied for some time but the composition of the solution at the root-soil interface is largely unknown. However, the processes governing movement of minerals to the root surface are subjects of much current research activity. Two processes are involved—movement by convective flow in soil water to the root and diffusion in response to concentration gradients. Both are strongly influenced by soil properties (Olson and Kemper, 1968; Nye, 1969).

Barber (1962) has estimated the likely contribution of convective flow to movement of minerals to roots from the concentrations of minerals in the soil solution and the transpirational loss of water from the plant. On this basis, convective flow can account for all or most of the calcium, magnesium, sulphur, and boron found in plants growing on many soils. It can account for relatively little of the potassium, phosphorus, molybdenum, iron, manganese, and zinc; these minerals are thought to move to the root primarily by diffusion in response to a concentration gradient arising from their depletion from solution at the root surface. Calculations indicate that rates of diffusion of potassium and phosphate in soils can account for the amounts of these minerals absorbed by plants (Nye, 1969).

Whatever the process of movement to the root, the solution concentration of many minerals in soils is very low in relation to the amounts absorbed by plants: large quantities of these minerals must be released from the solid phase during plant growth. Release of nitrogen, sulphur, and phosphorus by decomposition of organic matter, of potassium from primary minerals, and of phosphate and of potassium by desorption from soil colloids have long been recognized as important processes in mineral availability to plants. Those minerals which react strongly with soils can move only slowly through the soil. It is possible that the plant also assists release and movement of some minerals by secreting hydrogen ions, carbon dioxide, or substances such as citric acid. Appreciable quantities of hydrogen ions must be released in exchange for cations associated with the highly ionized uronic acid residues of pectic substances being synthesized near root tips. This production of hydrogen ions may explain the association frequently encountered between concentrations of cations in plants and their cation exchange capacity (Butler *et al.*, 1962; Drake, 1964). Further from the root tip, in the region of mineral absorption for translocation to herbage, hydrogen ions may also be excreted as the result of absorption of cations exceeding anions. More usually in soils, where nitrate rather than ammonium ions provide the plant with mineral nitrogen, absorption of anions exceeds cations resulting in secretion of bicarbonate ions and an increase in pH at the root surface in this region. Substances such as citric acid could act directly by desorbing or chelating minerals, or indirectly by promoting microbial activity of a similar nature. Some such activity may be important in the desorption of phosphate and zinc, in increasing the mobility of copper, iron, and zinc, and in the reduction of manganese oxides prior to the absorption of these minerals from soils. Enzymes on or secreted from root surfaces may also play an important role in modifying soil minerals prior to their absorption. Thus the recent demonstration that iron is absorbed by plants in the divalent form seems to establish that reductive activity at the root surface is an essential prerequisite for iron absorption from soils (Chaney *et al.*, 1971).

These examples serve to illustrate the importance not only of soil in modifying mineral supply to root but also of soil-root interactions. There can be no doubt too that, as Jenny has long maintained (see Jenny and Grossenbacher, 1963), contact between soil and root—its proximity, its extent, and its distribution—must also be important in the absorption of those minerals which can only move slowly through the soil.

III. MOVEMENT TO AND FROM LEAVES

Minerals absorbed into the xylem sap pass to the leaves in the transpiration stream. During passage to the leaves they may move rapidly into and out of cells of the stem. Cations are also adsorbed readily on to exchange sites, presumably pectin, within the xylem. Exchange adsorption can create difficulties in interpretation of short-term experiments with radioisotopes, but it does not interfere significantly with net movement of minerals from root to herbage.

Transpiration has variable effects on mineral translocation to plant tops. At low mineral supply, transpiration has little if any effect: apparently the rate of mineral absorption rather than transport limits mineral accumulation in herbage under these conditions. However, at high mineral supply, high transpiration enhances mineral transport to herbage. Under these conditions high transpiration rates may accelerate absorption by sweeping away minerals accumulating in the xylem sap, thus decreasing the rate of the back reaction from xylem into root symplasm (Russell and Barber, 1960; Brouwer, 1965; Weatherley, 1969). Recent experiments with potassium absorption suggest an alternative explanation for the effects of transpiration. Two pathways of absorption are proposed. The first which operates only at low concentrations of potassium ($<$ 1 mM) is not affected by water movement and possibly involves symplastic movement. The second which operates over a wide range of potassium concentrations follows water movement, moving along and within cell walls in the cortex and stele except at the endodermis where it must pass through the cytoplasm (Minchin and Baker, 1970).

Within the leaves, minerals are believed to move from the xylem sap into the leaf cells by the same mechanisms which govern their absorption by root cells. The characteristics of mineral absorption by leaf discs floating in solutions seem to be identical with those by excised roots (Epstein, 1966).

Minerals may move out of leaves by transport in the phloem or by leaching with rain, fog, or dew. Phosphorus, potassium, and nitrogen move readily in the phloem to roots and other organs. By contrast, calcium, strontium, boron, and iron do not move at all from leaves to other organs unless subjected to unusual experimental conditions. Other minerals behave in an intermediate manner, some like zinc tending towards the extreme immobility of calcium and others like sulphur tending towards the high mobility of nitrogen. The high mobility of nitrogen and phosphorus is associated with their incorporation into organic compounds in the phloem: in the case of phosphorus, phosphorylation of sugars is probably an important mechanism governing the movement of sugars into and out of phloem tissues. The reasons for the failure of elements to move out of leaves are not known—they could arise from strong retention by compounds within leaf cells, as appears to be the case for sulphur in protein compounds, or from failure to cross cytoplasmic boundaries into the phloem, as may be the case for calcium (Biddulph, 1959; Milthorpe and Moorby, 1969).

Phloem-mobile minerals generally occur at high concentrations in young, actively growing and metabolizing organs and at lower concentrations in old leaves. By contrast, the phloem-immobile minerals are generally present at low concentrations in roots, in stems, and in young organs and accumulate to high concentrations in old organs. Exceptions to these patterns develop when minerals are supplied at excessively high levels; in the case of some elements this leads to very high, sometimes toxic, concentrations in old leaves (as for example with phosphorus) while in the case of the minerals of the first transition metal series it may lead to excessively high concentrations in roots. Leaching of minerals from old leaves may also modify the pattern of accumulation.

Minerals which have accumulated in plant organs to concentrations in excess of their metabolic requirements appear to leach easily from leaves and stems in rain, fog, or dew. Consequently minerals leach rapidly from plants given an excessively high supply and from organs at maturity. The nature of the mineral seems less important since leaching can remove large amounts of both phloem-mobile and phloem-immobile elements. For example, leaching has removed up to 80% of both potassium and boron from leaves and more than 50% of calcium. Much of the loss may occur in guttation fluid which is subsequently washed from leaves. However, minerals may also be washed out of the leaves themselves (Tukey, 1970).

The mechanism of loss is not clear. Possibly most of the minerals lost from leaves are present in extracellular spaces having been transported to leaves in greater quantities than the leaf cells can absorb. This appears to be the case for boron, which has received most critical study (Oertli, 1968). Under conditions of high transpiration and high boron supply, leaves may accumulate very high concentrations of boron of which more than 80% can be lost in guttation fluid or by flushing water through the leaf xylem. By contrast, leaves which have not accumulated high concentrations of boron lose little or no boron by leaching. The behaviour of boron may provide a useful guide to the behaviour of other minerals.

Some minerals also leach from leaves which do not contain excessively high concentrations. In these cases the losses of minerals are much smaller but may still be appreciable. Thus leaves of rye grass containing around 1% potassium on a dry weight basis lost 2% of their potassium content in a single day of light, simulated rain; this rate of loss approximated the rate of potassium absorption. Rye grass lost appreciably more potassium than other species leading to the suggestion that leaching might result in large effects of climate on the comparative potassium nutrition of pasture species (Clement *et al.*, 1971). Much more critical work needs to be done before the effect of leaching on mineral content of herbage can be satisfactorily assessed.

IV. GROWTH AND ITS EFFECTS ON MINERAL COMPOSITION

The processes already discussed determine the content of minerals in herbage. Their effects on the concentrations of minerals in plants depend upon the rates at which they change mineral contents relative to the rate of change in the amount of plant matter, i.e. growth (Loneragan, 1968).

The effects of growth in diluting minerals in plants are well known in studies of nutrient responses in crops but they are sometimes overlooked in comparisons of the mineral composition of plant species and cultivars. The effects of leaching have also been neglected and there has been a tendency to look upon mineral absorption as the sole determinant of mineral composition.

Absorption of minerals is clearly an important process in all situations and in many it is dominant. However, it is clear that leaching may be extremely important in determining the concentrations of potassium and sodium and possibly of some other minerals of importance in animal nutrition. It also seems that under some conditions, as for example in tropical climates, growth is so fast that many plants do not maintain minerals at satisfactory levels for animal growth. All these possibilities should be considered in any programme designed to improve by selection, by breeding, or by manipulation the mineral status of herbage for animal nutrition.

REFERENCES

Arisz, W. H. (1969). *Acta Bot. Neerl.* **18**, 14-38.
Barber, D. A. (1968). *A. Rev. Pl. Physiol.* **19**, 71-88.
Barber, D. A. and Loughman, B. C. (1967). *J. exp. Bot.* **18**, 170-176.
Barber, S. A. (1962). *Soil Sci.* **93**, 39-49.
Biddulph, O. (1959). *In* "Plant Physiology" (F. C. Steward, ed.) Vol. 2, pp. 553-603. Academic Press, New York and London.

Bowen, G. D. and Rovira, A. D. (1969). *In* "Root Growth" (W. J. Whittington, ed.), pp. 170-198. Butterworths, London.

Briggs, G. E., Hope, A. B. and Robertson, R. N. (1961). "Electrolytes and Plant Cells". Blackwell, Oxford.

Brouwer, R. (1965). *A. Rev. Pl. Physiol.* 16, 241-266.

Butler, G. W., Barclay, P. C. and Glenday, A. C. (1962). *Pl. Soil* 16, 214-228.

Chaney, R. L., Brown, J. C. and Tiffin, L. O. (1972). *Pl. Physiol., Lancaster,* 47, 50, 208-213.

Chaudhry, F. M. and Loneragan, J. F. (1972). *Proc. Soil Sci. Soc. Am.* 36, 323-327.

Clarkson, D. T. and Sanderson, J. (1971). *A.R.C. Ann. Rep. Letcombe Lab.* 1970: 16-25.

Clement, C. R., Jones, L. H. P. and Hopper, M. J. (1972). *J. appl. Ecol.* 9, 249-260.

Clowes, F. A. L. and Juniper, B. E. (1968). "Plant Cells". Blackwell, Oxford.

Cook, G. M. W. (1971). *A. Rev. Pl. Physiol.* 22, 97-120.

Crafts, A. S. and Broyer, T. C. (1938). *Am. J. Bot.* 25, 529-535.

Dainty, J. (1969). *In* "Physiology of Plant Growth and Development" (M. B. Wilkins, ed.), pp. 455-485. McGraw Hill, London.

Drake, M. (1964). *In* "Chemistry of the Soil" (F. E. Bear and M. Drake, eds), pp. 395-444. Am. Chem. Soc. Monogr. No. 160.

Dunlop, J. and Bowling, D. J. F. (1971). *J. exp. Bot.* 22, 453-464.

Epstein, E. (1966). *Nature, Lond.* 212, 1324-1327.

Epstein, E. and Jefferies, R. L. (1964). *A. Rev. Pl. Physiol.* 15, 169-184.

Epstein, E., Rains, D. W. and Elzam. O. E. (1963). *Proc. natn. Acad. Sci. U.S.A.* 49, 684-692.

Esau, K. (1965). "Plant Anatomy". 2nd Ed. John Wiley & Sons, New York.

Fried, M. and Broeshart, H. (1967). "The Soil-Plant System in Relation to Inorganic Nutrition". Academic Press, London and New York.

Guinn, G. and Joham, E. (1962). *Soil Sci.* 94, 220-223.

Hagen, C. E. and Hopkins, H. T. (1955). *Pl. Physiol., Lancaster* 30, 193-199.

Harley, J. L. (1969). "The Biology of Mycorrhiza". 2nd Ed. Leonard Hill Ltd., London.

Hayman, D. S. and Mosse, B. (1971). *New Phytol.* 70, 19-27.

Helder, R. J. (1967). *In* "Handb. Pflanzenphysiol" (W. Ruhland ed.), Vol. 13, pp. 30-43. Springer, Berlin.

Hodges, T. K. and Vaadia, Y. (1964). *Pl. Physiol., Lancaster* 39, 109-114.

Humphries, E. C. (1951). *J. exp. Bot.* 2, 344-379.

Jenny, H. and Grossenbacher, K. (1963). *Proc. Soil Sci. Soc. Am.* 27, 273-277.

Johansen, C. (1971). "Mechanisms of Potassium Absorption in high-salt Plants". Ph.D. Thesis, Univ. W. Austral.

Johansen, C., Edwards, D. G. and Loneragan, J. F. (1968). *Pl. Physiol., Lancaster* 43, 1722-1726.

Johansen, C., Edwards, D. G. and Loneragan, J. F. (1970). *Pl. Physiol., Lancaster* 45, 601-603.

Kramer, P. J. (1969). Plant and Soil Water Relationships: a modern Synthesis". McGraw-Hill, New York.

Laties, G. G. (1969). *A. Rev. Pl. Physiol.* 20, 89-116.

Loneragan, J. F. (1968). *9th Int. Congr. Soil Sci. Trans.* Vol. 2, 173-182.

Loughman, B. C. (1969). *In* "Ecological Aspects of the Mineral Nutrition of Plants" (I. H. Rorison, ed.), pp. 309-322. Blackwell, Oxford.

Macrobbie, E. A. C. (1970). *Quart. Rev. Biophys.* 3, 251-294.

Macrobbie, E. A. C. (1971). *A. Rev. Pl. Physiol.* 22, 75-96.

Milthorpe, F. L. and Moorby, J. (1969). *A. Rev. Pl. Physiol.* 20, 117-138.
Minchin, F. R. and Baker, D. A. (1970). *Planta* 94, 16-26.
Moore, D. P., Mason, B. J. and Maas, E. V. (1965). *Pl. Physiol.* 40, 641-644.
Mosse, B. (1963). *In* "Symbiotic Associations" (P. S. Nutman and B. Mosse, eds), pp. 145-170. *13th Symp. Soc. Gen. Microbiol.* Cambridge Univ. Press.
Nye, P. H. (1969). *In* "Ecological Aspects of the Mineral Nutrition of Plants" (I. H. Rorison, ed.), pp. 105-114. Blackwell, Oxford.
Oertli, J. J. (1968). *Agrochimica* 12, 461-469.
Olsen, S. R. and Kemper, W. D. (1968). *Adv. Agron.* 20, 91-151.
Packer, L., Marakami, S. and Mehard, C. W. (1970). *A. Rev. Pl. Physiol.* 21, 271-304.
Pitman, M. G., Courtice, A. C. and Lee, B. (1968). *Aust. J. biol. Sci.* 21, 871-881.
Rees, D. A. (1969). *Adv. Carbo. Chem. Bio.* 24, 267-332.
Robertson, R. N. (1958). *In* Handb. Pflanzenphysiol. (W. Ruhland, ed.), Vol. 4, pp. 243-279. Springer, Berlin.
Robertson, R. N. (1968). "Protons, Electrons, Phosphorylation and active Transport". University Press, Cambridge.
Robson, A. D., Edwards, D. G. and Loneragan, J. F. (1970). *Aust. J. agric. Res.* 21, 601-612.
Russell, R. S. and Barber, D. A. (1960). *A. Rev. Pl. Physiol.* 11, 127-140.
Sanders, F. E. T. and Tinker, P. B. (1971). *Nature, Lond.* 233, 278.
Scott, F. M., Hamner, K. C., Baker, E. and Bowler, E. (1958). *Am. J. Bot.* 45, 449-461.
Sutcliffe, J. F. (1962). "Mineral Salts Absorption in Plants". Pergamon Press, London.
Tukey, H. B. (1970). *A. Rev. Pl. Physiol.* 21, 305-324.
Weatherley, P. E. (1969). *In* "Ecological Aspects of the Mineral Nutrition of Plants" (I. H. Rorison, ed.), pp. 323-340. Blackwell, Oxford.

CHAPTER 19

Mineral Biochemistry of Herbage

G. W. BUTLER

*Applied Biochemistry Division, Department of Scientific
and Industrial Research, Palmerston North, New Zealand*

and

D. I. H. JONES

*Welsh Plant Breeding Station, Plas Gogerddan,
Aberystwyth, Wales*

I. INTRODUCTION

In this chapter aspects of the biochemistry of macro- and micro-elements in herbage will be discussed from two viewpoints, namely the role of minerals in the metabolism of herbage plants and secondly the relationships between herbage mineral composition and the nutritional quality of the feed for grazing animals.

Sixteen elements are recognized as being essential for the growth of higher plants—C, H, O, N, P, S, K, Na, Mg, Fe, Mn, Zn, Cu, Mo, Cl and B—i.e. these elements meet the three criteria of Arnon (1950) for essentiality for an organism. In addition, Nicholas (1961) proposed that the term "functional element" be used to include any mineral nutrient that may function in some precise way in plant metabolism, irrespective of whether or not its action is completely specific or indispensable. Several elements in this class are discussed.

Certain elements are essential for grazing animals, although their essentiality for plants is not recognized. Chromium, cobalt and selenium fall in this class and fluorine is a functional nutrient for higher animals. The levels and availabilities of these elements in forage are clearly of importance.

Finally, mineral toxicities to herbivores are considered; the cycling of some non-essential elements from plants to grazing animals is also included because of the possibilities of toxicities to the herbivores or in human nutrition e.g. lead, cadmium and mercury.

II. PLANT BIOCHEMICAL ASPECTS

A. Modes of Action of Essential and Functional Elements in Plants

There is an extensive literature on this subject (viz. Nason and McElroy, 1963; Evans and Sorger, 1966; Bollard and Butler, 1966). Table I summarizes knowledge of the modes of action of the essential elements, excepting C, H, O and N. Knowledge is undoubtedly still incomplete for many elements; in the case of boron the basic reason for its essentiality is still unknown and the link with carbohydrate transport and/or flavonoid metabolism is still presumptive.

Knowledge of the modes of action of nine functional elements is summarized in Table II; the subject is reviewed in detail by Bollard and Butler (1966).

B. States of Combination of Elements in Herbage

The various states of chemical combination of elements within herbage tissues are of course basic to our understanding of metabolic processes in herbage. In addition, such knowledge is essential from

TABLE I

Modes of action of essential elements

Element	Main functions or modes of action	References[a]
Phosphorus	Plant tissues characteristically maintain high inorganic phosphate in their tissues. Organic esters of phosphate occur in great variety in metabolites and macro-molecules. Phosphate plays an important role in energy transfer in both respiration and photosynthesis. A mobile nutrient.	Nason and McElroy (1963) Nicholas (1961)
Calcium	Complements potassium in maintaining cell organization, hydration and per-meability, thus indirectly affecting many enzyme systems. Co-factor for some enzymes (phospholipases, ATPases). Constituent of cell wall middle lamella (calcium pectate). Relatively immobile, not readily redistributed.	Nason and McElroy (1963)
Magnesium	Approximately 10% of total leaf magnesium is present in chlorophyll. Magnesium ions function as metal activators of enzymes involving the group transfer of phosphate. Relatively mobile. Diurnal magnesium fluctuations in chloroplasts may represent photosynthesis control mechanism.	Nason and McElroy (1963) Nicholas (1961) Lin and Nobel (1971)
Potassium	Highly mobile. Functions as a univalent cation activator for a wide variety of important enzymes.	Evans and Sorger (1966)
Sulphur	Component of proteins as a constituent of amino-acids cystine, cysteine and methionine and a constituent of other vital cellular constituents—biotin, thiamine, coenzyme A and glutathione. Relatively immobile.	Thompson (1967)

TABLE 1—*continued*

Element	Main functions or modes of action	References[a]
Chlorine	Required in the photoproduction of oxygen in chloroplasts.	Cheniae (1970)
Zinc	Essential constituent of glutamic dehydrogenase, carbonic anhydrase.	Nicholas (1961)
Copper	Component of enzymes mediating substrate oxidation by atmospheric oxygen (polyphenol oxidase, monophenol oxidase, laccase and ascorbic acid oxidase). Component of plastocyanin, an electron carrier protein in photochemical system of photosynthesis.	Mason (1965) Peisach (1965) Bishop (1966)
Iron	Constituent of hemo-proteins (cytochromes) and also of non-heme iron proteins involved in photosynthesis, nitrogen fixation and respiratory-linked dehydrogenases.	Nason and McElroy (1963) Malkin and Rabinowitz (1967)
Molybdenum	Constituent of nitrate reductase and also of component 1 of nitrogenase.	Beevers and Hageman (1969) Klucas et al. (1968)
Manganese	Metal activator for many enzyme systems. Required in the photoproduction of oxygen in chloroplasts.	Nason and McElroy (1963) Cheniae (1970)
Boron	Indirect evidence for involvement in carbohydrate transport and/or in flavonoid synthesis.	Nicholas (1961) Nason and McElroy (1963) Rajaratnam et al. (1971)

[a] Most, but not all, of the references are to reviews.

TABLE II

Element	Main functions or modes of action	References[a]
Sodium	Essential for at least some higher plants, e.g. the halophyte *Atriplex vesicaria*. Often exerts a beneficial effect because of a sparing effect on potassium requirements. Shows marked genetic variability in amount absorbed.	Brownell (1965) Evans and Sorger (1966) Bollard and Butler (1966)
Rubidium	Readily absorbed and exerts a sparing effect on potassium requirements.	Evans and Sorger (1966) Bollard and Butler (1966)
Strontium	Exerts a sparing effect on calcium requirements.	Bollard and Butler (1966)
Fluorine	Some species contain organic fluorine compounds. Others are accumulators of fluoride.	See p. 142, Peters (1957) Bollard and Butler (1966)
Bromide	Exerts a sparing effect on chloride requirements, including photoproduction of oxygen in chloroplasts.	Broyer *et al.* (1954) Johnson *et al.* (1957) Ozanne *et al.* (1957)
Silicon	Required for diatom growth, apparently as a structural component. Beneficial to higher plants especially plants with high silicon contents (e.g. rice and other grasses and horsetails).	Lewin and Reimann (1969)
Selenium	Incorporated mainly into selenomethionine and selenocystine (free and protein-bound) in many plants and into other non-protein seleno-amino acids in selenium accumulators.	Shrift (1969)
Iodine	Angiosperms have a slight ability to incorporate iodide into tyrosine and tyrosine derivatives.	Fowden (1959)
Cobalt	Required as cobamide coenzyme in symbiotic nitrogen fixation. An additional requirement is suggested in higher plants (e.g. growth response in non-nodulated subterranean clover).	de Hertogh *et al.* (1964) Cowles *et al.* (1969) Wilson and Hallsworth (1965)

[a] Most, but not all, of the references are to reviews.

the viewpoint of animal nutrition, since the states of chemical composition markedly influence the availability of the element to the animal. Such studies are often complicated by the formation of artefacts during extraction of tissues and considerable care is required in the choice of techniques and in the interpretation of results. For example, Bieleski (1964) has demonstrated the persistence of phosphatase activity in plant tissue extracts and the need for appropriate solvent extractions if reliable assessments of the phosphate ester distributions are to be made. In the case of cations such as calcium, it is doubtful whether sequences of extractions by various solvents and reagents yield results which are physiologically meaningful. Similarly the ready complexing of borate and silicate anions with poly-hydroxy compounds such as carbohydrates readily yields artefacts during extraction.

Potassium is present in plant tissues as the free ion or in readily exchangeable combination and is the most mobile of the essential elements. *Sodium* is also present in ionic or readily exchangeable form, but there is varying discrimination between species in the extent of absorption of sodium and translocation from root to shoot tissue (Collander, 1941; Bollard and Butler, 1966).

Calcium is believed to be present in cells as the soluble calcium salts of organic acids (except oxalate), as partially soluble calcium phosphate, and salts of other inorganic acids; as calcium pectinate; as calcium bound or absorbed to protein and as insoluble calcium oxalate. Unequivocal extraction procedures for all these fractions cannot be made because of the artefact problem. *Strontium* appears to fractionate in plant tissues in a closely comparable manner to calcium and these fractionations have been reviewed by Bollard and Butler (1966). *Magnesium* is present in fresh herbage from temperate grasses and clovers to the extent of about 10% in the acetone soluble fraction, mostly as chlorophyll (Todd, 1961, 1962). Water-soluble magnesium comprises about 50% of the total, with approximately one-third remaining in the residue.

The presence in normal plant tissues of high levels of inorganic phosphate is the most important aspect of the manifold states of combination of *phosphorus* from the viewpoints of both mobility of phosphorus within the plant and also animal nutrition.

Phosphate is also present in phosphate ester linkage in a large variety of cellular constituents, including nucleic acid polymers, phospholipids and a wide variety of phosphate esters of low molecular weight. These latter compounds are in large measure the currency of cellular intermediary metabolism. The papers of Bieleski and colleagues report the development of reliable analytical techniques and the manner in which levels of phosphate compounds fluctuate in higher plant tissues, particularly the duckweed *Spirodela* (Bieleski, 1968a, b). In general, changes in nutritional states and stage of growth do not result in major changes in the proportion of organophosphorus compounds in tissues. Exceptions occur with respect to particular plant organs, e.g. phytic acid is typically stored in seeds and is not present in vegetative herbage.

Sulphur is present in herbage as the inorganic sulphate ion and as a variety of organic sulphur compounds, most of which are present as sulphydryl compounds and their derivatives. Prominent amongst these are proteins which contain the sulphur amino acids methionine, cystine and cysteine. Sulphur is also present in sulphate ester linkage in sulpholipids and in a number of compounds which are prominent in intermediary metabolism, including coenzyme A, thiamine and biotin. In addition, some plant species contain relatively large amounts of sulphur-containing secondary metabolites, such as the glucosinolates (Chapter 9, Vol. 1).

Iron occurs in plant cells principally in the form of porphyrins (tetrapyrrole cyclic structures with the iron atom in coplanar arrangement), acting as the prosthetic or functional groups for cytochromes, peroxidases, catalase and (in legume root nodules) leghemoglobin. Non-porphyrin iron is also present in various important enzymes, including important electron-transfer proteins (ferredoxins) which have as a common structural feature the presence at the redox centre of two nonheme iron and two "inorganic" sulphur atoms (Malkin and Rabinowitz, 1967). The iron storage protein phytoferritin has been detected by electron microscopy in developing plastids and in mature chloroplasts operating under suboptimal conditions (low light, particular tissue, viral infection, fluctuating iron supply, etc.). Phytoferritin is thought to be the iron storehouse from which developing plastids draw during development of the photosynthetic apparatus (Hyde *et al.*, 1963; Robards and Robinson, 1968). In xylem exudates from several species, iron is present as anion complexes of the hydroxy acids citrate, malate and maleate; the actual acids present depend on the plant species (Brown, 1961).

Over 60% of the *manganese* in ryegrass could be removed by sequential extraction with aqueous ethanol and water and this appeared to be present in cationic, non-complexed form (Bremner and Knight, 1970). In the same study, more than 60% of the *zinc* in ryegrass was removed by sequential extraction with aqueous ethanol and water and appeared to be present in the extracts as anionic complexes of unknown composition; Peterson (1969) reported a similar distribution of zinc in *Agrostis* species. Only a small proportion of zinc (measured as [65]Zn after growing the grasses in nutrient solutions containing added [65]Zn ions) was associated with the soluble protein fraction, but in both studies an association of the residual zinc with the carbohydrates of the cell wall was obtained by entirely distinct experimental approaches (solubilization by reagents and hydrolytic enzymes). Additional support comes from Turner (1970), who observed a relatively high distribution of zinc-65 in the cell-wall fraction from *Festuca* prepared by differential centrifugation.

Bremner and Knight (1970) also observed the major proportion of *copper* in ryegrass herbage to be extracted by aqueous ethanol and water and to be present as anionic complexes of unknown composition. Copper is present in several metalloproteins in plants (Table I) but the proportion of the total copper which is protein-bound is relatively small.

Molybdenum is a component of essential enzymes (Table I); it is absorbed and probably translocated as the molybdate ion and concentrates in the interveinal areas of leaves (Stout and Meagher, 1948). Nothing is known of the binding or otherwise of *boron* within plant cells; borate readily forms complexes with polyhydroxy compounds such as soluble carbohydrates and wall polysaccharides and artefacts could readily be formed during extraction.

The states of combination of *cobalt* in plant tissues are largely unknown. Bowen *et al.* (1962) observed that 37% of ^{60}Co in tomato leaf tissue was soluble in 95% ethanol; on paper chromatography this fraction was shown to be complex and clearly distinct from both cobaltous ion and vitamin B_{12}.

Selenium is present in most plants predominantly as selenomethionine in proteins, with lesser amounts of Se-methylselenonium compound and selenocysteic acid as soluble amino-acids (Peterson and Butler, 1962; Butler and Peterson, 1967). It is translocated both as selenite and selenate ions; work with tomato plants (Asher and Butler, unpublished) showed that selenite ions moved into the xylem an order of magnitude slower than did selenate ions, but selenite ion is the dominant form of selenium in neutral and acid soils (Allaway, 1968). In species which accumulate selenium, nearly all of the element is present as soluble non-protein seleno-amino-acids (see Chapter 1, Vol. 1).

In higher plants, most of the *silicon* is present as *silica*, deposited in cell walls, with the distribution between organs varying with species. Thus, in *Trifolium incarnatum* the concentration of silicon in the roots was about eight times that in the tops, apparently because it is able to exclude monosilicic acid from its transpiration stream (Handreck and Jones, 1967) but in species with an overall high silicon content, the aerial parts of the plant contain most silicon (Lewin and Reimann, 1969). Mechanical damage can also result in extensive deposition of silica in grasses and this is relevant to partial defoliation under grazing (Parry and Smithson, 1963). The silica is in the form of amorphous silica, as a silica gel. In rice plants silica constitutes 90-95% of the total silicon, orthosilicic acid constitutes 0·5-8·0% and colloidal silicic acid 0·0-3·3% (Yoshida *et al.*, 1962). There is no convincing evidence for the presence of silicon in organic combination, since silicic acid could readily complex with many substances during the extraction process (Lewin and Reimann, 1969). There is, however, evidence for the existence of plant silica in at least two forms, based on solubility criteria (van Soest, 1970); the more soluble phase has a solubility of 430 ppm in boiling water, compared with 0·4-1 ppm for the less soluble phase.

The states of combination of halogen elements have been reviewed by Bollard and Butler (1966). *Chlorine* is always present in herbage as the chloride ion and is required for the photoproduction of oxygen in chloroplasts; *bromide* exhibits a sparing effect for at least some functions normally met by chloride. *Fluorine* is generally present as inorganic fluoride in amounts of 0·1 to 10 mg per kg dry weight. Some species contain unusually high amounts of inorganic fluoride (e.g. camellias and the commercial tea plant). Certain African and Australian shrubby species contain organic compounds of fluorine, (see p. 142). *Iodine* is predominantly present in higher plants as inorganic iodide, but there is some evidence for the presence of free and protein-bound amino-acids in small amounts (3,5-di-iodothryonine, 3,5-di-iodotyrosine; Fowden, 1959).

Studies of the states of combination of chromium-51 in plants have been made by Lyon (1969). Red clover plants grown in nutrient solution were able to absorb and translocate ^{51}Cr to the leaves when the ^{51}Cr was supplied as either sodium chromate or chromic chloride. Aqueous ethanol extracts of various parts of such plants contained two unidentified compounds which were anionic at pH 5.3

in addition to chromate ion which was present in small amounts. The proportion of ^{51}Cr associated with proteins was negligible, but an appreciable quantity (7% of the total) was associated with RNA.

C. Assimilation of Nitrate, Phosphate and Sulphate in Herbage

The general principles governing the uptake of these ions are considered in the chapters by Loneragan and Dijkshoorn; factors governing assimilation of these nutrients into organic compounds are summarized here. Nitrate is assimilated in higher plants by the following pathway:

$$(NO_3^-)_{\text{soil solution}} \xrightarrow[\text{system}]{\text{Nitrate uptake}} (NO_3^-)_{\text{inside}} \xrightarrow[\text{reductase}]{\text{Nitrate}} NO_2^-$$

$$NO_2^- \xrightarrow[\text{reductase}]{\text{Nitrite}} NH_3 \longrightarrow \text{amino acids}$$

There is no evidence for the free existence of intermediate compounds such as hydroxylamine or hyponitrite in the tissues of higher plants. Both nitrate and nitrite reductases have been demonstrated in the roots and leaves of several plant species, but the relative importance of nitrate reduction in roots varies considerably between species. In leaves, nitrite reductase is located within the chloroplasts; in both leaves and roots nitrate reductase appears to be located in the general cytoplasm.

Nitrate reductase has not been fully purified from higher plants, but it is reported to be a flavoprotein of molecular weight 500,000-600,000, containing molybdenum, Fe^{2+} and FAD and capable of accepting electrons directly from NADH. The enzyme is adaptive in that the activity is inducible by nitrate and by molybdenum. It is unstable both *in vitro* and *in vivo* (e.g. exhibits a half-life of 4 h in detached corn leaves) (Beevers and Hageman, 1969; Hageman and Hucklesby, 1971).

Nitrite reductase has not been so extensively studied, but it appears to receive electrons directly from ferredoxin when the latter has been reduced by illuminated chloroplasts or by NADPH and a diaphorase enzyme. The 1300-fold purified enzyme from vegetable marrow leaves has an approximate molecular weight of 61,000-63,000; it appears that one protein mediates the transfer of

6 electrons and that free hydroxylamine or hyponitrite is not an intermediate in the reaction (Hucklesby and Hewitt, 1970). The enzyme is induced by light and appears to be more stable *in vivo* than nitrate reductase (Losada and Paneque, 1971).

A wide range of environmental and physiological conditions influence nitrate assimilation in higher plants, including light, drought, mineral nutrition, hormonal treatment, plant age and genetic composition. In most of these cases, it appears that control of nitrate reduction is mediated through a regulation of the enzyme nitrate reductase.

The importance of light in nitrate reduction is well-documented (Hewitt, 1971; Beevers and Hageman, 1969). The level of nitrate reductase in leaf tissue varies diurnally and is also influenced by intensity of illumination. The light effect is a complex one, probably arising from a general stimulation of protein synthesis together with the more generally reduced state of enzyme cofactors in leaves following illumination.

Nitrate accumulation or impaired reduction is "associated with deficiencies of several elements including calcium, potassium, magnesium, iron, manganese and molybdenum" (Hewitt, 1971). The dependence on molybdenum is of course to be expected because it is a constituent of nitrate reductase and because synthesis of nitrate reductase is induced by molybdenum in molybdenum-deficient plants grown in the presence of nitrate. The essentiality of manganese in photosystem II of chloroplasts is probably related to the activity of nitrite reductase. Since iron is a constituent of both nitrate reductase and ferredoxin, iron deficiency can plausibly result in nitrate accumulation. The effects of other mineral deficiencies in favouring nitrate accumulation appear to be more indirect.'

Nitrate induces the development of nitrate and nitrite reductases in a wide range of plant species, and in a few instances ammonia and amino-acids have been shown to inhibit the development of these enzyme activities (Filner *et al.*, 1969). Thus, there is evidence that both enzymes are regulated by the substrate of the pathway and by its end-products.

There is a considerable literature on the effects of plant herbicides such as 2,4-D-(2-4-dichlorophenoxy-acetate) and simazin-(2-chloro-4,6-bis-ethylamino-S-triazine) on nitrate assimilation. The effects are variable and probably indirect, arising from the regulatory functions of both nitrate reductase and the plant growth regulators on overall metabolism (Beevers and Hageman, 1969).

The level of nitrate reductase activity in various plants appears to be under relatively simple genetic control, e.g. corn (Hageman, Leng and Dudley, 1967), wheat (Beevers and Hageman, 1969) and ryegrass (Bowerman and Goodman, 1971). Further, there is evidence that nitrate reductase activity is positively correlated with dry matter or protein yield—selection of genotypes having high nitrate reductase activity may therefore have practical significance.

Finally, recent work with cultured tobacco cells shows that nitrate specifically induces the development of an active nitrate uptake system; the uptake system is subject to end-product regulation by ammonia and amino-acids (Heimer and Filner, 1971).

In contrast to the situation with nitrate, phosphate assimilation proceeds readily into a variety of phosphorylated metabolites and there is thus no simple control mechanism based on regulation of any particular phosphorylation. It is characteristic of plant tissues, however, that they accumulate high concentrations of inorganic phosphate in a non-metabolic pool, presumably in the vacuole. With the duckweed *Spirodela* Bieleski (1968b) found that under normal phosphate nutrition 88% of the total inorganic phosphate was in a non-metabolic pool, while the remaining 12% was in a metabolic pool. It appeared that growth normally occurred at the expense of external phosphate, but with deficiency of phosphate supply growth became dependent on release of phosphate from the non-metabolic pool. The relative proportions of various phosphate esters and phospholipids were not markedly affected by phosphate deficiency, indicating that the plant cannot respond to phosphate deficiency by making some phosphorus compounds at the expense of others. Reid and Bieleski (1970) observed a 50-fold increase of phosphatase activity of cell extracts from *Spirodela* with onset of phosphate deficiency; the evidence favoured *de novo* synthesis of the enzyme. It is thought that this enzyme might function in the release of inorganic phosphate from senescent tissue into the external medium for re-utilization in growing tissues. The possible ecological significance of acid phosphatases of plant roots has been considered by Woolhouse (1969).

The pathway of sulphate assimilation is known in considerable detail for various microorganisms (Thompson, 1967). The first two steps involve the formation of sulphate esters as follows:

$$SO_4^{2-} + ATP \xrightarrow[\text{sulphurylase}]{\text{ATP}} APS + AMP + PPi \quad (1)$$

$$\text{APS} + \text{ATP} \underset{}{\overset{\text{APS kinase}}{\rightleftharpoons}} \text{PAPS} + \text{ADP}$$

   (1)   (2)

ATP sulphurylase = ATP-sulphate adenyltransferase (E.C. 2.7.7.4)

APS kinase = ATP-adenylsulphate-3′-phosphotransferase (E.C. 2.7.1.25)

adenosine—O—P—O—S—O⁻ (APS, with O and O above each P and S, O and O below)

APS (1)

adenine—C—C—C—C—C—O—P—O—S—O⁻ (PAPS, with phosphate group O⁻—P—O⁻ below the chain)

PAPS (2)

Sulphite is then formed by reduction of PAPS by a hydrogen transporting protein; finally a six-electron oxido-reduction takes place converting sulphite to sulphide.

With higher plants, however, sulphate assimilation is very incompletely understood, but appears to differ in some respects from the pathway in microorganisms. Ellis (1969) has shown for several plant species that ATP-sulphurylase activity is present in supernatant fractions. However, APS-kinase has not been detected in extracts of any plant tissue; it seems possible that APS is reduced directly, rather than PAPS. The control of sulphate reduction also appears to differ, since the synthesis of ATP-sulphurylase by turnip, lettuce, tomato and *Lemna* plants grown under aseptic conditions is neither induced by sulphate (cf. nitrate) or repressed by cystine as in microorganisms.

D. Biochemical Diagnoses of Mineral Deficiency States

Total analyses for tissue mineral contents may give erroneous results with respect to the adequacy of the particular element for optimal plant growth, because of unavailability of much of the mineral present. Some investigations have therefore been made of the use of enzymic assays as indicators of particular mineral deficiencies, as exemplified in Table III. Such studies show that the method is applicable under particular conditions, but caution is required because of interactions between elements and because the activity of certain enzymes in plants under stress is very much dependent upon species.

The enzymic assays may be performed directly upon the deficient tissue. A modification of this approach is to measure early metabolic changes following removal of nutrient stress in a plant deficient in that nutrient. Thus Shaked and Bar-Akiva (1967) observed that the rate of increase in nitrate reductase activity in response to Mo infiltration into citrus leaf segments was negatively correlated with the leaf Mo level, and Randall (1970) observed similar responses for

TABLE III

Enzyme assays of mineral deficiency states

Deficient Element	Enzyme assayed	Plant	References
Mn	Peroxidase	Citrus	Bar-Akiva (1965)
Fe	Peroxidase		O'Sullivan et al. (1969)
N	Nitrate reductase	Citrus	Bar-Akiva and Sternbaum (1965)
Mo	Nitrate reductase	Citrus	Shaked and Bar-Akiva (1967)
		Wheat	Randall (1970)
Zn	Aldolase	Ryegrass, clover, oats, tomatoes, corn	O'Sullivan (1970) Clarke (1966)

wheat. Randall also noted increases in response to incubation with sulphate of protein nitrogen in young detached leaves from sulphur-deficient subterranean clover plants.

E. Biochemical Bases for Tolerance of Pasture Plants to Mineralized Environments

Pasture and rangeland plants are frequently to be observed growing on soils which have unfavourable mineral composition, apparently as a result of physiological and biochemical adaptations which mitigate against toxic effects. The varying nature of such adaptations will be discussed here.

Tolerance of herbaceous plants to saline conditions is a wide subject involving many biochemical and physiological factors. The reader is referred to Bernstein and Hayward (1958), Sutcliffe (1962) and Strogonov (1962). Tolerant plants often exhibit a capacity for retention of sodium and chloride ions in the roots with relatively little transport of these ions to the aerial parts.

Adaptations of plants to low and high soil pH are in large measure a result of changes in selectivity of absorption and translocation of particular ions. Hallsworth *et al.* (1957) summarized the adverse effects on plant growth of soil acidity as being attributable to (a) a deficienty of soil calcium or magnesium; (b) deficiency of molybdenum caused by lowered availability; (c) an excess of aluminium or manganese (and possibly nickel or chromium); (d) deficiency of available phosphate because of high anion-exchange capacity produced by association of hydroxyl ions, liberated mainly from the sesquioxides and to a lesser extent from clay minerals; and (e) a direct effect of pH on growth. Conversely, the failure of some plants to grow in calcareous soils is thought to be the result of a decrease in availability of some elements at high soil pH (such as potassium, phosphate and iron).

The mechanisms whereby some plants have adapted to tolerate "toxic" levels of elements and thereby to grow on soils which are highly mineralized or of high or low pH is of considerable interest. The general proposition can be stated that specific mechanisms for coping with the particular element(s) are present in order to avoid toxic effects.

For example, there may be specific exclusion mechanisms (lead and molybdenum); the element may be confined to the cell wall (zinc and copper); the element may be metabolized further to possible inactive components (selenium and fluorine) or metabolized by binding with another compound (aluminium and chromium) see Peterson (1971).

Mitchell and Reith (1966) reported that on soils in north-east Scotland derived from quite diverse parent materials, the normal lead content of rotational mixed pasture herbage during the period of active growth is between 0·3 and 1·5 ppm in the dry matter. In autumn with cessation of active growth, the lead content of herbage rises, reaching 10 ppm in late autumn and 30-40 ppm in late winter. It is not known whether this increase is caused by movement of lead which had accumulated in the roots during the period of active growth or by breakdown of a mechanism which excluded lead from the plants during active growth. Brooks and Lyon (1966) have proposed partial exclusion of molybdenum by the shrub *Olearia rari* to explain the low Mo levels found in this plant growing on mineralized areas.

Immobilization of zinc within root cell walls has been demonstrated for zinc-tolerant *Agrostis tenuis* and *A. stolonifera* plants growing on toxic mine wastes. Both by chemical fractionation of roots from plants growing in ^{65}Zn-containing nutrient solutions (Peterson, 1969) and also by studies of the zinc distribution in fractions from differential centrifugation of root homogenates (Turner, 1970; Turner and Marshall, 1971) evidence was obtained for high Zn levels in the cell walls. The ^{65}Zn studies revealed that the plants remove large amounts of zinc by binding to cation exchange sites in the pectin/uronic acid component of the cell walls.

With both fluorine and selenium, a small proportion of plant species accumulate the elements to relatively high levels, but metabolize them to unusual compounds. In most plant species the relatively abundant fluoride ion is discriminated against, but in some species fluoride is metabolized to organic compounds such as fluoracetate and ω-fluoro-oleic acid e.g. *Dichapetalum cymosum* (Marais, 1944), *Acacia georginae* (Oelrich and McEwan, 1961), *Gastrolobium grandiflorum* (McEwan, 1964). Fluoracetate is the major fluoro-component in these plants and can reach 120 mg/kg dry weight in *G. grandiflorum*; these species are toxic to grazing animals on unimproved Australian and South African rangelands because of the metabolism of fluoracetate to fluorocitrate by Krebs cycle enzymes after ingestion (Peters, 1957). Fluorocitrate is a potent inhibitor of the enzyme aconitate hydratase. It has now been shown that fluoride can be metabolized to a small extent by many plant tissues (including soyabean (*Glycine max*) and alfalfa) to fluoracetate and fluorocitrate. That this is not a result of associated microflora has been shown in the case of soyabean single-cell cultures (Peters and Shorthouse, 1972).

Similarly, with selenium most pasture species contain relatively low levels, over the range 0·1-2 ppm, but a small number of selenium-accumulating plants (usually confined to alkaline seleniferous soils) may contain up to 15,000 ppm. In the selenium accumulators, the major portion of the absorbed selenium is metabolized to non-protein seleno-amino acids (see Chapter 1, Vol. 1) and it is believed that this constitutes a detoxication mechanism for these plant species (Peterson and Butler, 1967). Whereas in most plants selenium is incorporated into plant protein as selenium analogues of methionine and cystine, the large amounts of selenium in these species are channelled into non-protein amino-acids.

Other cases are known where the element is bound with another compound. For example, Jones (1961) considered the tolerance to aluminium of plants such as clover growing on alkaline fly-ash from coal power stations and suggested that organic acids in plants may act as chelating agents for toxic amounts of aluminium in these plants. In the case of the shrub *Leptospermum scoparium*, a stable organic acid complex of chromium, trioxalatochromate (III) ion, has been identified in extracts (Lyon *et al.*, 1969); anionic chromium complexes of unknown structure were also observed in aqueous ethanol extracts of red clover plants (Lyon, 1969).

It should be pointed out in conclusion that knowledge of such tolerance mechanisms is being applied successfully in the choice of plant species and development of suitable varieties to colonize unfavourable mineralized environments (Bradshaw, 1969). Also, the information is used in the technique of geobotanical prospecting for various elements, especially heavy metals, where herbaceous plant species are often employed (Cannon, 1960; Malyuga, 1964).

III. ANIMAL NUTRITIONAL ASPECTS

A. *Relation of Mineral Contents of Herbage plants to the Requirements of the Herbivore*

The essentiality of several elements has, at present, only been demonstrated in experiments with laboratory animals maintained under dietary regimes deficient in the particular element. It is, however, generally assumed that if an element can be shown to be essential for species such as the rat, it is essential for all mammals. At the present time some fifteen elements (in addition to C, H, O and N) are regarded as essential viz. Ca, P, K, Na, Cl, Mg, S, I, Fe, Cu, Mn, Co, Zn, Se, Mo. It seems likely that Cr and Sn should also be added to this list (Mertz, 1969; Schwarz *et al.*, 1970) and current literature suggests the further addition of V, Ni and Si. The distribution and function of these elements in animal tissues and the evidence for their essentiality have been the subject of recent reviews by Gallagher (1964), Underwood (1966, 1971), McDonald (1968), Schwartz (1970) and O'Dell and Campbell (1970).

As pointed out earlier, elements essential to animals differ significantly from those essential to plants, e.g. the requirements of animals for iodine, cobalt, selenium and sodium. Many deficiencies in grazing animals are associated with these elements since normal pasture growth can occur on deficient soils.

Ranges of herbage contents of mineral nutrients are considered in relation to estimated requirements in Table IV. This data gives only a general indication of the complex interrelationships involved in adequate mineral nutrition of livestock. It should be considered in conjunction with the information on herbage mineral levels given in Chapter 12, Vol. 1; the undoubted importance for the mineral nutrition of grazing animals of direct soil ingestion should also be taken

TABLE IV

The mineral requirement of ruminants in relation
to the mineral content of pasture herbage

Element	Herbage Content[a]		Desirable Pasture Content[b]	
	Range reported in pasture	Normal content in pasture	Fattening sheep	Milking cow
All values % of D.M.				
Calcium	0·04—6·00	0·2—1·0	0·50	0·52
Phosphorus	0·03—0·68	0·2—0·5	0·25	0·42
Sodium	0·002—2·12	0·05—1·0	0·07	0·15
Chlorine	0·02—2·05	0·1—2·0	0·09	0·19
Magnesium	0·03—0·75	0·1—0·4	0·06	0·15
All values p.p.m. of D.M.				
Iodine	0·07—5·0	0·2—0·8	0·12	0·80
Iron	21—1,000	50—300	30	30
Cobalt	0·02—4·7	0·05—0·3	0·10	0·10
Copper	1·1—29·0	2—15	5	10
Manganese	9—2,400	25—1,000	40	40
Zinc	1—112	15—60	50	50
Selenium	0·01—4,000	0·03—0·15	>0·03	>0·03

[a]Whitehead (1966) and unpublished data.

[b] Agricultural Research Council (1965). Data calculated for 40 kg sheep gaining 200 g per day consuming 1·36 kg DM and for 500 kg cow yielding 20 kg milk per day and consuming 14·3 kg DM.

into account, as in Chapter 13, Vol. 1; and finally the problem of availability of minerals is considered below.

Diseases of livestock arising from inadequacies of inorganic elements are widespread and, unless corrected, make stock rearing and production uneconomic in many areas of the world.

Table V summarizes the occurrence of a number of such conditions in sheep and cattle.

Deficiencies of economic importance are mainly associated with phosphorus and magnesium, and the trace elements iodine, cobalt,

TABLE V

Disorders in ruminants arising from
mineral deficiencies in herbage[a]

Disorder	Country	Animal	Conclusion
Bush sickness	New Zealand	Sheep	
Morton Mains disease	New Zealand	Lambs	
Enzootic marasmum	Australia	Sheep and cattle	
Nakuruitis	Kenya	Sheep and cattle	
Pining	Great Britain	Sheep and cattle	Simple deficiency
Bodmin Moor sickness	Ireland	Sheep	of cobalt —
Grand traverse disease	United States	Cattle	herbage cobalt
Sukhotka, mossjuka	U.S.S.R.	Cattle and sheep	generally below
Vosk	Denmark	Cattle and sheep	0·07 ppm
Hinsch	Germany	Cattle	
Likzucht	Netherlands	Cattle and sheep	
Coast disease	Australia	Cattle and sheep	Dual deficiency of
Salt sick	United States	Cattle and sheep	copper and cobalt
Enzootic ataxia	Australia	Lambs	Simple deficiency
Falling disease	Australia	Cattle	of copper
Swayback	Great Britain	Lambs	Induced deficiency of copper — cause unknown
Pasture diarrhoea	Netherlands	Cattle	Induced deficiency
Teart Scours	Great Britain	Cattle	of copper due to
Peat Scours	New Zealand	Cattle	excess molybdenum
Grass staggers, tetany, lactation tetany	Europe New Zealand etc.	Cattle and sheep	Induced magnesium deficiency
Bone chewing and wasting diseases	World wide	Cattle and sheep etc.	Simple deficiency of phosphorus
Simple goitre	World wide	All animals	Simple or conditioned iodine deficiency
White muscle disease, unthriftiness, infertility	Widespread, especially in New Zealand and United States	Cattle and sheep	Deficiency of selenium and/or vitamin E

[a]Derived from the data of Russell and Duncan (1956).

copper and selenium. The distribution, aetiology and control of these diseases have been reviewed by several authors (e.g. Russell and Duncan, 1956; Underwood, 1966). The present discussion is, therefore, confined to consideration of the relationship between deficiency and herbage composition.

Underwood (1966) considers phosphorus deficiency to be the most widespread and economically important mineral deficiency which affects livestock. The deficiency, aphosphorosis, was first recognized in South Africa where the phosphorus content of pasture falls to 0·05-0.07% during the dry winter months. Losses of stock are usually accentuated by deficiencies of protein and energy. The disease appears to be confined to areas where a low soil phosphorus status is combined with the occurrence of a dry season. Under these conditions the only pasture available to stock is herbage low in protein and phosphorus at a mature stage of growth.

Hypomagnesemia (hypomagnesemic tetany, "grass staggers") occurs widely in Europe, North America and Australasia, particularly in dairy cows, and is economically important because of its high mortality rate. Although the disorder is associated with a deficiency of dietary magnesium, the relationship between the incidence of hypomagnesemia and the magnesium content of the pasture is by no means simple. For example, the incidence of hypomagnesemia tends to increase when pastures are improved by the application of nitrogen and potassium fertilizers. While the disorder is less likely to occur if the herbage contains over 0.2% magnesium, the balance of other constituents such as potassium, sodium and total nitrogen also influence the incidence. Environmental and physiological factors are also important (see review by Allcroft and Burns, 1968).

The most widespread trace element deficiency of livestock is probably goitre, arising from a simple or induced deficiency of iodine. The incidence of goitre is generally associated with soil type, and direct soil ingestion of iodine is probably an important factor (see Chapter 13, Vol. 1). The iodine contents of plant species and varieties growing on the same soil also differ considerably (Johnson and Butler, 1957; Alderman and Jones, 1967).

In addition, several plant species contain glucosinolates and/or cyanoglucosides, which give rise to goitrogenic substances when ingested (see Chapter 9, Vol. 1). Such substances either inhibit the synthesis of organically bound iodine (primarily thyroxine) or, alternatively, inhibit the selective uptake of iodine by the thyroid.

Cobalt deficiency in growing ruminants occurs in most countries of the world (variously described as bush sickness, wasting disease, pining, etc.). These diseases are associated with an acute deficiency of cobalt and occur in animals of all ages and on diverse pastures. It has been well established that the deficiency is due to the inability of rumen microorganisms to synthesize sufficient vitamin B_{12} for ruminant requirements unless the diet contains sufficient cobalt. The occurrence of deficiency is usually well related to the cobalt content of the pasture, being unlikely if this exceeds 0·08 ppm.

Copper deficiency often occurs as a dual deficiency with cobalt and is widespread in many countries. A simple deficiency of copper is rare and its occurrence is usually conditioned by various additional factors which influence copper utilization in the animal. A high molybdenum content in pasture is a frequent cause of copper deficiency (e.g. teart in Britain and peat scours in New Zealand) and high sulphate levels also reduce copper utilization (Dick, 1969). Copper deficiency of sheep in Britain, swayback, occurs on pastures relatively high in copper content and appears to be caused by unknown factors influencing copper utilization.

Naturally occurring deficiencies of selenium have been demonstrated in herbivores in many countries. Deficiencies occur on soils low in plant-available selenium (Allaway, 1968) and the herbage generally contains less that 0·05 ppm selenium. Other factors are also involved, notably the vitamin E status of the animals and probably unidentified herbage factors.

B. *Mineral Requirements of Rumen Microorganisms*

Apart from the specific requirements of the ruminant itself for an adequate supply of dietary minerals, the rumen microorganisms also require an optimum supply of minerals and other nutrients for normal growth and function. Since the major proportion of herbage feeds are digested as a result of rumen fermentation, an efficient rumen function is clearly of importance in relation to herbage digestion.

Barnett and Reid (1961) have reviewed in some detail *in vitro* studies on the optimum and toxic levels of mineral elements for the efficient growth of rumen bacteria. Several workers have shown the requirement of rumen bacteria for iron and the simulating effect of its addition on cellulose digestion. Copper and cobalt have been found to inhibit digestion at concentrations above 2·5 ppm in the medium, while selenium was inhibitory at 0·3 ppm. The stimulating effect of alfalfa ash on fibre digestion has been extensively investigated by American workers both *in vivo* and *in vitro* (Burroughs *et al.*, 1950, 1951). The addition of molybdenum (Ellis *et al.*, 1958; Varela *et al.*, 1970) and cobalt (Jones, 1970) have also been shown to increase fibre digestibility *in vivo*.

C. *Availability of Herbage Minerals for the Herbivore*

Several instances have been cited earlier of mineral deficiencies occurring in ruminants despite apparently adequate levels of the minerals concerned being present in the herbage consumed. It must be concluded in these instances, that the mineral elements are present in the herbage in a form unavailable to the animal or that availability has been impaired through interaction with other dietary constituents.

Considerable information exists on the content and distribution of most inorganic elements in herbage plants when expressed as a total content after destruction of organic matter. The availability of herbage minerals has, however, been little studied largely due to the considerable difficulties inherent to its assessment. Availability may be defined as the proportion of the dietary intake of an element which can be used by the animal to make good endogenous loss or promote storage (A.R.C. 1965). The determination of "apparent" availability presents few problems and only requires the measurement of faecal excretion of an element and its intake viz.

$$\text{Apparent availability} = \frac{(\text{Intake} - \text{faecal excretion}) \times 100}{\text{Intake}}$$

The determination of "true" availability, however, requires data on the proportion of the element voided which is of endogenous origin. This may be estimated from excretion on diets free of the element under study or by radioisotope techniques where the metabolic pool is labelled allowing a partition of the element in the faeces between that of indigestible and endogenous origin. Various techniques for the assessment of mineral availability have been reviewed by Thompson (1965).

Many of the estimates for availability quoted in the literature are of "apparent" availability and, since they ignore endogenous losses, give values considerably lower than the true availability. Lofgreen and Kleiber (1953), for example, found 91% of the phosphorus in alfalfa to be available to lambs when endogenous losses were accounted for but an apparent digestibility of only 22%. Care (1960), similarly, gives figures of 30% and 17% for the true and apparent availability of the magnesium in meadow hay to sheep.

Even when true availability is determined, the data requires careful interpretation. Hill (1962) points out that the availability of calcium and phosphorus are influenced by level of feeding and the values may only therefore apply to the feeding conditions of the particular experiment and bear little or no relation to the inherent properties of the feed. Considerable variation in availability of minerals occurs between animals and in the same animal from day to day (Wright, 1955). Availability also appears to be related to age of the animal; Braithwaite and Riazuddin (1971), for example, show dietary calcium to be less available to mature than to growing sheep and conclude that mature sheep only absorb sufficient calcium to maintain equilibrium. In these circumstances it is not surprising that

TABLE VI

Estimates of the availability of herbage Ca, P and Mg to the ruminant

Element	Animal	Feed	% availability	Reference
Ca	Sheep	Orchard grass and concentrate	45	Shroder & Hansard (1958)
	Cattle	Alfalfa hay	31	Hansard et al. (1957)
		Lespedeza hay	36	
		Orchard grass hay	39	
P	Sheep	Alfalfa hay	95	Lofgreen and Kleiber (1954)
	Sheep	Ryegrass/clover hay and concentrate	69	Wright (1955)
	Cattle	Alfalfa hay and concentrate	57	Kleiber et al. (1951)
Mg	Sheep	Meadow hay	30	Care (1960)
	Cattle	Alfalfa hay and grain	28	Simensen et al. (1962)

little progress has been made in comparing the availability of minerals from different herbages. Using retention in the rat as a criterion, however, Armstrong and Thomas (1952) and Armstrong et al. (1953, 1957), were able to show differences in the availability of Ca in different herbages ranging from 70% for cocksfoot to 95% for plantain, values for other grasses and legumes being intermediate. These values are of course not applicable to the ruminant but may indicate potential differences in availability between herbage species.

While similar comparisons have not been made using ruminant animals some of the estimates of true or net availability made using the radioisotope technique suggest interesting differences in the availability of different inorganic elements (Table VI). Phosphorus in herbage, for example, appears to be more available than calcium while magnesium is poorly available to both cattle and sheep. Sodium in herbage is highly available, Kemp (1964) giving a mean value of 85% for apparent availability.

Although there is a lack of quantitative information on the availability of herbage minerals, the various factors which may influence availability have been intensively studied for elements such as magnesium, where nutritional disorders are known to be poorly related to dietary intake. Table VII summarizes some of the factors

TABLE VII

Some factors considered to influence the availability of herbage magnesium and its utilization by the ruminant

Factor	Example
Mg present in herbage in unavailable form.	Mg in chlorophyll—an indigestible constituent. Irvin *et al.* (1953).
Interaction of Mg with other dietary constituents in the digestive tract leading to formation of unavailable compounds or chelates.	Reaction of Mg with higher fatty acids to form insoluble soaps. Kemp *et al.* (1966).
Interference with Mg absorption by other ions.	Mg absorption increased by high Na and decreased by high K in gut. Care *et al.* (1967).
Other dietary constituents modifying conditions in the rumen or gut.	High N content of diet. Underwood (1966).
Factors reducing the energy content or intake of feed.	Utilization of Mg reduced when animal is in negative energy balance. Swan and Jamieson (1956 a, b).

known to influence either the absorption of dietary magnesium by the ruminant or its subsequent utilization. Transaconitic acid has also been suggested to affect Mg utilization (Burau and Stout, 1965; Stout *et al.*, 1967) but this has not been supported by dosing experiments in animals (Kennedy, 1968; Wright and Wolff, 1969). Elements such as calcium and phosphorus have been less well investigated but dietary constituents are known to influence their availability (Hill, 1962). The availability of calcium is, for example, reduced by high concentrations of oxalate due to the formation of insoluble oxalate which is excreted in the faeces. Phytate phosphorus, which occurs in grain in substantial quantities, is poorly available to non-ruminants and it may reduce the availability of calcium and other ions such as zinc. Phytate phosphorus appears to be well utilized, however, in the ruminant (Tillman and Brethour, 1958).

A variety of dietary constituents influence the absorption, utilization or retention of trace elements by the ruminant. These have recently been summarized by Mills and Williams (1971) and Table VIII is extracted from their data. Many of the constituents listed only influence trace element metabolism when present in high concentrations in the diets; with respect to iron, deficiency is unlikely in herbage fed animals. The variety of factors influencing

TABLE VIII

Dietary constituents known to modify trace element availability, utilization or requirement

Trace element	Dietary constituent
Co	None known.
Cu	Mo, organic and inorganic S compounds, Fe, Zn, unidentified components.
I	Thiooxazolidones, Co, thiocyanate.
Fe	Ca, P, Cu, Zn, carbonates.
Mn	P, unidentified components.
Se	Organic and inorganic S compounds, unidentified components.
Sn	None known.
Zn	Ca, Cu, unidentified components.

(After Mills and Williams, 1971)

copper and iodine absorption and utilization are, however, of particular economic importance as shown in the previous section. While no specific constituents have been shown to impair cobalt absorption or utilization, some evidence exists of differences in availability between herbages (Patil and Jones, 1970) while Andrews et al. (1970) have recently noted an association between cobalt deficiency and therapy and worm infestation in sheep.

Considerable interest is at present focused on the formation of complexes between trace elements and organic ligands in the digestive tract of the ruminant (Bremner, 1970). Formation of these complexes is clearly important in relation to the absorption of trace elements and, consequently, their availability (O'Dell, 1962). The effect of molybdenum in reducing copper availability, for example, has been attributed by Matrone (1970) to the formation of a Cu-Mo compound which is unavailable and excreted in the faeces. Molloy and Richards (1971) have recently reported that a considerable proportion of Ca^{2+} and Mg^{2+} may be complexed by lignin and organic acids in solutions of similar cation content to duodenal fluid. Complex formation does not necessarily imply, however, a reduction in availability. Mills (1956), for example, found organic complexes of grassland herbage to be readily available to sheep while Kirchgessner and Grassmann (1970) found copper-amino acid complexes to be more rapidly absorbed in rats than copper sulphate, larger complexes were, however, unavailable. Van Campen (1970) suggests that complex formation may be an integral part of the absorption process and that the antagonism between copper and zinc may result from their competing for amino acids, proteins or other potential carriers.

Further information on the formation of complexes involving inorganic elements in the digestive tract and the availability of these complexes is clearly needed for a better understanding of the factors influencing the availability of herbage minerals to the ruminant. To complement this, the nature of the compounds containing inorganic elements in herbage needs to be elucidated together with their stability during the digestive process. Such information would allow the potential availability of herbage minerals to be assessed more accurately.

D. Accumulation of Toxic Levels of Minerals in Herbage

Although many minerals are toxic to the ruminant when ingested at high concentrations in the diet, only selenium, copper and molybdenum occur in toxic quantities in pastures under natural conditions. Toxic amounts of nitrate generally arise from use of high levels of nitrogen fertilizer although this may also occur under certain environmental conditions in unfertilized pastures. Toxic levels of other minerals, notably fluorine and some heavy metals, arise mainly from contamination of pastures with airborne dusts e.g. from industrial plants.

The well recognized disorders associated with toxic levels of minerals in herbage are summarized in Table IX derived from the reviews of Russell and Duncan (1956) and Underwood (1971). Toxicity from contaminated pastures has been excluded since this arises primarily from elements present on the surface of the herbage.

TABLE IX

Disorders in ruminants arising from mineral excesses in herbage[a]

Disease	Country	Animal	Conclusion
Scouring on teart pastures	Great Britain	Cattle	Excess Mo causing induced Cu deficiency.
Peat scours	New Zealand		
Chronic copper poisoning	Australia	Sheep	Poisoning due to excess Cu in pasture.
Alkali disease— blind staggers	N. America, limited areas in Ireland, Israel, Australia, U.S.S.R.	Cattle, Sheep, etc.	Poisoning due to excess Se in pasture.
Oat hay poisoning	United States	Cattle	Poisoning due to excess nitrate in herbage.

[a] Derived from the data of Russell and Duncan (1956).

1. Molybdenum

The induced copper deficiency which occurs in ruminants as a result of ingesting high levels of molybdenum has already been referred to. The disorder was originally recognized as scouring on teart pastures in Britain, the herbage containing 20-100 ppm of Mo and 11-18 ppm Cu compared to the normal content of less than 3 ppm Mo (Ferguson *et al.*, 1940, 1943). Although sheep are susceptible, dairy cows and beef cattle are more susceptible. The condition may be prevented or cured by administration of copper and may therefore be regarded as a copper deficiency resulting from the effect of high molybdenum ingestion on copper storage in the animal. Many of the symptoms, e.g. depigmentation of hair, are similar to those of simple copper deficiency.

A similar condition, peat scours, occurs in cattle grazing reclaimed swamps and peaty soils in New Zealand. The molybdenum contents of the herbage range up to 17 ppm while copper contents tend to be low (Cunningham 1944, 1954). Molybdenum toxicity has also been reported in Ireland and in N. America. Barshad (1948) in the United States noted a wide variation in uptake of molybdenum by different plant species, values ranging in September-October from 9-10 ppm for orchard grass to 72-93 ppm for *Lotus corniculatus* and 78 ppm for *Melilotus alba*. An interesting observation was that young herbage was more dangerous to stock than mature herbage although it contained less total molybdenum, possibly inferring a difference in availability of molybdenum.

Molybdenosis has also resulted from aerial contamination of pastures from factories. Buxton and Allcroft (1955) for example, found levels of up to 126 ppm Mo in pastures in industrial areas.

2. Copper

Chronic copper poisoning of sheep, enzootic jaundice or "yellows", is common in certain areas of Australia and is characterized by a sudden rise in blood copper following the storage of abnormally high amounts in the liver. It occurs on cupriferous soils where certain herbage plants may contain over 60 ppm Cu (Bull, 1951). It also occurs in other areas where soil copper is not unduly high but where conditions favour growth of herbage with a high ratio of Cu to Mo.

Subterranean clover is often implicated in poisoning, containing a fairly high copper (11-18 ppm) but very low molybdenum content (less than 0·4 ppm). It is recommended that subterranean clover is not grown in these areas except in association with a high proportion of grass (Toxaemic Jaundice Investigation Committee, 1947, 1949).

Chronic copper poisoning of sheep also occurs in Australia in association with ingestion of *Heliotropium europeum* which contains hepatoxic alkaloids causing liver damage (see Chapter 8, Vol. 1). Copper is retained in abnormally high amounts in the damaged livers and copper poisoning results (Bull and Dick, 1959).

Underwood (1971) quotes several instances of copper poisoning in sheep resulting from ingestion of contaminated herbage, e.g. vineyards and orchards sprayed with fungicide, pastures sprayed with molluscicide, and from free access to trace element mixtures while at pasture. It should be noted that sheep are particularly susceptible to copper toxicity with very little margin between requirement and excess.

3. Selenium

Selenium toxicity occurs extensively in the Great Plains area of N. America both in the chronic form, alkali disease, and acute form, blind staggers. It also occurs in limited areas of other countries (Moxon and Rhian, 1943; Underwood, 1971). The marginal level for toxicity appears to be around 5 ppm Se but depends on the form of selenium ingested, the duration of feeding and other dietary factors such as protein intake.

Certain weed species, notably those of the *Astralagus, Stanleya, Aplopappus* and *Aster* species accumulate selenium readily from soils in seleniferous areas to concentrations of over 4000 ppm. These plants are rare or absent on adjacent non-seleniferous soils and are, therefore, referred to as 'indicator' plants. Stock normally avoid ingesting these plants except in times of drought while stock new to the area may consume them readily (Byers, 1935; Russell and Duncan, 1956).

Acute poisoning results from the ingestion of large quantities of seleniferous weeds, a complicating factor being the fact that some of these weeds appear to be toxic even in the absence of selenium. The chronic condition arises from the daily consumption of small amounts of selenium, about 4 ppm Se appears to be as high a level as livestock can tolerate (Williams *et al.*, 1941). In Ireland, chronic selenium poisoning occurred on farms when herbage contained 150-500 ppm Se (Walsh *et al.*, 1951).

4. Nitrate

Wright and Davison (1964) have reviewed the factors influencing the accumulation of toxic levels of nitrate in herbage and its toxicity to the ruminant. Under practical conditions nitrate accumulates primarily as a result of using high levels of nitrogen fertilizers, particular under unfavourable growing conditions (Butler, 1959). High levels of nitrate accumulate in Britain in spring and autumn when growth is limited by factors other than nitrogen supply (Jones *et al.*, 1961).

Although nitrate itself is relatively non-toxic it may be reduced to nitrite by soil microorganisms in cut herbage or hay stored under moist conditions (Olson and Moxon, 1942; Jones and ap Griffith, 1965). It is also readily reduced to nitrite by rumen microorganisms and high levels of nitrite may accumulate in the rumen under certain conditions. Absorption of nitrite into the bloodstream results in the conversion of haemoglobin into methaemoglobin due to the oxidation of Fe^{2+} to Fe^{3+} by nitrite. Methaemoglobin prevents the transport and release of oxygen by the blood and conversion of a substantial proportion of haemoglobin therefore results in internal asphyxiation.

The toxicity of nitrate was recognized as a result of investigations of oat hay poisoning in the United States (Bradley *et al.*, 1939). As a result of these and subsequent drenching experiments the toxic level of nitrate in feeds was considered to be of the order of 0·15-0·2% nitrate N in the dry matter. More recent work has shown that animals can ingest considerably higher levels of nitrate without apparent ill effect. Kennedy and Crawford (1960), for example, fed oat hay containing 0·27% nitrate N to pregnant heifers for 5 weeks with no ill effect. While 45 g nitrate per 100 lb body weight could be consumed with no ill effect when incorporated into the feed a third of this quantity proved fatal as a drench. Investigations in Britain have supported the fact that calves (Hodgson and Spedding, 1966) and sheep (Sinclair and Jones, 1964a, b) show no ill effect from ingestion of levels of nitrate considerably above 0·2% nitrate N.

Sub-clinical toxicities of nitrate may be of economic importance. Muhrer *et al.* (1956) found milk yield and reproduction in cows to be influenced by low levels of feed nitrate. Other workers have found nitrate injection to influence vitamin A storage and thyroid function.

5. Fluorine

Fluorosis in animals has been the subject of several reviews (Mitchell and Eldman, 1952; Phillips *et al.*, 1960). In grazing animals it arises from ingestion of herbage contaminated with fluorine. It occurs in N. Africa as a result of contamination of pasture from airborne dust from phosphate deposits and in many areas from drinking water high in fluorine, this often being derived from low lying rock formations. Plants appear to have a limited ability to absorb fluorine from soil although fluorine accumulators, notably the tea and camellia plants, are known (see also p. 142).

An important cause of fluorosis in industrial areas is the aerial contamination of pastures from factories and smelting plants (Burns and Allcroft, 1964).

6. Lead

While toxicity of lead to ruminants has been studied (Allcroft and Blaxter, 1950), there appears to be little danger of poisoning occurring under natural grazing conditions. It may arise, however, in localized areas adjacent to old lead mines or near lead smelters (Rains, 1971). Stewart and Allcroft (1956), Goodman and Roberts (1971) have recently shown high levels of lead, zinc, cadmium, nickel and copper in herbage in the industrial area of S. Wales. They suggest that the heavy metal concentration was related to the occurrence of sporadic livestock disorders in the area. As stated on p. 142,

Mitchell and Reith (1966) found high lead levels in herbage growing on north-east Scotland soils in late autumn and winter.

7. Other Elements

The toxicity of other minerals appears to be of minor significance for herbage fed ruminants. Elements which are relatively non-toxic may however be important in certain conditions. Matrone et al. (1959), for example, found as little as 45 ppm manganese to reduce haemoglobin formation in anaemic lambs. High levels of salt may be detrimental to sheep (Pierce, 1957). Wilson (1966) found the ingestion of saltbush and bluebush containing 3·2-8·2% Na, to be dependent on the availability of fresh water. Zinc appears to be relatively non-toxic to ruminants but high levels were found by Ott et al. (1966) to be toxic to rumen microorganisms.

An interesting effect of a trace element in stock disorders occurs in relation to *Phalaris* staggers in Australia. On cobalt deficient or marginally deficient soils a neurotoxic substance present in *Phalaris* induces staggers, unless they are treated with cobalt (see Chapter 8, Vol. 1). On soils of higher cobalt content sufficient cobalt is available in the herbage for normal nutritional requirements and to detoxify the toxin in *Phalaris* (Underwood, 1971).

The cycling of other elements into herbivore tissues is receiving increasing attention from the viewpoint of possible toxicity in human food chains, for example cadmium, vanadium and mercury levels. Particularly in highly industrialized communities employing intensive agricultural practices, these problems are likely to continue to be of concern. Such issues are reviewed by Allaway (1968).

IV. CONCLUSIONS

Although the mineral biochemistry of herbaceous plants has been studied extensively, the complexity of the interactions which are possible means that the subject is still at a relatively unsophisticated level. From the viewpoint of plant production, there remain many unanswered questions on the modes of action of essential and functional elements and their states of combination and pathways of assimilation. From the viewpoint of animal nutrition, the requirements of herbivores in relation to the availability of particular minerals is an extremely complex subject. From both points of view, multiple interactions between elements increase the complexity. With the development of automated techniques for multiple analysis and data processing, elucidation of multiple interactions between minerals in both plant and animal nutrition is becoming practicable.

REFERENCES

Alderman, G. and Jones, D. I. H. (1967). *J. Sci. Fd Agric.* **18**, 197-199.

Allaway, W. H. (1968). *Adv. Agron.* **20**, 235-274.

Agricultural Research Council (1965). "The Nutrient Requirements of Farm Livestock. No. 2, Ruminants". ARC, London.

Allcroft, R. and Blaxter, K. L. (1950). *J. comp. Path. Ther.* **60**, 209-218.

Allcroft, R. and Burns, K. N. (1968). *N.Z. Vet. J.* **16**, 109-128.

Andrews, E. D., Hogan, K. G., Stephenson, B. J., White, D. A. and Elliott, D. C. (1970). *N.Z. Jl agric. Res.* **13**, 950-964.

Armstrong, R. H. and Thomas, B. (1952). *J. agric. Sci., Camb.* **42**, 454-460.

Armstrong, R. H., Thomas, B. and Horner, K. (1953). *J. agric. Sci., Camb.* **43**, 337-342.

Armstrong, R. H., Thomas, B. and Armstrong, D. G. (1957). *J. agric. Sci., Camb.* **49**, 446-453.

Arnon, D. I. (1950), *In* "Trace Elements in Plant Physiology" (T. Wallace, ed.), pp. 31-39, Chronica Botanica, Waltham, Mass.

Bar-Akiva, A. (1965). *Øyton* **22**, 131-136.

Bar-Akiva, A. and Sternbaum, J. (1965). *Pl. Cell Physiol., Tokyo* **6**, 575-577.

Barnett, A. J. G. and Reid, R. L. (1961). "Reactions in the rumen", 170-207, Arnold, London.

Barshad, I. (1948). *Soil Sci.* **66**, 187-195.

Beevers, L. and Hageman, R. H. (1969). *A. Rev. Pl. Physiol.* **20**, 495-522.

Bernstein, L. and Hayward, H. E. (1958). *A. Rev. Pl. Physiol.* **9**, 25-46.

Bieleski, R. L. (1964). *Anal. Biochem.* **9**, 431-442.

Bieleski, R. L. (1968a). *P. Physiol., Lancaster* **43**, 1297-1308.

Bieleski, R. L. (1968b). *Pl. Physiol., Lancaster* **43**, 1309-1316.

Bishop, I. (1966). *A. Rev. Pl. Physiol.* **17**, 185-208.

Bollard, E. G. and Butler, G. W. (1966). *A. Rev. Pl. Physiol.* **17**, 77-112.

Bowen, H. J. M., Cawse, P. A. and Thick, J. (1962). *J. exp. Bot.* **13**, 257-267.

Bowerman, A. and Goodman, P. J. (1971). *Ann. Bot.* **35**, 353-366.

Bradley, W. B., Eppson, H. F. and Beath, O. A. (1939). *J. Amer. vet. med. Ass.* **94**, 541-542.

Bradshaw, A. D. (1969). *In* "Ecological Aspects of the Mineral Nutrition of Plants" (I. H. Rorison, ed.), pp. 415-427, Blackwell, Oxford and Edinburgh.

Braithwaite, G. D. and Riazuddin, S. (1971). *Br. J. Nutr.* **26**, 215-225.

Bremner, I. (1970). *In* "Trace Element Metabolism in Animals" (C. F. Mills, ed.), pp. 366-369, S. & L. Livingston, Edinburgh and London.

Bremner, I. and Knight, A. H. (1970). *Br. J. Nutr.* **24**, 270-289.

Brooks, R. R. and Lyon, G. L. (1966). *N.Z. Jl Sci.* **9**, 706-707.

Brown, J. C. (1961). *Adv. Agron.* **13**, 329-369.

Brownell, P. F. (1965). *Pl. Physiol., Lancaster* **40**, 460-468.

Broyer, T. C., Carlton, A. B., Johnson, C. M. and Stout, P. R. (1954). *Pl. Physiol., Lancaster* **29**, 526-532.

Bull, L. B. (1951). Proc. Spec. Conf. in Agric., Australia 1949, 300-310.

Bull, L. B. and Dick, A. T. (1959). *J. Path. Bact.* **78**, 483-502.

Burau, R. G. and Stout, P. R. (1965). *Science, N.Y.* **150**, 766-767.

Burns, K. N. and Allcroft, R. (1964). Anim. Disease Survey Report No. 2, Part 1, H.M.S.O., London.

Burroughs, W., Gerlaugh, P. and Bethke, R. M. (1950). *J. Anim. Sci.* 9, 207-213.

Burroughs, W., Latona, A., De Paul, P., Gerlaugh, P. and Bethke, R. M. (1951). *J. Anim. Sci.* 10, 693-705.

Butler, G. W. (1959). *Proc. N.Z. Soc. Anim. Prod.* 19, 99-110.

Butler, G. W. and Peterson, P. J. (1967). *Aust. J. biol. Sci.* 20, 77-86.

Buxton, J. C. and Allcroft, R. (1955). *Vet. Rec.* 67, 273.

Byers, H. G. (1935) U.S.D.A. Tech. Comm. No. 482.

Cannon, H. L. (1960). *Science, N.Y.* 132, 591-598.

Care, A. D. (1960). *Res. vet. Sci.* 1, 338-349.

Care, A. D., Vowles, L. E., Mann, S. O. and Ross, D. B. (1967). *J. agric. Sci., Camb.* 68, 195-204.

Cheniae, G. M. (1970). *A. Rev. Pl. Physiol.* 21, 467-498.

Clarke, R. B. (1966). *Crop. Sci.* 6, 593-596.

Collander, R. (1941). *Pl. Physiol., Lancaster* 16, 691-720.

Cowles, J. R., Evans, H. J. and Russell, S. A. (1969). *J. Bact.* 97, 1460-1465.

Cunningham, I. J. (1944), *N.Z. Jl Agric.* 69, 559-569.

Cunningham, I. J. (1954). *N.Z. vet. J.* 2, 29-36.

De Hertogh, A. A., Mayeux, P. A. and Evans, H. J. (1964). *J. biol. Chem.* 239, 2446-2453.

Dick, A. T. (1969). *Outl. Agric.* 6, 14-18.

Ellis, R. J. (1969). *Planta* 93, 34-42.

Ellis, W. C., Pfander, W. H., Muhrer, M. E. and Pickett, E. E. (1958). *J. Anim. Sci.* 17, 180-188.

Evans, H. J. and Sorger, G. J. (1966). *A. Rev. Pl. Physiol.* 17, 47-76.

Ferguson, W. S., Lewis, A. H. and Watson, S. J. (1940). Jealott's Hill Res. Stat. Bull. No. 1.

Ferguson, W. S., Lewis, A. H. and Watson, S. J. (1943). *J. agric. Sci., Camb.* 33, 44-51.

Filner, P., Wray, J. L. and Varner, J. E. (1969). *Science, N.Y.* 165, 358-367.

Fowden, L. (1959). *Physiologia Pl.* 12, 657-664.

Gallagher, C. H. (1964). "Nutritional Factors and Enzymological Disturbances in Animals". Crosby Lockwood and Son Ltd., London.

Goodman, G. T. and Roberts, T. M. (1971). *Nature, Lond.* 231, 287-292.

Hageman, R. H. and Hucklesby, D. P. (1971). *In* "Methods in Enzymology" (A. San Pietro, ed.), 23A, 491-503, Academic Press, New York and London.

Hageman, R. H., Leng, E. R. and Dudley, J. W. (1967). *Adv. Agron.* 19, 45-86.

Hallsworth, E. G., Greenwood, E. A. N. and Auden, J. (1957). *J. Sci. Fd Agric.* 8, 60-65.

Handreck, K. A. and Jones, L. H. P. (1967). *Aust. J. biol. Sci.* 20, 483-485.

Hansard, S. L., Crowder, H. M. and Lyke, W. A. (1957). *J. Anim. Sci.* 16, 437-443.

Heimer, Y. M. and Filner, P. (1971). *Biochem. biophys. Acta* 230, 360-372.

Hewitt, E. J. (1971). *In* "Nitrogen Nutrition of the Plant" (E. A. Kirkby, ed.), pp. 78-103, The Waverley Press, Leeds.

Hill, R. (1962). *Wld Rev. Nutr. Diet.* 3, 129-148.

Hodgson, J. and Spedding, C. R. W. (1966). *J. agric. Sci., Camb.* 67, 155-167.

Hucklesby, D. P. and Hewitt, E. J. (1970). *Biochem. J.* 119, 615-627.

Hyde, B. B., Hodge, A. J., Kohn, A. and Birnstel, M. L. (1963). *J. Ultrastruct. Res.* 9, 248-258.

Irvin, H. M., Wiseman, H. G., Shaw, J. C. and Moore, L. A. (1953). *J. Anim. Sci.* 12, 541-551.

Johnson, J. M. and Butler, G. W. (1957). *Physiologia Pl.* **10**, 100-111.
Johnson, C. M., Stout, P. R., Broyer, T. C. and Carlton, A. B. (1957). *Pl. Soil* **8**, 337-353.
Jones, D. I. H. (1970). *Nature, Lond.* **226**, 772.
Jones, D. I. H., ap Griffith, G. and Walters, R. J. K. (1961). *J. Br. Grassld Soc.* **16**, 272-275.
Jones, D. I. H. and ap. Griffith, G. (1965). *J. Sci. Fd Agric.* **16**, 721-724.
Jones, L. H. P. (1961). *Pl. Soil* **13**, 297-310.
Kemp, A. (1964). *Neth. J. agric. Sci.* **12**, 263-280.
Kemp, A., Deijs, W. B. and Kluvers, E. (1966). *Neth. J. agric. Sci.* **14**, 290-295.
Kennedy, G. S. (1968). *Aust. J. biol. Sci.* **19**, 529-538.
Kennedy, W. K. and Crawford, R. F. (1960). Proc. Cornell Nutr. Conf. 30-34.
Kirchgessner, M. and Grassmann, E. (1970). *In* "Trace Element Metabolism in Animals" (C. F. Mill, ed.), pp. 277-285, E. & S. Livingston, Edinburgh and London.
Kleiber, M., Smith, A. H., Ralston, N. P. and Black, A. L. (1951). *J. Nutr.* **45**, 253-263.
Klucas, R. V., Koch, B., Russell, S. A. and Evans, H. J. (1968). *Pl. Physiol., Lancaster* **43**, 1906-1912.
Lewin, J. and Reimann, B. E. F. (1969). *A. Rev. Pl. Physiol.* **20**, 289-304.
Lin, D. C. and Nobel, P. S. (1971). *Archs Biochem. Biophys.* **145**, 622-632.
Lofgreen, G. P. and Kleiber, M. (1953). *J. Anim. Sci.* **12**, 366-371.
Lofgreen, G. P. and Kleiber, M. (1954). *J. Anim. Sci.* **13**, 285-264.
Losada, M. and Paneque, L. (1971). *In* "Methods in Enzymology" (A. San Pietro, ed.), **23A**, 487-491, Academic Press, New York and London.
Lyon, G. L. (1969). Ph.D. thesis, Massey University, Palmerston North, New Zealand.
Lyon, G. L., Peterson, P. J. and Brooks, R. R. (1969). *Planta* **88**, 282-287.
McDonald, I. W. (1968). *Nutr. Abstr. Rev.* **38**, 381-400.
Malkin, R. and Rabinowitz, J. C. (1967). *A. Rev. Biochem.* **36**, 113-148.
Malyuga, D. P. (1964). "Biogeochemical Methods of Prospecting" (Acad. Sci. Press, Moscow, 1963, Translated Consultants Bureau, N.Y.), 205 pp.
Matrone, G. (1970). *In* "Trace Element Metabolism in Animals" (C. F. Mills, ed.), pp. 354-361, E. and S. Livingston, Edinburgh and London.
Matrone, G., Hartman, R. M. and Clawson, A. J. (1959). *J. Nutr.* **67**, 309-317.
Marais, J. S. C. (1944). *Onderstepoort J. vet. Sci. Anim. Ind.* **20**, 67-73.
Mason, H. S. (1965). *A. Rev. Biochem.* **34**, 595-634.
Mertz, W. (1969). *Physiol. Rev.* **49**, 163-239.
McEwan, T. (1964). *Nature, Lond.* **201**, 827.
Mills, C. F. (1956). *Biochem. J.* **63**, 190-193.
Mills, C. F. and Williams, R. B. (1971). *Proc. Nutr. Soc.* **30**, 83-90.
Mitchell, H. H. and Eldman, M. (1952). *Nutr. Abstr. Rev.* **21**, 787-804.
Mitchell, R. L. and Reith, J. W. S. (1966). *J. Sci. Fd Agric.* **17**, 437-440.
Molloy, L. F. and Richards, E. L. (1971). *J. Sci. Fd Agric.* **22**, 397-402.
Moxon, A. L. and Rhian, M. (1943). *Physiol. Rev.* **23**, 305-337.
Muhrer, M. E., Garner, G. B., Pfander, W. H. and O'Dell, B. L. (1956). *J. Anim. Sci.* **15**, 1291-1292 (Abstr.).
Nason, A. and McElroy, W. D. (1963). *In* Plant Physiology (Steward, F. C., ed.), Vol. 3, pp. 451-536, Academic Press, New York and London.
Nicholas, D. J. D. (1961). *A. Rev. Pl. Physiol.* **12**, 63-90.
O'Dell, B. L. (1962). Proc. Cornell Nutr. Conf. 77-83.
O'Dell, B. L. and Campbell, B. J. (1970). *In* "Comprehensive Biochemistry. Vol. 21. Metabolism of Vitamins and Trace Elements" (M. Florkin and E. H. Stotz, eds), pp. 179-266, Elsevier, Amsterdam.

Oelrich, P. B. and McEwan, T. (1961). *Nature, Lond.* **190**, 808-809.
Olson, O. E. and Moxon, A. L. (1942). *J. Am. vet. med. Ass.* **100**, 403-406.
O'Sullivan, M. (1970). *J. Sci. Fd Agric.* **21**, 607-609.
O'Sullivan, M., Flynn, M. J. and Codd, F. J. (1969). *Ir. J. agric. Res.* **8**, 111-119.
Ott, E. A., Smith, W. H., Harrington, R. B., Stob, M., Parker, H. E. and Beeson, W. M. (1966). *J. Anim. Sci.* **25**, 432-438.
Ozanne, P. G., Woolley, J. T. and Broyer, T. C. (1957). *Aust. J. Biol. Sci.* **10**, 66-79.
Parry, D. W. and Smithson, F. (1963). *Nature, Lond.* **199**, 925-926.
Patil, B. D. and Jones, D. I. H. (1970). *Proc. XIth Int. Grassld Congr.*, Brisbane, Australia, 726-730.
Peisach, J. (1965). Ed. "The Biochemistry of Copper", Academic Press, New York and London.
Peters, R. A., (1957). *Adv. Enzymol.* **18**, 113-159.
Peters, R. A. and Shorthouse, M. (1972). *Phytochemistry* **11**, 1339.
Peterson, P. J. (1969). *J. exp. Bot.* **20**, 863-875.
Peterson, P. J. (1971). *Sci. Prog., Lond.* **59**, 505-526.
Peterson, P. J. and Butler, G. W. (1962). *Aust. J. biol. Sci.* **15**, 126-146.
Peterson, P. J. and Butler, G. W. (1967). *Nature, Lond.* **213**, 599-600.
Pierce, A. W. (1957). *Aust. J. agric. Res.* **8**, 711-722.
Phillips, P. H., Greenwood, D. A., Hobbs, C. S., Huffman, C. F. and Spencer, G. R. (1960). Publ. Nat. Res. Council, Washington, No. 824.
Rains, D. W. (1971). *Nature, Lond.* **233**, 210-211.
Rajaratnam, J. A., Lowry, J. B., Avadheni, P. N. and Corley, R. H. V. (1971). *Science, N.Y.* **172**, 1142-1143.
Randall, P. J. (1970). *Proc. XI Int. Grassld Congr.*, Brisbane, Australia, 338-341.
Reid, M. S. and Bieleski, R. L. (1970). *Planta* **94**, 273-281.
Robards, A. W. and Robinson, C. L. (1968). *Planta* **82**, 179-188.
Russell, F. C. and Duncan, D. L. (1956): "Minerals in Pasture", Tech. Comm. 15, 2nd Ed., Comm. Bureau of Anim. Nutr., Aberdeen.
Schwartz, K. (1970). *In* "Trace Element Metabolism in Animals" (C. F. Mills, ed.), pp. 25-37, E. & S. Livingston, Edinburgh and London.
Schwartz, K., Milne, D. B. and Vinyard, E. (1970). *Biochem. biophys. Res. Commun.* **40**, 22-29.
Shaked, A. and Bar-Akiva, A. (1967). *Phytochemistry* **6**, 347-350.
Shrift, A. (1969). *A. Rev. Pl. Physiol.* **20**, 475-494.
Shroder, J. D. and Hansard, S. L. (1958). *J. Anim. Sci.* **17**, 343-352.
Simensen, M. G., Lunaas, T., Rogers, T. A. and Luick, J. R. (1962). *Acta vet. scand.* **3**, 175-184.
Sinclair, K. B. and Jones, D. I. H. (1964a). *J. Sci. Fd Agric.* **15**, 717-721.
Sinclair, K. B. and Jones, D. I. H. (1964b). *Br. vet. J.* **120**, 78-86.
Stewart, W. L. and Allcroft, R. (1956). *Vet. Rec.* **68**, 723-728.
Stout, P. R. and Meagher, W. R. (1948). *Science, N.Y.* **108**, 471-473.
Stout, P. R., Brownell, J. R. and Burau, R. G. (1967). *Agron. J.* **59**, 21-24.
Strogonov, B. P. (1962). "Physiological Basis of Salt Tolerance of Plants." Moscow (translated Israel Programme for Scientific Translations, Jerusalem, 1964).
Sutcliffe, J. F. (1962). "Mineral Salts Absorption in Plants," Pergamon Press, London.
Swan, J. B. and Jamieson, N. D. (1956a). *N.Z. Jl Sci. Technol.* (A) **38**, 137-151.
Swan, J. B. and Jamieson, N. D. (1956b). *N.Z. Jl Sci. Technol.* (A) **38**, 316-325.
Thompson, A. (1965). *Proc. Nutr. Soc.* **24**, 81-88.
Thompson, J. R. (1967). *A. Rev. Pl. Physiol.* **18**, 59-84.

Tillman, A. D. and Brethour, J. R. (1958). *J. Anim. Sci.* **17**, 104-112.

Todd, J. R. (1961). *J. agric. Sci., Camb.* **57**, 35-38.

Todd, J. R. (1962). *J. agric. Sci., Camb.* **58**, 277-279.

Toxaemic Jaundice Investigation Committee (1947). *Aust. vet. J.* **23**, 38-50; 253-261.

Toxaemic Jaundice Investigation Committee (1949). *Aust. vet. J.* **25**, 202-208.

Turner, R. G. (1970). *New Phytol.* **69**, 725-731.

Turner, R. G. and Marshall, C. (1971). *New Phytol.* **70**, 539-545.

Underwood, E. J. (1966). "The Mineral Nutrition of Livestock", F.A.O. and C.A.B.

Underwood, E. J. (1971). "Trace Elements in Human and Animal Nutrition," Academic Press, New York and London.

van Campen, D. (1970). *In* "Trace Element Metabolism in Animals" (C. F. Mills, ed.), pp. 287-296, E. & S. Livingston, Edinburgh and London.

van Soest, P. J. (1970). *Proc. Cornell Nutr. Conf.* 103-109.

Varela, G., Escriva, J. J. and Boza, J. (1970). *Proc. Nutr. Soc.* **29**, 38A-39A.

Walsh, T., Fleming, G. A., O'Connor, R. and Sweeney, A. (1951). *Nature, Lond.* **168**, 881.

Whitehead, D. C. (1966). "Nutrient Minerals in Grassland Herbage," Comm. Agric. Bur. Publ. No. 1/1966.

Williams, K. T., Lakin, H. W. and Byers, H. G. (1941). U.S.D.A., Tech. Bull. No. 758.

Wilson, A. D. (1966). *Aust. J. agric. Res.* **17**, 155-163.

Wilson, S. B. and Hallsworth, E. G. (1965). *Pl. Soil* **22**, 260-279.

Woolhouse, H. W. (1969). *In* "Ecological Aspects of the Mineral Nutrition of Plants" (I. H. Rorison, ed.), pp. 357-380, Blackwell, Oxford and Edinburgh.

Wright, E., (1955). *N.Z. Jl Sci. Technol.* (A) **37**, 332-348.

Wright, M. J. and Davison, K. L. (1964). *Adv. Agron.* **16**, 197-247.

Wright, D. E. and Wolff, J. E. (1969). *N.Z. Jl Agric. Res.* **12**, 287-292.

Yoshida, S., Ohnishi, Y. and Kitagishi, K. (1962). *Soil Sci. and Plant Nutrition, Tokyo* **8**, 107-113.

CHAPTER 20

Organic Acids, and their Role in Ion Uptake

W. DIJKSHOORN

Institute for Biological and Chemical
Research on Field Crops and Herbage,
Wageningen, the Netherlands

I. INTRODUCTION

The main aim of this chapter is to review those aspects of organic acid accumulation which are applicable to the interpretation of plant analysis in relation to plant growth and nutrition. It should be read in relation to Chapter 18. Reference to plant species other than pasture species has been necessary; only thus can the material be viewed in perspective.

The synthesis of organic acids in plants will be considered in relation to the balance of uptake and metabolism of the major nutrient ions. Hypotheses will be introduced only if they add anything to the convenience of our formulations. The possible role

163

of organic acids in the general pattern of plant nutrition will also be considered. The term "carboxylates" will be used to designate salts of organic acids throughout this chapter.

Accumulation of the salts of organic acids as an end product in ion metabolism is in itself no measure of their importance. One might even go further and say that their accumulation is a measure of their uselessness to the plant producing them. Indeed, evidence forces us to conclude that graminaceous plants lose much, and the unicellular alga *Scenedesmus* practically all of the carboxylates formed as a result of assimilation of absorbed nitrate.

But even if we are not directly concerned with their role in intermediary metabolism, the level of carboxylates may serve as a convenient index for the state of ion metabolism and accumulation.

A range of treatments can be selected that will be bound to show differences in the level of carboxylates and growth. If in addition, all the other nutrient ions are given due consideration in evaluating results, evidence can be obtained to suggest that the level of carboxylates should be considered in any system for evaluating balanced nutrition for optimal growth (de Wit *et al.,* 1963; Van Tuil, 1965; Kostić *et al.,* 1967).

Because of emphasis upon their direct relation to accumulation and metabolism of nutrient ions, ion equivalent weights will be used as the units to express quantities of the anions of the organic acids.

II. MECHANISMS OF ACCUMULATION

A. *Uptake of Cations*

1. *Ammonium*

An acid is defined as any substance which is capable of liberating free H^+ ions in solution. Likewise, a base is any substance capable of eliminating free H^+ ions from solution.

When plants are grown in a nutrient solution containing ammonium ions as the source of nitrogen, the solution gradually becomes more acid. With nitrate as the source of nitrogen, the solution will become more alkaline with growth of, for instance, graminaceous plants.

In accordance with these definitions, it can be said that in the presence of ammonium ions the plant acts as an acid on the medium, whereas with nitrate it may exhibit the properties of a base.

A simple theory to explain this behaviour may first be presented. For simplicity of exposition, consider a spherical plant body with a central chlorophyllous part surrounded by a colourless zone. Let the outer zone provide a path which permits ions to pass from the medium into the central metabolic region.

If the organism is immersed in distilled water, any synthesis of organic acid will result in acidification of the tissue. But experiment has shown that most plant tissues do not significantly change their acid content if there is no uptake of salt ions from the medium (Ulrich, 1942).

On placing in a solution of ammonium chloride, the organism will be surrounded by NH_4^+ and Cl^- ions coming from the salt, and by H^+ and HCO_3^- ions coming from the dissociation of water in the presence of carbon dioxide. If the plant normally contains 2 mequiv. of nitrogen per gram dry matter, the new growth will absorb 2 meq NH_4^+ for each gram dry matter produced. This transfer will proceed along with 2 meq (Cl^- + HCO_3^-) by reason of electroneutrality.

Analysis for N and chloride in the whole plant material reveals the quantity of NH_4^+ ion supplied for assimilation in balance with chloride and bicarbonate ions.

Whereas the quantity of NH_4^+ ions absorbed is controlled by the growth and requirement for nitrogen, the proportion of chloride and bicarbonate will be controlled by the selectivity of the outer surface.

If the membrane is impermeable to chloride, all the NH_4^+ will be assimilated along with HCO_3^- according to:

$$NH_4^+ + HCO_3^- \rightarrow (NH_3) + H_2O + CO_2$$

The parenthesized term denotes organic N which in all forms represents substituted ammonia and is non-ionic. Its quantity is conveniently expressed as the ion-equivalent ammonium (or nitrate) consumed for its synthesis.

There will be left in the medium an excess of Cl^- over NH_4^+ which is now matched by H^+ originating from the dissociation of H_2CO_3 and the removal of HCO_3^- by uptake in place of Cl^-. Consequently, part of the weak carbonic acid is replaced in the medium by the strong hydrochloric acid.

Were there no absorption of HCO_3^-, all the NH_4^+ would be supplied to metabolism along with Cl^-, and metabolized according to:

$$NH_4^+ + Cl^- \rightarrow (NH_3) + H^+ + Cl^-$$

Except for the removal of some ammonium chloride the medium would not change in acidity. But the metabolic tissue would furnish 2 meq HCl for each gram dry matter produced. If the amount of tissue sap were of the order of 6 grams per gram dry matter, its titratable acidity would increase to 0·3 N. Since in the normal region of pH the buffer capacity is only about 30 meq/1 per unit change in pH (Leuthart, 1927; Hurd-Karrer, 1930; Ulrich, 1941) this would create intolerable acidity.

Experiments have shown that with ammonium ions in the nutrient solution, tissue pH remains unchanged (Arnon, 1939) or is only slightly lowered (Keyssner, 1931). From this it would appear that ammonium ions are absorbed and assimilated in balance with bicarbonate ions. The possibility of excretion of acid as fast as it is formed, or its destruction by bicarbonate released from a pool of carboxylates within the tissue through decarboxylation, are not supported by physiological evidence to be explained later.

Hence, the acidity in the medium originates from differential uptake of cations and anions of the nutrient salt, rendered possible by the participation of the ions supplied by the dissociation of water or carbonic acid. The extent to which H^+ or HCO_3^- participate can be deduced from the balance-sheet of uptake of cations and anions other than hydrogen or bicarbonate ions.

The above considerations show that uptake and utilization of ammonium ions yields organic nitrogen and mineral acid, but no carboxylates.

2. Metal Cations

On placing the organism in a solution of potassium sulphate, K^+ will be absorbed greatly in excess of $SO_4^=$ when the uptake system exhibits the common type of selectivity. Analysis for potassium and sulphate in the plant material can be applied to determine the potassium absorbed in association with bicarbonate. If the increments were 2 and 0·5 meq/g dry matter, respectively, the tissue would contain 1·5 meq of potassium bicarbonate per gram dry matter.

If potassium chloride were applied, the gain in bicarbonate might have been smaller, e.g. 0·5 meq/g dry matter. Since chloride is absorbed much faster than sulphate, the participation of bicarbonate will be considerably reduced (Jacobson, 1955; Hiatt, 1967).

3. Carboxylation

In plants where the uptake of bicarbonate is appreciable, adequate methods of analysis show that it is still undetectable within the tissue, but that the expected numerical value agrees with an increase in the anions of organic acids such as malate, citrate, oxalate, etc. (Böhning and Böhning-Seubert, 1932; Overstreet et al., 1940).

Hurd (1958) has shown that radiocarbon from bicarbonate in the medium accumulates in beet disks mainly in the form of malate. Similar results were obtained by Bedri et al. (1960) for roots of intact plants, by Jackson and Coleman (1959) for excised bean roots, by Hiatt (1967) for barley roots, and by other workers for a variety of plant material.

It therefore appears that the bicarbonate absorbed along with the metal cations is converted into carboxylate by carboxylation:

$$K^+ + HCO_3^- + RH \rightarrow K^+ + RCOO^- + H_2O$$

where RH stands for some metabolite, for instance pyruvate, which is subject to carboxylation, and $RCOO^-$ denotes the anion of the various organic acids common in plants.

It is often presumed that the entry of bicarbonate ions will stimulate enzyme activity involved in the synthesis of organic acids, e.g. Hiatt and Hendricks (1967). If this were true the tissue would synthesize organic acids first which, on dissociation, yield H^+ ions to neutralize the entering HCO_3^- ions. The result will be the same; the absorbed metal cations will accumulate within the tissue counterbalanced by carboxylates.

Even with more acid media the presumption of available HCO_3^- is the more convenient one. However, there remains the possibility that the tissue creates organic acid first and that the H^+ supplied by its dissociation is released to the medium in exchange for K^+ (Hiatt and Hendricks, 1967).

A spontaneous temporary increase in organic acid content has been observed in excised leaves of crassulacean plants placed in the dark (Gustafson, 1925; Vickery, 1956). During the first day the level of free organic acid (mainly malic) is raised by about 0·2 meq/g fresh weight, and there is a significant fall in pH of the tissue. The level of their salts, the carboxylates, remains unchanged because there is no uptake of salt. This fact is often overlooked, doubtless due to the emphasis which has been placed on organic acid metabolism.

In this context, the finding by Ulrich (1941), Butler (1953), and Neirinckx (1967) that root tissue begins to absorb the cation in excess of the anion in the supplied salt for some hours, and that this is followed by a relative increase in uptake of the salt anion, may be of interest.

The tendency of HCO_3^- to combine with any free H^+ irrespective of its source is greater than that of the carboxylate anions; HCO_3^- is a stronger base than $RCOO^-$. Its conversion therefore prevents the sap from becoming more alkaline.

Thus the uptake of metal cations unmatched by non-basic nutrient anions such as chloride, nitrate, phosphate and sulphate, generates carboxylic anions within the tissue.

B. Metabolism of Anions

1. Sulphate

The sulphate ion $(S^{6+}O_4^=)^=$ is assimilated in the form $(S^=H_2^+)$ since the majority of organic sulphur substances in tissues are in a chemical sense substituted hydrogen sulphide.

The ionic equation for this reductive conversion is:

$$(S^{6+}O_4^=) + 8H \rightarrow (S^=H_2^+) + 2H_2O + 2OH^-$$

The hydrogen for hydrogenation supplied by metabolism is represented for brevity as atomic. Consequent on the formation of non-ionic organic S, the sulphate anion is replaced by its equivalent of hydroxyl anion.

If sulphate anions were taken up along with H^+ ions and assimilated, the OH^- ions would be neutralized to form water, and the only result would be a gain in organic S. If sulphate anions were taken up along with metal cations, the hydroxyl ions will combine with CO_2 to give bicarbonate ions, and this will be converted into carboxylates:

$$2K^+ + SO_4^= + 8H + 2CO_2 + 2RH \rightarrow (SH_2) + 2K^+ + 2RCOO^- + 4H_2O$$

The latter equation expresses the electroneutrality of anion assimilation. If the absorbed anions are assimilated, carboxylates substitute for assimilated anions in amounts equivalent to match the metal cations.

2. Nitrate

For nitrate assimilation the reduction equation is:

$$(N^{5+}O_3^=)^- + 8H \rightarrow (N^{3-}H_3^+) + 2H_2O + OH^-$$

The parenthesized term at the right of the arrow denotes the various organic nitrogen assimilation products which are all substituted ammonia. The exact pathway of assimilation does not affect the argument; the present formulation merely illustrates the principle involved in carboxylate synthesis.

Were the nitrate anion transferred to the metabolic tissue along with H^+ ion the OH^- ion would be neutralized to form water, and the only detectable result would be a gain in organic N. But were K^+, or any other metal cation, the partnering ion, the tissue would be left positive and the formation of carboxylate from OH^-, CO_2, and some metabolite RH subject to carboxylation would be required:

$$K^+ + NO_3^- + 8H + RH \rightarrow (NH_3) + 3H_2O + K^+ + RCOO^-$$

It is important to note that what is metabolized is an anion, and that carboxylate anion is synthesized in amounts sufficient to match the absorbed metal cations.

Sulphate and nitrate metabolism thus require the synthesis of carboxylate for a different reason than the previous type in that the bicarbonate for synthesis is generated within the tissue where nitrate and sulphate are assimilated.

C. Disparity Between Carboxylate and Nitrogen Level

Suppose the plant body were suspended in a solution of potassium nitrate containing K^+, H^+, NO_3^- and HCO_3^-. The ions will enter in proportions determined by the selectivity of the membrane, but $K^+ + H^+$ will be equal to $NO_3^- + HCO_3^-$ by reason of electroneutrality of ion transfer.

If 2 meq NO_3^- were absorbed along with 1 meq K^+ then these ions would have passed along with 1 meq H^+ through the outer zone to the metabolic zone. Then, metabolism would produce 2 meq organic N, but only 1 meq potassium carboxylate.

There will be removed from the medium 2 meq NO_3^- and 1 meq K^+, so that 1 meq K^+ will be left in balance with HCO_3^-. Thus part of the nitrate is replaced by bicarbonate and the medium will become more alkaline. This is in harmony with the situation in graminaceous plants. The deficit in carboxylate is matched by the gain in bicarbonate in the medium. In ryegrass organic N accumulates at around 2 meq/g dry matter, and carboxylates are at a level of 1 meq/g dry matter so that these plants exhibit a very distinct external alkaline effect when exposed to nitrate containing media. In the unicellular alga *Scenedesmus* the organic N content is 5 meq/g dry matter, and the carboxylate level must be very low when judged from the data of Krauss and Thomas (1954) so that the external alkaline effect will be considerable: nearly 1 equivalent of titratable alkali per equivalent of nitrate consumed by the plants.

In young guayule (Cooil, 1948), sugar beet (Van Egmond and Houba, 1970), potato (Dijkshoorn, 1970) and tomato plants (Kirkby and Mengel, 1967) organic N and carboxylates accumulate in about equivalent quantities with nitrate in the medium so that the external alkaline effect, if detectable, must be very small.

These data refer to plants with a complete supply of nutrients where sulphate metabolism too contributes to the generation of carboxylates. However, there is much evidence that sulphate metabolism proceeds at the very low rate of only 0·06 equivalents per 1 equivalent of organic N formed within the plant (Dijkshoorn and van Wijk, 1967).

If a deficit in carboxylate content were due to transfer of H^+ to the metabolic tissue, the amount of titratable acidity among the ions en route should be a value of significant magnitude relative to nitrate. It is of interest therefore to consider in the following section the ionic composition of transport liquids.

D. Site of Synthesis, Transfer and Degradation of Carboxylates in Higher Plants

1. Leaves as Organs of Synthesis of Carboxylates: the Nitrate System

It is appropriate at this stage to reshape the model in accordance with the organization of the higher plant. Replace the outer membrane by the roots, the outer zone by the transport tissues xylem and phloem, and the inner metabolic region by the leaves.

The proportion of ions entering the transport system is controlled by the uptake system in the roots. The capacity for uptake is determined by the growth, for instance, 2 meq of the nitrogenous ion per gram dry matter produced.

The absorbed salts are released to the xylem and conducted to the leaves where nitrate is metabolized to generate organic N and carboxylate. In addition, a much smaller amount of carboxylate is created by sulphate metabolism which will be neglected in this discussion.

If a medium is prepared containing the nitrates, chlorides, di-hydrogen phosphates and sulphates of potassium, sodium, mag-nesium and calcium, the sum of cation equivalents (C) is equal to the sum of anion equivalents (A). In such a medium the value for (C-A) is zero, and the concentrations of H^+ and HCO_3^- are low and equal. The capacity of the medium to supply the latter two ions is unrestricted, because of further dissociation of water in the presence of carbon dioxide.

If some acid of the anions A were added, the solution would contain H^+ and (C-A) would be negative to the extent of the H^+ concentration. Likewise, if some bicarbonate of the cations C were added, the solution would be positive in (C-A).

If the external alkaline effect and its cause or effect, the deficit in carboxylate, were due to H^+ absorption, the xylem liquid would exhibit a negative value for (C-A).

Investigations into the ionic composition of xylem exudate from maize grown with nitrate in the medium have revealed that (C-A) in the xylem liquid is zero under these conditions. Since no detectable H^+ migrates with the ions, all the nitrate and other anions A are supplied for assimilation along with metal cations C. This implies that carboxylates are generated by the leaves in an amount equiva-lent to that of organic nitrogen.

The volume of the conducting elements is so small that whole plant analysis reflects the composition of the accumulating tissues. Sequential harvests of replicates have shown that carboxylates accumulate in amounts of about one-half the organic nitrogen for a long series of stages with maintained levels of supply. The combined evidence suggests that the deficit in carboxylates must be due to a continuous removal from the plant body.

The only pathway by which carboxylates can escape is their release from the roots to the medium. Since the medium became alkaline, and titration curves of the partially exhausted nutrient solution revealed an amount of bicarbonate equal to the deficit in carboxylates in the plant grown on it, the missing carboxylates must have passed into the medium as bicarbonate. As the carboxylates arise from nitrate metabolism in the leaves, the portion lost by the plant must have moved downward via the phloem along with other assimilates to the roots. Analysis of honey dew from aphids feeding on the plants showed the presence of a variety of potassium carboxylates. Phloem sap delivered by severed stylets of aphids feeding on willow was found to contain potassium salts of organic acids as the main electrolyte constituents (Peel and Weatherley, 1959).

From this it would appear that carboxylates move down the phloem as such, and that decarboxylation takes place in the root tissues:

$$K^+ + RCOO^- + H_2O \rightarrow K^+ + HCO_3^- + H_2O + RH$$

The bicarbonate is then released to the medium in exchange for extra nitrate from the medium (Hoagland and Broyer, 1936).

There is no sure way of measuring the quantity transferred by the phloem. However, judging from data on xylem sap composition and the nature of the external alkaline effect, the above picture seems fairly well established.

2. Roots as Organs of Synthesis of Carboxylates: the Bicarbonate System

Usually more nitrate and sulphate are absorbed than can be metabolized and this excess accumulates unchanged. If the nitrate supply is discontinued by transfer to a medium in which all the nitrate is replaced by sulphate, further growth consumes the nitrate in the tissues and the plant becomes progressively starved of nitrate. Within a few days after transfer, carboxylates can be detected in the xylem liquid, and with the further advance of depletion nitrate disappears and is completely replaced by carboxylates. The effect can be reversed by subsequent restoration of nitrate to a depleted culture. Then the level of nitrate is raised from zero to a final level of about 15 meq per litre xylem exudate, and the carboxylate concentration falls from around this level to practically zero (Dijkshoorn, 1970). Chromatographic analysis showed that for maize all the (C-A) constituents in the xylem liquid were in the form of the carboxylates of the non-volatile organic acids, and that malate was the major component. Thus when the medium is replaced by one containing additional sulphate in place of nitrate, maize plants continue to absorb nutrient ions, but now bicarbonate substitutes for nitrate in anion uptake. The roots begin to generate carboxylates from the absorbed bicarbonates, and these are released to the transpiration stream in the xylem, transferred to the shoot, and further distributed via the phloem.

In this way bicarbonate absorption provides another bulk source of carboxylates to all the organs of the plant. In maize it is not until the nitrate in the tissues is exhausted that appreciable quantities of carboxylates can be detected in the xylem liquid. Here, the bicarbonate system operates only in the absence of nitrate. By contrast buckwheat (*Polygonum fagopyrum*), with a full supply of nitrate and the other nutrients, accumulates carboxylates in excess of organic N. From this it would seem that both systems operate together.

A typical case is the feeding with ammonium in place of nitrate ion. Here, whole plant analysis accounts for bicarbonate as the partnering anion for all the ammonium, and for some part of the metal cations C. The former bicarbonate fraction is consumed together with the ammonium ions to form organic nitrogen, the latter bicarbonate fraction is converted into carboxylates of the metal cations which accumulate as such within the tissues. The balance-sheet of ion analysis shows that the synthesis of organic acids proceeds at a slow rate and, for this reason, the level of organic acids is always lower when the plants are grown with ammonium instead of nitrate.

Thus both chemical considerations and physiological evidence invite us to consider the nitrate and the bicarbonate systems as the principal stimuli for the synthesis of organic acids, which accumulate in the form of their salts, the carboxylates, as a consequence of ion uptake, ion utilization, and growth. Emphasis may be placed on their common aspect of absorption of metal cations in balance with anions consumed by metabolism; and on their distribution within the plant, with the nitrate system operating in the leaves and the bicarbonate system in the roots.

III. PLANT COMPOSITION

A. The Value (C-A) in Plant Analysis

1. The Ionic Balance

The carboxylates are the quantity thought of in conversation when we speak of organic acids. It is unfortunate that this mode of expression from the parlance of the laboratory is in use, because it incorrectly suggests titratable acidity (Willaman *et al.*, 1919). Based on titratable acidity of plant sap the findings of Vickery on the decrease in "organic acid" content of tobacco following substitution of ammonium for nitrate in the medium were recently questioned by Dougall and Birch (1967). There is every reason to expect an increase, if any, in the acidity of the sap, but consideration of the ionic balance predicts a fall in the carboxylate level, and this appears to have been widely established.

From the static point of view the "organic acids" are merely the complement to accumulation of inorganic nutrient ions left unchanged by metabolism. The carboxylate content is numerically fixed by the excess of metal cations over the inorganic anions. Plant analysis for K^+, Na^+, Mg^{2+}, Ca^{2+} gives the sum C. Plant analysis for NO_3^-, Cl^-, $H_2PO_4^-$, $SO_4^=$, all referring to the unchanged ionic form in the tissues, gives the sum A. This quantity should include total phosphorus since organophosphate compounds in tissues generally carry the same charge as monovalent phosphate ion. The quantities of other inorganic ions in the plant are usually so small that their omission will make little difference to the additive values C and A. From the results of analysis for the inorganic ions we can thus calculate (C-A), the numerical value for the total of carboxylates in the plant material, expressed in milliequivalents.

2. Ash Alkalinity

A crude but nevertheless successful method for quick measurements of (C-A) is to measure the alkalinity of the ash. If the powdered dry plant material is gently ashed at 500°C all the carboxylates and nitrates are converted into oxides of the metal cations C. Addition of excess standard acid and titration to pH 5 gives the ash-alkalinity. To obtain (C-A) the value is corrected for nitrate determined in another subsample of the material. Advantage of this can be taken in survey work on small samples; the ash can be further analyzed for the cations and phosphate. This procedure gives results agreeing reasonably well with those obtained from analysis for the inorganic ions (Van Tuil et al., 1964; Kirkby and Mengel, 1967).

3. Resolution of (C-A): Organic Acid Composition

The carboxylates from plant material are usually extracted in the form of their acids, and isolated by means of ion exchange resins. For their further resolution and quantitative determination partition chromatography with silica gel as the supporting medium, dilute sulphuric acid as the stationary phase, and butanol-chloroform mixtures as the mobile phase is often applied (Bulen *et al.*, 1952; Dijkshoorn and Lampe, 1962).

Where plants with a more acidic cell sap may contain some free organic acids which do not contribute to (C-A), these acids are included. On the other hand, some 10 to 20% of the value for (C-A) consists of the salts of polyuronic acids which are insoluble and should be measured by a further analysis of the plant material left after decationization and extraction. This correction will only be approximate unless allowance is made for the uronic acid groups which are esterified. With material high in calcium special precautions are often necessary to attain complete recovery of oxalic acid (Houba *et al.*, 1971).

During the isolation of the water soluble acids evaporation to dryness is usually carried out and the volatile acids are lost. Resolution of the non-volatile organic acids from ryegrass has shown that malic and citric are dominant, with smaller quantities of oxalic, fumaric, succinic, malonic, and quinic acid. The total recovered accounted for about 90% of the value for (C-A) in the herbage (Dijkshoorn, 1962; Van Tuil, 1965).

Studies on the availability of calcium from grasslands herbs involving feeding experiments with rats have given evidence that a higher oxalic acid content in the herbage reduces the percent absorption of calcium from the food (Armstrong *et al.*, 1953, 1957). In grass species the level of oxalic acid is low and equivalent to about one-third (cocksfoot) to one-fifth (ryegrass) of the calcium in the herbage (Dijkshoorn *et al.*, 1959). In sorrel the level of oxalic acid is high and the oxalic acid : calcium ratio in the foliage was found to vary between 2 and 3 depending on fertilizer treatment. Oxalic acid is toxic to animals grazing herbage if it is present in sufficient amounts. Toxic levels of oxalic acids are normally present in the leaves of plants such as *Chenopodium*, *Oxalis* and *Amaranthus* species which are not usually regarded as herbage. Quite high levels can, however, occur in some tropical, panicoid grasses and deaths in cattle grazing the tropical grass *Setaria sphacelata* were attributed to the high levels (5-7% of dry weight) of oxalic acid present (Jones *et al.*, 1970). Oxalic acid levels recorded in several studies of tropical grasses are listed in Table I. The variations in amounts of oxalic acid in the several varieties of *Panicum maximum* and strains of *S. sphacelata* listed in Table I indicate considerable intraspecific variation. Detailed studies on oxalate levels in *S. sphacelata* indicate that

TABLE I

Oxalic acid levels in tropical (panicoid and eragrostoid) grasses grown and sampled under similar conditions

Common Name	Species	Oxalic Acid (% of total plant dry weight)
	1. Garcia-Rivera and Morris (1955)	
Cuban grass	*Andropogon caricosus*	0·25
Carpet grass	*Axonopus compressus*	0·02
Mexican bluegrass	*Chloris inflata*	0·43
Bermuda grass	*Cynodon dactylon* (2 var.)	0·02, 0·16
Star grass	*C. plectostachyus*	0·09
Pangola grass	*Digitaria decumbens*	0·89
Malojilla	*Eriochloa polystachya*	0·22
Molasses grass	*Melinis minutiflora*	0·41
Guinea grass	*Panicum maximum* (5 var.)	1·10, 1·05, 1·65, 2·01, 2·26
Para grass	*P. purpurascens*	1·24
Venezuela grass	*Paspalum fasciculatum*	0·02
Sweet grass	*P. plicatulum*	0·02
Buffel grass	*Pennisetum ciliare*	0·83
Merker grass	*P. purpureum var. merkerii*	2·48
Elephant grass	*P. purpureum*	2·57
Cerrillo	*Sporobulus indicus*	0·22
Beach grass	*S. virginicus*	0·12
San Augustine	*Stenotaphrum secundatum*	1·20
	2. Lal *et al.* (1966)	
Napier grass	*Pennisetum purpureum*	3·31–3·57
Bajra grass	*P. typhoides*	2·95–1·54
Pusa giant	*P. purpureum* ×	
Napier grass	*P. typhoides*	3·04–1·32
	3. Jones *et al.* (1970)	
	Setaria sphacelata (6 var.)	2·78–5·13[a]
	S. trinervia	4·16
	S. splendida	3·97

[a] Values up to 7·80% were recorded in some samples.

they are highest in the leaves and are increased by nitrogen fertilization (Jones and Ford, 1972).

Stout *et al.* (1967) have shown that transaconitate is a normal constituent of herbage plants and accumulates to higher concentrations in some graminaceous species. They suggest that their finding may be related to the possible action of carboxylates in hypomagnesia induction in grazing cattle.

From the point of view of plant physiology there is little use in a more complete record of data on the various types of organic acids in plants. Attention is better directed to an account (given below) of their distribution within selected plant organs etc. where malate and citrate exhibit some typical features. Nevertheless these acids can represent for the ruminant a significant part of the readily digested carbon in the herbage. For this reason data obtained from several major herbage species for levels of total and individual organic acids during the growing season are given in Tables II-V. On the whole the results indicate that from 3-10% of the dry matter of herbage can be organic acids and most, but not all, of the fall in organic acid content during the growing season is due to an increase in the proportion of stem tissue. Malonic acid has been detected in many but not all legume leaves (Bentley, 1952) and appears to be more important in this herbage than in grasses.

B. Partition Between Plant Organs

1. Potato tuber and fruit

Absorbed ions move in the xylem, in proportions determined by the supply and the selective properties of the uptake system in the roots, to the assimilatory tissues in the leaves. The phloem transfers the carboxylates released by metabolism of nitrate (and sulphate) along with the other assimilates to the heterotrophic organs of the plant. The main ionic constituent of phloem sap is potassium carboxylate with some phosphate and magnesium, but there is a marked absence of calcium compared with xylem sap and foliage (Ziegler, 1963). Apparently, transfer to the conducting elements of the phloem involves a second selective system with a marked preference for potassium.

The expanding potato tuber is only a special case of phloem import of assimilates into a growing organ. Here, the cation regime is mainly delegated to potassium, and experiment shows that there is a striking constancy of the concentration of potassium and the sum of cations C throughout the time of development.

The nutritional system of the developing fruit is another case of phloem import of potassium carboxylates—hence the close correlation between potassium and citrate in the orange fruit (Smith and Rasmussen, 1964). In the leaves there is little change in the level of citric acid with the increase in potassium, but here oxalate and malate are dominant and the levels of these acids vary with that of calcium (Rasmussen and Smith, 1961). In the tomato fruit the

TABLE II

Changes in levels of individual organic acids in total herbage in festucoid grasses during the growing season

Grass	Number of cuts	Organic Acids (% of plant dry weight)							References
		Shikimic	Quinic	Succinic	Malic	Citric	Fumaric	Total	
Ryegrass (Lolium perenne)	3	0·25–0·05[a]	1·51–0·14	0·30–0·36	1·74–1·28	0·70–0·51	0·04–0·06	4·54–2·34	Jones and Barnes (1967)
	5	0·30–0·07	0·61–0·05	0·05–0·03	0·82–0·07	0·50–0·04	—	2·28–0·73	Martin (1970)
	4							7·22–2·3[b]	Fauconneau (1958)
Cocksfoot (Dactylis glomerata)	3	0·54–0·20	1·24–0·74	0·34–0·25	0·74–0·71	0·30–0·22	0·08–0·04	3·24–2·16	Jones and Barnes (1967)
	4	—	0·68–0·36	0·08–0·06	1·29–1·24	0·32–0·12	—	2·36–1·72	Fauconneau (1959)
Meadow Fescue (Festuca pratensis)	4	—	0·92–0·31	—	2·45–1·34	0·39–0·23	—	3·75–1·89	Fauconneau (1959)

[a] Values given are for first and last cuts.
[b] Levels of malic, citric and quinic acids also given as % of the total acid values.

TABLE III

Levels of organic acids in total herbage from festucoid grasses grown under the same conditions

Species	Organic acids (% of plant dry weight)							
	Shikimic	Quinic	Succinic	Malic	Citric	Fumaric	Trans-aconitic	Total
1. Jones and Barnes (1967)								
Meadow fescue (Festuca pratensis)	0·12	0·62	0·25	1·10	0·37	—	—	2·46
Timothy (Phleum pratense)	0·18	0·56	0·22	0·82	0·42	0·02	—	2·22
Perennial ryegrass (Lolium perenne)	0·12	0·30	0·24	0·88	0·45	0·02	—	1·99
Cocksfoot S143 (Dactylis glomerata)	0·14	0·44	0·28	0·81	0·32	—	—	1·99
Cocksfoot S37	0·37	0·66	0·32	0·94	0·27	0·02	—	2·58
N.Z. Agrostis (Agrostis tenuis)	0·05	0·14	0·30	0·56	0·24	0·01	—	1·30
Red fescue (Festuca rubra)	0·22	0·24	0·28	1·30	0·28	—	—	2·32
2. Molloy (1969)[a]								
Perennial ryegrass							0—0·02	
Italian ryegrass (Lolium multiflorum)							0—0·02	
Cocksfoot							0—0·02	
Yorkshire fog (Holcus lanatus)							0·36—1·69	
Barley grass (Hordeum murinum)							0·50—1·42	
Prairie grass (Bromus unioloides)							1·84—3·40	

[a] Ranges of values obtained from 4-20 samples for each species.

TABLE IV

Levels of organic acids in individual plant parts of festucoid grasses

| Plant part | Organic acids (% of plant dry weight) | | | | | | | References |
	Shikimic	Quinic	Succinic	Malic	Citric	Fumaric	Total	
1. Ryegrass (*Lolium perenne*)								
Leaf	0·07–0·30[a]	0·40–0·61	0·05–0·03	0·82–1·32	0·50–0·65	—	2·01–2·50	Martin (1970);
Stem	0·01	0·10	0·01	0·57	0·38	—	1·08	Fauconneau (1958)
Sheath	—	—	—	—	—	—	2·06–2·83	
2. Cockfoot (*Dactylis glomerata*)								
Leaf	—	0·77–0·90	0·09–0·12	1·47–2·27	0·17–0·38	—	2·75–3·35	Fauconneau
Sheath	—	0·21–0·51	—	0·86–1·19	0·12–0·19	—	1·52–1·68	(1959)
Stem	—	0·19–0·32	—	1·22–1·02	0·09–0·13	—	1·30–1·67	
3. Fescue (*Festuca pratensis*)								
Leaf	—	0·68–1·07	—	2·64–2·94	0·42–0·67	—	3·83–4·53	Fauconneau
Sheath	—	0·32–0·50	—	0·98–1·36	0·12–0·18	—	1·63–1·86	(1969)
Stem	—	0·24–0·28	—	1·34–1·35	0·09–0·15	—	1·68–1·79	

[a]Results given as range measured.

TABLE V

Levels of organic acids in herbage species other than grasses

Species	Plant Part	Malic	Citric	Malonic	Quinic	Succinic	Total	References
Lucerne (*Medicago sativa*)	Whole plant	6·33—2·60[a]	0·88—1·28	1·45—1·86	1·13—0·40	—	9·80—5·84	Fauconneau (1958)
	Leaf	7·22—3·63	0·86—2·21	1·57—3·15	0·34—1·02	—	10·3—10·74	
	Stem	3·85—2·20	0·85—0·48	1·05—1·37	2·77—0·21	—	8·50—4·46	
Red clover (*Trifolium repens*)	Whole plant	—	—	—	—	—	2·80—3·80	Mangan and Johns (1957)
White Clover (*T. pratense*)	Whole plant						3·00—3·50	Mangan and Johns (1957)
Chou Moellier (*Brassica oleracea*)	Leaf blade	6·94	1·43	—	0·31[b]	0·07	8·75	Fauconneau (1958)
	Petiole and leaf midrib	2·43	0·37	—	0·10[b]	0·08	2·98	
	Stem	1·54	0·42	—	0·23[b]	0·08	2·27	

[a] Results given as range from first to last cuts; values for individual acids calculated from quoted percentage of each acid in total acids.
[b] Quinic + shikimic acid.

relation between potassium and carboxylate has been demonstrated by Bradley (1964) and Davies (1964).

The mother tuber of potato exports its material to the growing plant and this export takes place in the phloem. Here again, there is little change in the concentration of potassium in the dried material with the progressive fall in dry weight. But the tuber receives continuously ions from the medium by uptake and, among these, calcium is not reexported by the phloem so that the dry substance is left rich in calcium. The loss in dry material and potassium, and the gain in calcium proceed in such a manner that there is a continuous increase in the concentration of total cations and carboxylates (C-A), in the dry matter with the further progress of exhaustion by the growing plant. This change goes along with a progressive substitution of malate for citrate as the dominant carboxylate (Hagemann, 1961; Jolivot, 1959; Dijkshoorn, 1970).

2. Ageing of Leaves

If we compare the levels of calcium, (C-A), and malate in leaves of different age the analogy with the situation in potato tuber will become manifest.

In early life the leaf depends on import of assimilates from older leaves via the phloem. When grown to a final size it in turn exports assimilates to the other organs (Webb and Gorham, 1964). During the life of the leaf there is an increasing supply of nutrient ions from the transpiration stream in the xylem. Substitutions similar to those already described for the mother tuber of potato would produce a marked change towards increasing levels of calcium, carboxylates (C-A), and malate, when during ageing the leaf is directed towards export of assimilates. Indeed the levels of calcium, (C-A), and malate were found to increase by a factor of 10 from the younger top leaves to the older bottom leaves of plant species that absorb calcium readily, such as guayule (Cooil, 1948), tobacco (Vickery, 1961), Brussels sprouts (Kirkby and DeKock, 1965) and cabbage (DeKock and Morrison, 1958).

3. Role of Potassium

When the uptake system is capable of absorbing calcium readily enough, absence of K in the medium may stimulate Ca uptake to such an extent that the average level of carboxylates (C-A), in the whole plant is raised to a value higher than the normal. In tobacco deficient in potassium there is an increased level of carboxylates compared with that in normal plants (Böhning and Böhning-Seubert, 1932; Chouteau, 1960).

In graminaceous plants the uptake of calcium is delegated to a system of low capacity and non-competitive with respect to potassium (de Wit *et al.*, 1963), so that calcium does not significantly substitute for potassium when the latter ion is omitted. Here, shortage of potassium creates a deficit in the cations C, and the level of (C-A) is low compared with plants grown under normal conditions (Kostić *et al.*, 1967).

The uptake of potassium by graminaceous plants is delegated to a system of high capacity, and potassium contributes to a considerable extent in supplying the plant with metal cations. Although its function in plants is still a subject for speculation, it seems that its particular role in plant nutrition concerns the distribution of carboxylates by its ready release to the conducting system of the phloem. It is only in this latter property that it differs for all plants from the other metal cations. Its unrestrained mobility suggests a regulatory function in the traffic and accumulation of carboxylates.

C. Time Course of Accumulation in Ryegrass

1. General Considerations

The concentration (milliequivalents per gram) of a constituent within plants is the integrated result of accumulation (milliequivalents per plant, per pot, per unit surface) and production of dry matter (grams per plant, per pot, per unit surface).

If the concentration remains constant throughout the life of the plant the data on accumulation and dry weight should give a straight line for a series of stages when charted as in Fig. 1, and this line, when extrapolated, should pass through the origin.

Disproportional slopes measure change in concentration: if the extrapolated line intercepts the milliequivalent axis the concentration falls (a), otherwise it rises (b) with the increase in dry weight. If the two curves intercept, this point refers to identity in concentration and dry weight but the plants may differ considerably in rate of accumulation relative to dry weight production as shown by the slopes of the tangents to the curves.

Identity in concentration may be an incidental matter valid for one stage of growth. In efforts to relate growth to composition for foliar diagnosis, difficulties may arise if such distinctly different ways of establishment of concentrations are not recognized.

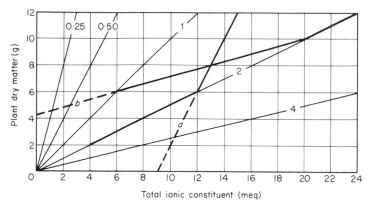

Fig. 1. Schematic representation of the accumulation of dry matter and ionic constituents by a growing plant. The slopes of the lines drawn correspond to ionic constituent concentrations of 4, 2, 1, 0·5, 0·25 meq/g DM respectively. At the level of 6 g dry material per pot the medium is changed from the deficient state to adequate nutrition (b), and from adequate supply to shortage (a). Where the heavy lines intercept, plant weight and concentration of nutrient within the plants are identical. If the growth is influenced by treatment this stage will be reached at different times of growth. Changes with time may be recorded separately by extending the graph to the left and to below with time scales. The thin straight lines through the origin are those for constant concentration at the indicated levels. Insertion of grid lines should accord with the precision of the experiment, generally about 5%.

Experiments with sequential harvests of replicates have led to still incomplete but instructive results as to establishment of the level of carboxylates in ryegrass under varying conditions of nutrition (Dijkshoorn et al., 1968). These will be summarized below on a "per pot" basis.

2. The Nitrate System

For ryegrass, the value of (C-A) is only a constant by grace of the maintenance of a full supply of nitrate, potassium, and other nutrients. If during growth the medium is replaced by one in which all the potassium nitrate is replaced by calcium sulphate, the plants lacking a supply of potassium and nitrate will reduce their level of (C-A) considerably. If potassium nitrate is restored to the depleted culture, the plant resumes a rapid incorporation of nitrogen in the starved tissues, and accumulation of organic nitrogen proceeds for a time at a much faster rate than dry matter production (Fig. 2). Parallel to the consumed nitrate there is a rapid accumulation of carboxylates but the curve for (C-A) levels off at 1 meq/g dry matter and this normal level is maintained during the further increase in dry weight. Subsequent to this stage, the difference in slope between (C-A) and organic nitrogen gives the overproduction of

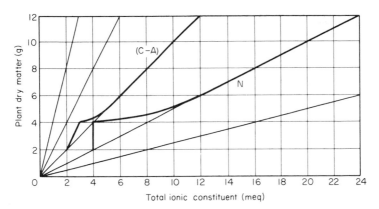

Fig. 2. Accumulation of carboxylates (C-A) and organic nitrogen N by ryegrass, in relation to dry matter production. The plant production lines are drawn as in Fig. 1 and the heavy curves refer to (C-A) and organic nitrogen N. At the level of 2 g dry matter per pot, the medium is replaced by one containing additional sulphate in place of nitrate, and at 4 g dry matter nitrate is restored to the depleted culture.

carboxylates giving rise to the external alkaline effect (as outlined earlier), which is here constant at 1 meq/g dry matter produced. In experiments of longer duration, the organic nitrogen level is usually reduced with age of the plant and, with maintenance of (C-A) at 1 meq/g, the external alkaline effect is diminished.

3. The Bicarbonate System

With nitrate and an adequate supply of all nutrients, ryegrass accumulates carboxylates at a constant level of 1 mequiv. (C-A)/g dry matter. When transferred to a nutrient solution with all the nitrate replaced by additional sulphate, accumulation of carboxylates is retarded to 0·5 meq/g dry matter produced, and further growth dilutes the concentration of the whole plant to values lower than the normal. After transfer the bicarbonate system acts as the generator of carboxylates and since in ryegrass it proceeds at a lower rate, the level of carboxylates is reduced by nitrate shortage (Fig. 3).

Ryegrass must suffer drastic treatment to completely repress the production of carboxylates. Experiment showed that this repression can be attained by transfer to a medium in which all the potassium nitrate is replaced by ammonium chloride. Ammonium can readily supply the nitrogen requirements in balance with bicarbonate. But omission of potassium and addition of the readily absorbed chloride resulted in all the metal cation C entering the plant in balance with chloride, phosphate, and sulphate. Consequently, there was no further supply of carboxylates, the unchanged pool of carboxylates

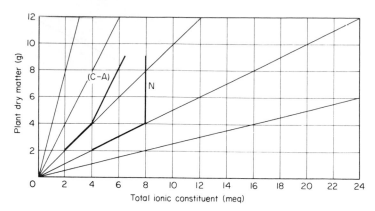

Fig. 3. Carboxylates, (C-A), and organic nitrogen in ryegrass transferred at the 4 g dry matter level from nitrate containing nutrient solution to one containing additional sulphate in place of nitrate. Drawn to the same conventions as Fig. 2. Carboxylate accumulation continues at a lower rate via the bicarbonate system; growth ceases at 9 g dry weight owing to shortage of nitrogen.

within the plant was diluted by further growth, and the level of carboxylates fell to one-third the normal value when the dry weight had risen to three times its value at transfer (Fig. 4).

Under normal conditions of ammonium supply with potassium and the other nutrients, and no chloride, bicarbonate was absorbed in excess of ammonium, and this excess balanced by metal cations, supplied the plant with carboxylates at a moderately low level of about 0·5 meq/g dry matter.

The results are very similar to those obtained with sugar beet by Houba *et al.* (1971).

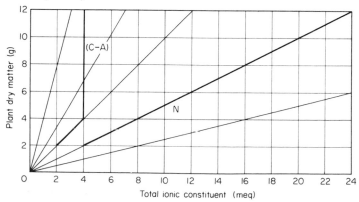

Fig. 4. Drawn to the same conventions as Figs 2 and 3. Complete cessation of carboxylate synthesis in ryegrass as a result of transfer at the 4 g dry matter per pot level from complete nutrient solution with nitrate to a medium with ammonium in place of nitrate containing chloride and no potassium.

Figures 2, 3 and 4 are schematic representations of results from Dijkshoorn *et al.* (1968).

REFERENCES

Armstrong, R. H., Thomas, B. and Horner, K. (1953). *J. agric. Sci., Camb.* **43**, 337-341.
Armstrong, R. H., Thomas, B. and Armstrong, D. G. (1957). *J. agric. Sci., Camb.* **47**, 446-453.
Arnon, D. I. (1939). *Soil Sci.* **48**, 295-307.
Bedri, A. A., Wallace, A. and Rhoads, W. A. (1960). *Soil Sci.* **89**, 257-263.
Bentley, L. E. (1952). *Nature, Lond.* **170**, 847-848.
Böhning, K. and Böhning-Seubert, E. (1932). *Bioch. Z.* **247**, 35-67.
Bradley, D. B. (1964). *J. agric. Fd Chem.* **12**, 213-216.
Bulen, W. A., Varner, J. E. and Burrell, R. C. (1952). *Analyt. Chem.* **24**, 187-190.
Butler, G. W. (1953). *Physiologia Pl.* **6**, 617-635.
Chouteau, J. (1960). Ph.D. Thesis, University of Bordeaux.
Cooil, B. J. (1948). *Pl. Physiol., Lancaster* **23**, 403-424.
Davies, J. N. (1964). *J. Sci. Fd Agric.* **15**, 665-673.
De Kock, P. C. and Morrison, R. I. (1958). *Bioch. J.* **70**, 272-277.
Dijkshoorn, W. (1962). *Nature, Lond.* **194**, 165-167.
Dijkshoorn, W. (1970). Proc. 6th Intern. Colloq. Plant Analysis and Fertilizer Problems, Tel-Aviv (in press).
Dijkshoorn, W. and Lampe, J. E. M. (1962). *Jaarb. Inst. biol. scheik. Onderz. LandbGewass.* 135-148.
Dijkshoorn, W. and van Wijk, A. L. (1967). *Pl. Soil* **26**, 129-157.
Dijkshoorn, W., Lampe, J. E. M. and Said, I. M. (1959). *Jaarb. Inst. biol. scheik. Onderz. LandbGewass.* 155-157.
Dijkshoorn, W., Lathwell, D. J. and de Wit, C. T. (1968). *Pl. Soil* **29**, 369-390.
Dougall, H. W. and Birch, H. F. (1967). *Pl. Soil* **26**, 85-98.
Fauconneau, G. (1958). *Annls agron.* suppl. 1, 1-13.
Fauconneau, G. (1959). *Annls. Physiol. veg., Paris* **2**, 179-187.
Garcia-Rivera, J. and Morris, M. P. (1955). *Science, N.Y.* **122**, 1089.
Gustafson, F. B. (1925). *J. gen. Physiol.* **7**, 719-728.
Hagemann, C. (1961). *Kühn-Arch.* **78**, 225-258.
Hiatt, A. J. (1967). *Pl. Physiol., Lancaster* **42** 294-298.
Hiatt, A. J. and Hendricks, S. B. (1967). *Z. Pflphysiol.* **56**, 220-232.
Hoagland, D. R. and Broyer, T. C. (1936). *Pl. Physiol., Lancaster* **11**, 471-507.
Houba, V. J. G., van Egmond, F. and Wittich, E. M. (1971). *Neth. J. agric. Sci.* **19**, 39-47.
Hurd, R. G. (1958). *J. exp. Bot.* **9**, 159-174.
Hurd-Karrer, A. M. (1930). *Pl. Physiol., Lancaster* **5**, 307-328.
Jackson, W. A. and Coleman, N. T. (1959). *Soil Sci.* **87**, 311-319.
Jacobson, L. (1955). *Pl. Physiol., Lancaster* **30**, 264-269.
Jolivot, M. E. (1959). *C.r. hebd. Séanc. Acad. Sci., Paris* **248**, 3208-3210.
Jones, E. C. and Barnes, R. J. (1967). *J. Sci. Fd Agric.* **18**, 321-324.
Jones, R. J. and Ford, C. W. (1972). *Aust. J. expt. Agric. anim. Husb.*, **37**, 400-406.
Jones, R. J., Seawright, A. A. and Little, D. A. (1970). *J. Aust. Inst. agric. Sci.* **36**, 41-43.
Keyssner, E. (1931). *Planta* **12**, 575-587.
Kirkby, E. A. and deKock, P. C. (1965). *Z. PflErnähr. Düeng. Bodenk* **111**, 197-203.
Kirkby, E. A. and Mengel, K. (1967). *Pl. Physiol., Lancaster* **42**, 6-14.
Kostić, M., Dijkshoorn, W. and de Wit, C. T. (1967). *Neth. J. agric. Sci.* **15**, 267-280.

188   CHEMISTRY AND BIOCHEMISTRY OF HERBAGE

Krauss, R. W. and Thomas, W. H. (1954). *Pl. Physiol., Lancaster* 29, 205 214.
Lal, B. M., Johari, R. P. and Mehta, R. K. (1966). *Curr. Sci.* 35, 125-126.
Leuthardt, F. (1927). *Kolloid-chem Beih.* 25, 1-68.
Mangan, J. L. and Johns, A. T. (1957). *N.Z. Jl Sci. Tech.* 38A, 956-965.
Martin, A. K. (1970). *J. Sci. Fd Agric.* 21, 496-501.
Neirinckx, L. J. A. (1967). *Annls Physiol. vég. Brux.* 12, 13-38.
Overstreet, R., Ruben, S. and Broyer, T. C. (1940). *Proc. natn. Acad. Sci. U.S.A.* 26, 688-695.
Peel, A. J. and Weatherley, P. E. (1959). *Nature, Lond.* 184, 1955-1956.
Rasmussen, G. K. and Smith, P. F. (1961). *Pl. Physiol., Lancaster* 36, 99-101.
Smith, P. F. and Rasmussen, G. K. (1964). *Proc. Am. Soc. hort. Sci.* 74, 261-265.
Stout, P. R., Brownell, J. and Burau, R. G. (1967). *Agron. J.* 59, 21-24.
Ulrich, A. (1941). *Am. J. Bot.* 28, 526-537.
Ulrich, A. (1942). *Am. J. Bot.* 29, 220-227.
Van Egmond, F. and Houba, V. J. G. (1970). *Neth. J. agric. Sci.* 18, 182-187.
Van Tuil, H. D. W. (1965). *Agric. Res. Reports Wageningen* 657, 1-83.
Van Tuil, H. D. W., Lampe, J. E. M. and Dijkshoorn, W. (1964). *Jaarb. Inst. biol. scheik. Onderz. LandbGewass.* 157-160.
Vickery, H. B. (1956). *Pl. Physiol., Lancaster,* 31, 455-464.
Vickery, H. B. (1961). *Conn. agric. Expt. Sta. New Haven Bull.* 640, 1-42.
Webb, J. A. and Gorham, P. R. (1964). *Pl. Physiol., Lancaster* 39, 663-672.
Willaman, J. J., West, R. M., Spriesterback, D. O. and Holm, G. E. (1919). *J. agric. Res.* 18, 1-31.
Wit, C. T., de, Dijkshoorn, W. and Noggle, J. C. (1963). *Verslag. Landbouwk. Onderzoek.* No. 69. 15, 68 pp.
Ziegler, H. (1963). *Planta* 60, 41-45.

CHAPTER 21

Symbiotic Nitrogen Fixation by Legumes

F. J. BERGERSEN

Division of Plant Industry,
Commonwealth Scientific and Industrial
Research Organization, Canberra, Australia

I. INTRODUCTION

There are at least 12,000 species of legumes widely distributed throughout tropical and temperate regions of the world. Herbaceous legumes provide a substantial proportion of the food intake of

189

grazing animals raised on sown pastures in many countries and also where domesticated or native animals graze natural vegetation (e.g. Table I). Even the foliage of woody species or their seedlings may be utilized. The range of species of forage legumes is very extensive although relatively few are exploited in agriculture. For example Foury (1954) describes 307 species as being of potential use in Morocco. Legume crops grown for grain production are also of importance. The great development of the soybean industry which has taken place in the past 30 years in the U.S.A. is well known but many other grain legumes are used elsewhere. For example 18 species of legumes are used in grain production for human and animal consumption in central and southern Africa and Madagascar (Stanton, 1966).

TABLE I

Percentage of soil surface covered by green vegetative plants in northern Negev in March 1965. The legumes comprised 33 species in 15 genera (Data from Hely and Ofer, 1971)

Soil Type	Legume cover	Total plant cover
Loessial arid brown	19·6	65·7
Sandy rogosols and arid brown	7·9	48·0
Loessial sierozem	6·9	29·1

It is very likely that both the present wide distribution of legumes in nature and their considerable exploitation in agriculture result from the same primary cause, viz. the ability of a large proportion of the species to bear root nodules which are capable of making atmospheric nitrogen available to the plants for their growth. This characteristic enables legumes to be largely independent of soil nitrogen compounds. These compounds frequently limit plant growth because of loss from the root zone by leaching and because of their susceptibility to gaseous loss resulting from microbial activity. The property of atmospheric nitrogen fixation enables legumes to function as pioneering plants in ecological successions on poor soils and as an important source of nitrogen input into ecosystems (see Chapter 22): in agriculture, it enables their use as a means of enriching the soil in ley systems of farming and as good sources of food protein for grazing animals and for human consumption. For these reasons, a discussion of nitrogen fixation by legumes finds a place in a book about the chemistry and biochemistry of herbage.

The property of nitrogen fixation appears to be limited to procaryotic microorganisms and the only plants which can utilize N_2 for their growth are those which have developed a symbiotic relationship with such microorganisms. Nitrogen-fixing symbioses are widely scattered among a number of genera of angiosperms and gymnosperms and involve bacteria, actinomycetes and blue-green algae as the symbionts. Non-leguminous symbiotic nitrogen fixation is important in many special ecosystems (e.g. Bergersen *et al.*, 1965; Lawrence *et al.*, 1967). However, only in the legumes is nitrogen fixation widely distributed among the genera and species and with a few exceptions, only legumes are exploited in agriculture in a N_2-fixing role. The annual contribution of nitrogen fixed by legumes has been recorded to be as high as 575 lb/acre for white clover but values of about 200 lb/acre are more usual for perennial pasture legumes. Of the nitrogen fixed, it is common that only a proportion is retained in the soil, being recorded to be in the range 40-100 lb/acre/annum. Enrichment of soil nitrogen by crop legumes is usually lower than by pasture legumes because a greater proportion of the fixed nitrogen is removed in the grain (review by Henzell and Norris, 1962).

The legume root nodule bacteria have been studied as components of the soil micro-flora. Environmental and physiological factors affecting the infection of legume roots, the development of nodules and nitrogen fixation, have been studied in a number of pasture and crop legumes. The biochemistry of nitrogen fixation however, has been studied in only a few legume species and most of the information in the literature has been gathered from studies of only one, the soybean. This plant conveniently produces large volumes of nodule tissue under glasshouse conditions and has certain structural advantages in the preparation of the large amounts of uniform material which are necessary for biochemical study. There is no reason to suppose that great differences will be found between the nitrogen fixation processes of different legumes because it has been found that the enzyme concerned in the primary reaction, nitrogenase, is very similar in all nitrogen fixing systems irrespective of whether these are found in free-living or symbiotic microorganisms. However, important differences in ancillary metabolism may occur between different legumes.

The fixation of N_2 by legume root nodules only occurs at the completion of a complex series of development steps (Vincent, 1967) each of which has its own distinct chemistry. The following sections will give an outline of what is known about these steps so that N_2-fixation in this system may be seen in its proper context. It will be apparent that much interesting chemistry remains to be investigated. The material given is by no means exhaustive but it is the author's intention that a broad impression is given of this very complex topic. References are given to publications which will allow the interested reader to find his way more fully into the literature.

II. NODULE BACTERIA IN THE SOIL AND IN CULTURE

Legume root nodule bacteria belong to the genus *Rhizobium* according to current bacterial nomenclature and species names are mostly derived from the names of the legumes upon which the various groups of bacteria produce nodules. This classification is not very satisfactory for bacteriologists, although it is convenient for agriculturalists. Many cultures are unclassifiable upon this basis and a number of proposals for taxonomic revision have been made (e.g. Lange, 1961; Graham, 1964a). This chapter is not the place to discuss the relative merits of such proposals but it is interesting to observe that chemical methods such as DNA base composition (DeLey and Rassel, 1965) and other biochemical considerations, are playing an increasingly important part. The modern approach to the taxonomy of *Rhizobium* species is inevitably related to the study of the origin of this group of bacteria and the evolution of the symbiosis. Two authors, Norris (1956) and Parker (1968) have been the chief contributors to these aspects and they present two opposing views.

A. Occurrence and Detection in Soil

Legume root nodule bacteria occur in soils as part of the heterotrophic microflora, requiring hexoses, polyols, organic acids or amino acids as energy sources. They are usually more numerous in soils which bear the appropriate host plants but they may be numerous in soils devoid of them. They are rhizosphere bacteria; that is, they are usually more numerous in the soil adjacent to the roots of their hosts than in the soil away from the roots. Occasionally they are associated in high numbers with the roots of non-legumes (Rovira, 1961). The chemical components of root exudates which are responsible for the rhizosphere effect with *Rhizobium* spp. have not been identified (Harris and Rovira, 1962) but they are presumably growth factors and/or compounds which are readily utilized as energy or nitrogen sources. The dimensions of the soil populations of rhizobia may vary from one organism in several grams of soil to several millions per gram. In addition to the effects of host roots, numbers of the bacteria in soil fluctuate in response to seasonal conditions (Brockwell, 1963). Soil pH and inorganic and organic soil constituents are also important. Because they are motile, the moisture content of soil plays an important part in the spread of nodule bacteria (Hamdi, 1971). Although no heat-resistant form equivalent to a spore is known in any species of *Rhizobium* (most workers do not agree with the assertions of Bisset, 1952), some species are able to survive high temperatures and prolonged desiccation in soil (e.g. Jensen, 1961; Brockwell and Phillips, 1965).

Rhizobium spp. have few cultural features distinguishing them from other soil bacteria and therefore direct culture from soil is usually impractical. Instead, cultures are prepared from surface-sterilized nodules produced on "trap" hosts which have been grown in the field or grown aseptically in the laboratory and inoculated with aliquots of soil (Vincent, 1970). Enumeration of nodule bacteria in the soil is usually derived from the nodulation pattern on a series of host plants grown under bacteriologically controlled conditions and inoculated from dilutions of soil suspensions (Vincent, 1970).

B. *Cultural Characteristics of* Rhizobium *spp.*

In culture, nodule bacteria are aerobic rods whose dimensions are 0·5-0·9 x 1·2-3·0 μm, growing optimally at 25-30°C and pH 6-7. They are Gram negative and motility in liquid culture is effected by polar, sub-polar or peritrichous flagella (Kleczkowska *et al.*, 1968). These organisms are able to utilize a very wide range of carbon compounds as energy sources for growth, but there are differences between species (Graham, 1964b). The biochemistry of carbohydrate metabolism has been investigated in only a few species. Katznelson and Zagallo (1957) observed the presence of enzymes of the glycolytic pathway of Embden, Myerhof and Parnas, the pentose phosphate pathway, and the Entner-Doudoroff pathway in strains of *R. phaseoli,* and *R. meliloti.* Keele *et al.* (1969), in a radiorespirometric study concluded that glucose was metabolized exclusively by the Entner-Doudoroff pathway and the tricarboxylic acid cycle in *R. japonicum*. Polyols such as D-mannitol are often preferred as carbon sources for the culture of *Rhizobium* species. Martinez de Drets and Arias (1970) showed that inducible NAD-polyol dehydrogenases were produced in cultures of *R. meliloti* grown on D-mannitol. The action of these enzymes was to oxidize D-arabinose to D-xylulose, and D-mannitol and D-sorbitol to D-fructose. A mannitol-induced fructokinase, phosphohexose isomerase and glucose-6-phosphate dehydrogenase were also detected but hexokinase was absent. These observations indicated that the pathway of mannitol metabolism in this organism is:

$$\text{D-mannitol} \rightarrow \text{D-fructose} \rightarrow \text{D-fructose-P} \rightarrow$$
$$\text{glucose-6-P} \rightarrow \text{6-phosphogluconate}$$

The subsequent metabolism was not investigated in these experiments.

In culture, nodule bacteria are not able to utilize N_2 as the nitrogen source for their growth; nitrogen fixation is a unique property of the symbiotic form of the organism (the bacteroid) which will be discussed later in this chapter. Nitrate and NH_4^+ may be utilized by most strains of *Rhizobium* in culture but often better growth is obtained with an amino acid such as glutamate (Bergersen, 1961). However, glycine, alanine, leucine and valine. have been reported to be selective for mutants which are less effective in nitrogen fixation and these amino acids should not be used as nitrogen sources in culture media (Vincent, 1970).

Many species and strains of *Rhizobium* require or are stimulated by the B vitamins biotin, thiamine and pantothenate. The requirements for these factors are variable and some strains of some species grow best without any of them; biotin may even be inhibitory (Fig. 1). Most ordinary growth media employ yeast extract as the source of both nitrogen and growth factors (Vincent, 1970).

Cultures of nodule bacteria generally fall into two classes in terms of growth on ordinary culture media. One group, the fast growers, produce abundant growth in 3-5 days at 25°C and include typically *R. trifolii, R meliloti* and *R. leguminosarum*. The slow growing group require 1-2 weeks at 25°C to produce abundant growth and includes most strains of *R. lupini, R. japonicum* and the cowpea-type rhizobia. A few fast-growing members of these species have been described. Fast growing nodule bacteria often produce an acid reaction in culture and the slow growers a neutral to alkaline reaction. Norris (1965) proposed that these characteristics are associated with adaptations to host species which are adapted to the acid soils of the tropics in the case of the slow growers and to host species adapted to the more alkaline soils of temperate regions in the case of the fast growers; the activities of the bacteria thus tending to adjust the pH of their microenvironment towards neutrality. Norris interpreted this as being related to the postulated evolution of the legumes and the symbiosis from an ancestral tropical source. Support

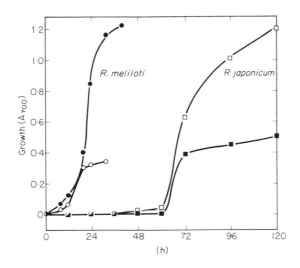

Fig. 1. Growth curves of examples of biotin-requiring and biotin-inhibited strains of rhizobia. The difference in the forms of the curves (a) for fast-growing types, typified by *R. meliloti* Rothamsted strain 146A (○,●) and (b) for slow-growing types, typified by *R. japonicum* strain U711 (□,■) is also illustrated. Biotin concentrations were 50 μg/1 (●,■), 0·5 μg/1 (○) and zero (□).

for some aspects of Norris' views has come from the work of Brockwell *et al.* (1966) who studied the reactions in culture of nodule bacteria isolated from acid, neutral and alkaline soils. They found that *R. trifolii* isolated from alkaline soils tended to produce more acid than cultures isolated from acid soils, although all produced terminal pH values of less than 7 in the media used. Some workers consider that the pH changes observed in the media in this type of experiment are not related to the evolutionary relationships of the legumes but are a consequence of the growth rate of the culture and the constitution of the medium; fast growing cultures tending to produce acid. The causes for the different rates of growth of the two classes of *Rhizobium* have not been elucidated but in some cases slow growers have been found to exhibit a long lag before growth commences. This is followed by quite a rapid growth which may be similar in rate to rates observed with fast growing cultures. Figure 1 shows such an effect with a slow growing culture of *R. japonicum,* compared with a fast growing *R. meliloti.*

Inorganic ions such as K^+, Mg^{2+}, PO_4^{3-}, and SO_4^{2-} are required for the growth of all bacteria. In addition to these, trace amounts of Ca (Vincent, 1962, Bergersen, 1961), Co (De Hertogh *et al.,* 1964) and probably Fe are required for growth of *Rhizobium* in culture. Calcium has been found to be involved in cell-wall structure (Humphrey and Vincent, 1962) and several metabolic functions in nucleotide metabolism, propionate metabolism and haem bio-synthesis have been found for Co-containing co-enzymes by Evans and co-workers (Cowles *et al.,* 1969; De Hertogh *et al.,* 1964; Jackson and Evans, 1966). The nodule bacteria use a cytochrome-containing terminal respiratory pathway and Fe will therefore be required although the author is not aware of any experimental work specific-ally demonstrating this.

Many species of *Rhizobium* produce one or both of two polymers in culture and during symbiotic growth. Poly-β-hydroxybutyric acid is found as prominent inclusions within the cells of many strains and in some, this material may account for a large proportion of the cell mass. Many cultures also produce large amounts of extracellular polysaccharides and these are sometimes present as true capsules (Dudman, 1968). Composition of these polysaccharides is variable among species but they usually contain glucose, and sometimes glucuronic acid and galactose as major components. A number of other sugars may be encountered as minor components. Recently, pyruvyl and acetyl groups have been found in *Rhizobium* poly-saccharides and their presence has been related to antigenic proper-ties (Amarger *et al.,* 1967; Bailey *et al.,* 1971; Dudman, 1964; Dudman and Heidelberger, 1969; Humphrey and Vincent, 1959).

The lipopolysaccharide constituents of the cell-walls of nodule bacteria, which are known as O-antigens, are important as major antigens for the serological identification of strains in field experi-ments (Vincent, 1970). The chemistry of the O-antigens of *R. trifolii* has been studied by Humphrey and Vincent (1969) who found that they contained firmly bound lipid, 2-keto-3-deoxyoctonate, glucose, mannose and fucose and sometimes a heptose. These antigens differ from the O-antigens of the Enterobacteriaceae in containing little phosphorus and in containing glucuronic acid.

III. INFECTION OF LEGUME ROOTS

A. *The Rhizosphere*

The growth of legume root nodule bacteria in the soil is greatly stimulated in the rhizosphere and rhizoplane of legumes but the chemistry of the stimulation is obscure (Nutman, 1963). Other soil microorganisms are also stimulated in the rhizosphere and some very complex microbial interactions result, some of which may affect the subsequent infection of the host plant by nodule-producing rhizobia. For example, mild fungal infections of developing roots may damage root hairs in the zone of the root in which infection by nodule bacteria would otherwise occur and in this way nodule development may be delayed or prevented.

The host plant sometimes exerts selective effects, stimulating those rhizobia which are capable of infecting it (Robinson, 1967); other soil microorganisms which are selectively stimulated in the rhizosphere may be antagonistic to the rhizobia (e.g. Hely *et al.*, 1957). In addition to these types of interactions, strains of the rhizobia interact with one another and many of these effects can be demonstrated in the rhizosphere. Competition between strains of rhizobia may be the result of differential nutritional requirements on growth rates but others are more directly due to antagonistic effects. Schwinghamer and Belkengren (1968) studied an antibiotic produced by a strain of *R. trifolii* which was active against strains of several species of *Rhizobium*. This antibiotic was a polypeptide of 1-2000 molecular weight, similar in action to basic polypeptide surface-active bacteriocides produced by other genera of bacteria. Other interactions involving other types of antibiotics, bacteriophages and bacteriocins active against rhizobia have been described and many of these are also of potential importance in the rhizosphere of legumes (Schwinghamer, 1971).

The rhizosphere of legumes also is a special environment in physical terms. Dart and Mercer (1964) showed that the bacteria were embedded in mucilaginous material on root surfaces and this must result in a zone in which special microenvironments can be established, especially in terms of chemical concentration gradients, aeration and ionic charge density.

An adequate rhizosphere population of suitable rhizobia is an absolute prerequisite for infection. The aim of legume seed inoculation is the rapid establishment of such a population in the rhizosphere of the developing seedling, so that prompt infection and nodule development may be achieved, thus ensuring that the seedling becomes supplied with fixed atmospheric nitrogen as soon as possible.

B. Susceptibility of the Host to Infection

As indicated earlier in this chapter, the legume root nodule bacteria have been grouped into species according to the host genera which they nodulate. In addition, there are several groups of miscellaneous host-rhizobial combinations which do not exhibit such specificity. The best known of these is the so-called "cowpea miscellany", in which several botanically diverse host genera may be nodulated by a single culture. The chemical basis for infection specificity is not known but it is under the control of genes in both host and bacteria. Nutman studied the nuclear and cytoplasmic inheritance of resistance to infection in red clover and other workers have studied non-nodulating lines of soybean and other legumes which have been bred for resistance to infection by certain strains of nodule bacteria (Nutman, 1969). Mutations in the rhizobia may also result in loss of infective ability. The mutations may occur naturally with considerable frequency in some strains, thus presenting a hazard in the maintenance of stock cultures. Schwinghamer (1968) found that mutants of R. leguminosarum which were resistant to structural analogues of L-amino acids were frequently unable to nodulate peas. In later work (Schwinghamer, 1969) it was found that some of these non-nodulating mutants were auxotrophic; that is, they had a specific nutritional requirement for growth in culture. Some prototrophic revertants of these auxotrophs (reverse mutants now again independent of the nutritional requirements) regained their infectiveness for peas and appeared to be identical with the parent cultures. Furthermore, some of the non-infective auxotrophs were enabled to infect their host plants when the substances required for their growth in culture were supplied to the rooting medium in which the plants were grown. However, the nodules which were produced did not fix N_2 although the parent culture was fully effective.

Legume roots are not uniformly susceptible to infection by the appropriate rhizobia. Nutman (1962) showed that zones of susceptibility occurred around initial foci as the main roots of 12 species of

Trifolium and *Vicia hirsuta* elongated. The further development of these zones of infection was arrested when nodules developed or when nitrate was added to the growth medium. Root temperature also exerts profound effects upon infection. In temperate legumes such as the clovers, infection is sparse and slow below 7°C and above 33°C. In tropical legumes, the range is much narrower being from 18-35°C (Roughley *et al.*, 1970; Gibson, 1967a, 1971). Treatments which delay infection also change the subsequent pattern of infection, because of the changing distribution of zones of susceptibility with time.

C. *Root-Hair Infection*

The infection of legume roots by the rhizobia has been recorded to be by way of deformed root-hairs in the Trifolieae, the Vicieae and some other groups. In a few special cases, such as the aquatic species *Neptunia oleracea* which has no root-hairs, other routes of infection are found but these are rare. In a few species, infection through wounds and points of emergence of lateral roots has been recorded (Nutman, 1956). Even in the latter case, root-hairs concealed within the partly emerged lateral root may be the infection route (Gibson, private communication). It may therefore be generalized that root-hair infection is the most common route for the infection of legume roots by the rhizobia.

The growth of the rhizobia in the rhizosphere of legumes produces a characteristic deformation of the root-hairs. This usually occurs whether the plant will be infected or not, but the phenomenon appears to be restricted to the legumes (Nutman, 1956). Furthermore, a specific type of root-hair curling associated with nodule-producing combinations of host-plant and rhizobia, can be distinguished (Yao and Vincent, 1969). The number of curled hairs is greatly in excess of the number of infections which occur. Infection of the root-hair is initiated at points of sharp folding of the hair surface, where stresses in the outer layers may expose the primary structure of the host cell-wall. The chemical basis of root-hair curling has been investigated in a number of laboratories. McCoy (1932) recorded that culture filtrates from the rhizobia were capable of producing a degree of root-hair deformation. Kefford *et al.* (1960) showed that tryptophane was among the amino acids excreted by subterranean clover roots and concluded that this was converted to indole-3-acetic acid (IAA) by the action of *R. trifolii* in the rhizosphere; it was considered that IAA may have in important role in infection. Experiments in which IAA has been added to clover roots have failed to reproduce root-hair curling although a subsidiary role for auxin produced by the bacteria seems likely at this and subsequent stages of infection and nodule development. In a search for other extracellular bacterial products which may be "curling factors", a soluble, non-dialyzable bacterial product has been found to produce a degree of root-hair deformation which was similar to, but distinguishable from, that produced by the bacteria. Relatively crude preparations of extracellular polysaccharide from cultures of *R. trifolii* were active, but these certainly contained other components and further purification of the "curling factor" has not been done (Yao and Vincent, 1969; Li and Hubbell, 1969; Hubbell, 1970). Thus, the chemistry of processes involved in the first step of infection remains obscure.

The next step in infection involves the penetration or invagination of the exposed primary wall of deformed root-hairs. The rhizobia do not themselves produce cellulases or pectinases and it has been postulated that infection is promoted by polygalacturonase (or perhaps pectin transeliminase) which is induced locally in the host by a product of the rhizobia on the root-hair surface and which plasticizes the primary cell wall of the hair. Evidence for this has come from the work of Ljunggren and Fahraeus (1961), but apart from the report of Munns (1969) there has been no published confirmation of these results.

Low pH and the presence of nitrate inhibit nodulation. Munns (1968a-d) found that nodulation of *M. sativa* in solution culture was reduced below pH 5·5 and prevented at pH 4·5 and that there were parallel effects upon root-hair curling, which was also prevented at the lower pH and appeared to coincide with the acid sensitive step of infection. Exposure of roots to these low pH values during the first few hours of contact with the rhizobia, followed by a return to mild pH, inhibited early root-hair curling and nodulation. In contrast, exposure to a mild pH for the first 12 h after inoculation allowed normal nodule development, even when the plants were then placed in acid solution. In a later paper (Munns, 1969) it was shown that these results could be explained in terms of the activity of a pectinase, whose production by inoculated roots was little affected by pH values from 4-6 but whose reaction rate increased with increasing pH above 4. This increase in rate closely paralleled the increasing nodule number, which resulted when infection was allowed to proceed over the same pH range. The effects of nitrate were found to be due primarily to an inhibition of infection, but they could not be localized in time to any particular phase. Experiments on the interactions of nitrate and auxin (IAA) have indicated that the auxin is involved in the extension of infection threads, because IAA at $10^{-9}-10^{-8}$ M doubled the number of threads reaching the cortex and doubled the number of nodules formed in the presence of nitrate (Valera and Alexander, 1965; Munns, 1968d).

D. Infection Threads

Although the details of events at the primary penetration of root-hairs are not known, the subsequent development of the infection is relatively well documented and has been studied by time-lapse photomicrography (Doncaster *et al.*, 1970). These and earlier studies from many laboratories have shown that the infection proceeds from the point of penetration of the root-hair by the rhizobia, by means of a developing tube whose inner surface is lined with cellulose and within which the bacteria lie end to end in a polysaccharide matrix. This structure is the so-called "infection thread" (Fig. 2). Nutman (1956) proposed that this thread could be regarded as an invagination of the root-hair wall and electron microscopy of infected root-hairs and developing nodules is in substantial agreement with this concept (Sahlman and Fahraeus, 1963; Goodchild and Bergersen, 1966). The continued growth of the infection thread within the root-hair apparently depends upon a close spatial relationship being maintained with the hair-cell nucleus; when the nucleus moves away from the tip of the developing infection thread during cyclosis of the hair cytoplasm, further growth of the thread ceases and the infection is said to abort. In successful infections, the thread penetrates from the root-hair cell into the cortical tissue of the root and then ramifies into the cells which ultimately develop into the nodule. In some types of nodules, such as those which are characteristic of the clovers, infection threads continue to develop within young cells developing from the nodule meristem, thus providing a continuing source of infection as the nodule increases in size. In other types of nodule such as those of lupins, soybeans and some tropical legumes, infection threads are only a transient feature of nodule development and increase in nodule size is accomplished by the division of a few infected cells into which bacteria have been released from infection threads. It was

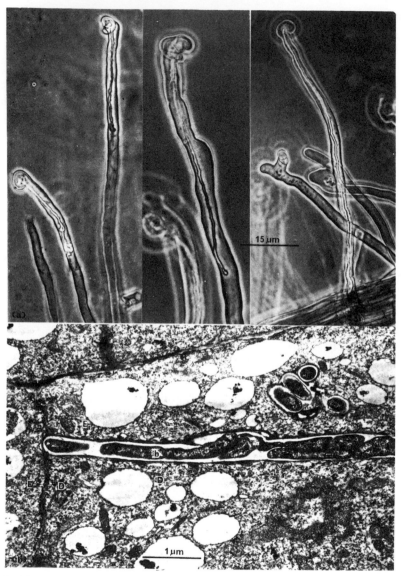

Fig. 2. (a) Infected root hairs of *M. sativa;* phase contrast photomicrographs. (Adapted from Munns, 1968c, with the permission of the author). (b) An electron micrograph of a thin section of an infection thread in the very early stages of the development of a soybean nodule. D, host cell membrane; E, host cell wall; D_1, infection thread membrane; E_1, infection thread cellulose wall; b, bacteria. (Adapted from Goodchild and Bergersen, 1966, with the permission of the authors).

originally thought that infection threads grew from the extending tip which would consist of naked host membrane, but electron micro-scopy has shown that the structures associated with the walls of the thread tube (Goodchild and Bergersen, 1966) resemble those associated with expanding plant cell walls. Extension growth of the thread may therefore be a general stretching of the wall, accompanied by

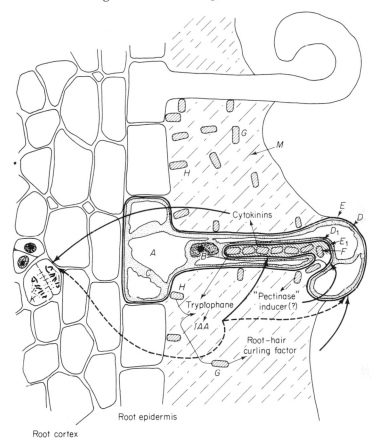

Fig. 3. A diagrammatic representation of root-hair infection in legumes. A, root-hair cell; B, root-hair cell nucleus; C, cytoplasm in active cyclosis; D, root-hair cell membrane; D_1, infection thread membrane; E, root-hair cellulose wall; E_1, infection thread cellulose wall; F, infection thread matrix; G, rhizosphere rhizobia; H, rhizoplane rhizobia; K, dividing cortical cells which will give rise to a nodule (these develop at a later stage of infection thread penetration than that shown here); M, rhizosphere mucilage. The activities shown are located at the origins of the fine arrows and their probable points of action are at the points indicated by the broad arrows.

the interpolative deposition of new cellulose fibres (Fig. 2b). Auxin production by the bacteria within the thread would probably play an important part in this type of growth. Figure 3 shows a schematic diagram of the known chemical and physiological events associated with infection.

IV. NODULE DEVELOPMENT AND STRUCTURE

A. Nodule Development

The details of the early stages of root nodule development have been studied in only a few legumes. In all of them, the infection thread penetrates from the initially infected root-hair cell, passing through and between the cells of the root cortex towards the stele. Certain cells of the root cortex, often near protoxylem points, are stimulated to divide by the approaching infection thread and from these cells the nodule structures develop. These nodule-initial cells mostly contain double the chromosome number of root apices and have been termed "disomatic". A few monosomatic cells also divide. The disomatic cells give rise to the central tissue of the nodule in which the rhizobia develop their functional intracellular symbiotic association with the host plant. The other cells give rise to the nodule cortex, to the vascular tissue which connects the nodule to the root stele and to other nodule structures. The literature does not agree as to whether the polyploid condition of the disomatic cells results from the infection or preexists in the root (Wipf and Cooper, 1938; 1940). Torrey (1961) reported that pea root segments treated with IAA and kinetin produced polyploid mitoses which occasionally showed diplochromosomes. These were presumed to arise following endomitosis in certain cells in the root cortex. In later work it was shown that similar effects could be found in the first few days of nodule development (Torrey and Barrios, 1969) and that cultures of *R. japonicum* produced cytokinins (Phillips and Torrey, 1970). It therefore seems reasonable to propose that the rhizobial infection leads to a cytokinin-auxin stimulation (auxin is also produced by rhizobia, see above), causing endomitotic cells already present in the root to undergo mitosis thus producing the disomatic cells which ultimately give rise to the infected nodule tissue (Fig. 3).

B. Nodule Structure

Further nodule development involves the organization of a meristem followed by the production of the various tissues. The infection of the cells forming the central tissue takes place when infection thread branches penetrate them. After approaching the nucleus, a bulge develops on the thread, which there becomes deficient in cellulose, allowing bacteria within the thread to become applied to the surface of the infection thread membrane. This area then folds inwards, enveloping the bacteria in an infection vacuole which passes into the cytoplasm (Goodchild and Bergersen, 1966). This process of endocytosis is the commonest way in which bacteria enter the cells of the higher plants and animals (c.f. phagocytosis). The bacteria continue to divide within the host cell following its infection, and the infection vacuoles appear to divide also, so that the entire host cell cytoplasm becomes filled with bacteria, each discrete within its own vacuole. The subsequent development of the infection is accompanied by hypertrophy of the host cell and degeneration of its nucleus. Two main paths of further bacterial development are found. (1) In some legumes, further division ceases and the bacteria then enlarge greatly and frequently assume very pleomorphic shapes. It was for these that the term "bacteroids" (bacteria-like bodies) was originally coined. (2) In other legumes, the bacteria undergo a few more divisions but apparently because of lack of space, these occur within the infection vacuole, resulting in membrane envelopes containing several bacteria. In this type of nodule gross enlargement of the bacteria does not usually occur but it is customary to refer to them also as bacteroids. The reason for this is that in both types of nodule active nitrogen fixation only occurs when they contain a central tissue composed of host cells packed with either of these two types of symbiotic bacteria and it is these host cells which contain the red pigment leghaemoglobin which is a characteristic feature of active legume nodules. The term bacteroid is thus a functional one. There are good biochemical reasons why bacteroids should be distinguished from the vegetative, free-living forms of the bacteria, as we shall see later. In addition, bacteroids have a very much reduced ability to produce cultures on ordinary media, compared with vegetative rhizobia (Almon, 1933; Bergersen, 1968). Figure 4 illustrates two different types of nodule and the distinctive structures which they contain. Fuller accounts of nodule structure are found in the work of Lechtova-Trnka (1931), Fred et al. (1932) and in the electron microscope studies of Dart and Mercer (1963) and Goodchild and Bergersen (1966); Bergersen (1971a, b) gives references to other recent work.

As nodules become older, in those which have a terminal meristem, the basal central tissue begins to decay; in spheroidal or "collar type" nodules, this usually begins from the centre of the tissue. Breakdown of this tissue occurs when the bacteroids lyse and

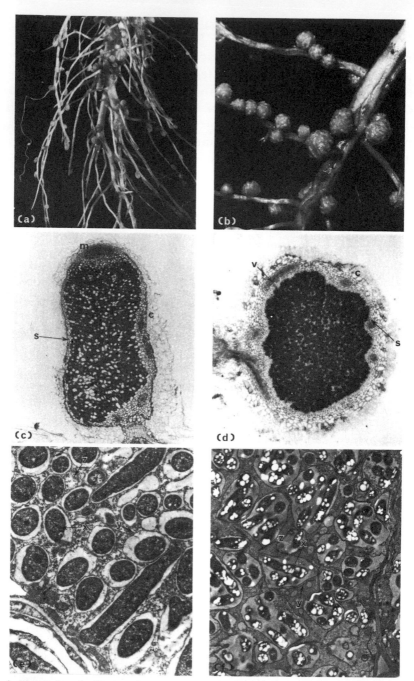

Fig. 4. Illustrating the form and structure of two types of nodules, (a, c, e), the terminal meristem type of clovers and (b, d, f), the spheroidal type of soybean. 1, gross nodule appearance; 2, longitudinal and maximum diameter sections; 3, the appearance of the bacteroids in the cells of the central tissue, c, cortical tissue; v, vascular strands; m, meristem; s, symbiotic tissue composed of bacteroid-containing cells; x, bacteroids; y, infection vacuole membrane envelope; z, host cytoplasm. (e and f are electron micrographs kindly supplied by C. Pankhurst and D. J. Goodchild).

residual vegetative bacteria appear to grow saprophytically upon the dying tissue which becomes first brown and then green in colour as the leghaemoglobin is transformed into bile pigment-like compounds. Finally, the tissue becomes grey-brown with phenolic materials. Tissue breakdown of this type frequently accompanies nodule growth but when the rate of breakdown exceeds the production of new tissue, the activity of the nodules declines and decay soon involves the entire structure.

The biochemistry of most of the processes involved in the development and breakdown of nodule tissue is virtually unknown and it has been described here in structural terms and in terms of elementary plant physiology only. There is great scope for biochemical and chemical experimentation in this field but it has proved very difficult so far because of the almost complete lack of suitable technology.

C. Environmental Factors and Nodule Development

The rates at which the various phases of nodule initiation and development proceed are strongly influenced by root temperature (e.g. Dart and Mercer, 1965a; Gibson, 1967a, b; 1971). Periods of temperature above 30-33°C for temperate species and above 35-40°C for tropical legumes cause rapid breakdown of nodule tissue. In this there is a strong influence of the bacterial strain forming the nodules, some strains being much more tolerant of supra-optimal temperatures than others. In contrast, periods of sub-optimal root temperature, although reducing nodule activity, do not induce gross structural changes although there may be extensive starch accumulation. Return to near optimal conditions allows rapid recovery of activity.

At certain times of the year when nitrification in the soil produces transient high nitrate levels, or in soils treated with nitrate fertilizers, gross cytological changes may be initiated in nodules of most legumes, leading ultimately to tissue breakdown (e.g. Dart and Mercer, 1965b). It has been found that some combinations of host plant and rhizobial strain are more resistant to these effects than others.

Prolonged cloudy weather, or artificial darkening of nodulated plants reduces activity and induces starch accumulation. If continued, low light intensity leads to nodule breakdown. Similar effects upon nodule structure are encountered when plants are defoliated and these therefore seem to be a consequence of deprivation of photosynthetic products. Some pasture plants are, however, able to maintain N_2-fixation when subjected to regular defoliation.

D. Ineffective Nodules

Not all legume root nodules fix N_2. In some well-documented cases, strains of rhizobia which produce active nodules (effective) on one host species, produce nodules on other related species (or cultivars of the same species) which do not fix N_2. These nodules are said to be symbiotically ineffective. They are often small, white and very numerous compared with effective ones and most fail to produce bacteroid-containing tissue or if they do, it does not persist for long enough to fix significant amounts of N_2 (Bergersen, 1957). Ineffective nodules grade through to various degrees of effectiveness when different host lines or species and different rhizobial strains are examined. In these grades, degrees of effectiveness are usually correlated with the amount and degree of persistence of the bacteroid-containing central tissue of the nodules.

In addition to the effects of host-strain specificities upon symbiotic effectiveness, genetic defects in the host or in the rhizobia may arise spontaneously or as a result of genetic manipulation (Nutman, 1969). The study of some of these may provide valuable information about the biochemistry of symbiosis. For example, Schwinghamer (1969) found that mutants of a strain of R. trifolii originally effective for red clover, were auxotrophic, requiring riboflavin for their growth. These mutants produced ineffective nodules on red clover unless riboflavin was supplied to the medium in which the plants were grown. Spontaneous or induced revertants to the proto-trophic condition simultaneously regained their symbiotic effectiveness. Further study (Pankhurst et al., 1972) showed that bacteroid development was arrested in nodules produced by the auxotroph, although most other structural features were normal. The addition of riboflavin allowed the development to continue and near-normal tissue resulted. These results indicated that bacteroid development has a higher requirement for riboflavin than other stages of nodule development and suggest that flavin-containing co-enzymes are pro-duced in quantity as a part of the changes in bacterial metabolism which accompany acquisition of the nitrogen-fixing system.

V. BIOCHEMISTRY OF NITROGEN FIXATION

A. Studies with Intact Plants

Until about 1952, all that was known about symbiotic N_2-fixation had been derived from studies with intact plants. These studies elicited a number of important facts, which were summarized by Wilson (1940). Chiefly these are the establishment of H_2 as a specific inhibitor of N_2 fixation by nodulated red clover, the establishment of the affinity of N_2, $Km(N_2)$, for nodulated red clover, and extensive studies of the soluble nitrogen compounds of N_2-fixing soybean plants. Extensive studies of the control of N_2-fixation exerted by combined N were also done and the theory of control based on carbohydrate-nitrogen ratio was developed and although some aspects of this have had to be modified in the light of new knowledge, much of it remains as a valid description of the behaviour of nodulated legumes.

Studies of intact nodulated plants have shown that requirements for several elements was greater when N_2 was the N-source rather than soil mineral N, and these results have influenced more detailed biochemical studies. Molybdenum (Anderson and Thomas, 1946) is now known to be a constituent of the nitrogenase complex of nodules (Klucas et al., 1968) as well as of the free-living bacteria (Mortenson et al., 1967). The increased requirements for phosphate by nodulated legumes (McLachlan and Norman, 1961) may be related to the production of ATP which is required for N_2-fixation (Koch et al., 1967). The specific requirement for Co by nodulated legumes (Ahmed and Evans, 1960) has been shown to be related to several functions in bacteroid metabolism (e.g. Jackson and Evans, 1966; Cowles et al., 1969). Nodulated legumes also have a specific Cu requirement (Hallsworth, 1958); so far this has not been related to any particular metabolic step in N_2-fixation or related metabolism.

Nodulated legumes require illumination for the support of nitrogen fixation. Darkened plants rapidly lose N_2-fixing ability (Virtanen et al., 1955). Photosynthetic products are rapidly translocated from leaves to nodules (Bach et al., 1958) and consist principally of sucrose, glucose and organic acids. These materials constitute the fuels used to supply energy and reducing power for N_2-fixation. Recently, nodule nitrogenase activity (measured as

acetylene-reducing activity) was found to be related to the illumination received by soybean plants during the preceding 2 h. Furthermore, nodulated roots were more active than detached nodules
during the day but the level of nitrogenase activity fell during
darkness to that characteristic of detached nodules. These results
showed that photosynthetic products, present in the roots, were
available to support the activity of nodules. The activity of detached
nodules is largely supported by endogenous reserve compounds and
is less affected by illumination prior to detaching the nodules from
the plants (Bergersen, 1970).

B. Detached Nodules

In 1952 Aprison and Burris reported for the first time, active
N_2-fixation in soybean nodules detached from the host plant. Using
the heavy isotope ^{15}N, they were able to show that activity
continued for several hours before declining. This enabled a series of
experiments to be done in which the symbiotic system could be
studied, isolated from the influence of the rest of the host plant.

1. The Products of N_2 Fixation in Nodules

Studies of detached soybean nodules incubated with $^{15}N_2$ showed
that most ^{15}N appeared first in glutamic acid followed by serine,
threonine or asparagine and NH_3. This pattern was consistent with
the entry of fixed N into assimilatory pathways by way of NH_3
(Aprison et al., 1954). Later experiments with improved ^{15}N
techniques established that fixed N in soybean nodules was almost
completely in the soluble fraction of the nodules (Bergersen, 1960)
and with short term experiments it was clearly shown that the first
free product was NH_3 which was rapidly assimilated into α-amino
compounds (Bergersen, 1965). Similar results have been obtained
with nodules of Ornithopus sativa (serradella) by Kennedy
(1966a, b). The primary assimilatory product in these nodules was
glutamic acid and trans-amination reactions followed. Some assimilatory reactions are probably mediated by plant enzymes and some
by the bacteroids (Bergersen, 1971a).

2. Effects of pO_2 and pN_2

Experiments with growing red clover plants in atmospheres containing 150 mm Hg O_2 and varying concentrations of N_2 indicated
that N_2 concentration was not limiting from 152 mm Hg upwards

and the apparent Michaelis constant (K_m (N_2)), was 38 mm Hg (Wilson, 1940). With a pO_2 of 140 mm Hg, the K_m (N_2) of detached soybean nodules was 50 mm Hg. Further, N_2-fixation was not at a maximum until a pO_2 of 280-350 mm was reached. Above this partial pressure, increasing pO_2 inhibited N_2 fixation sharply (Fig. 5), in a manner which was close to being competitive. Measurements of nodule respiration over the pO_2 ranges showed that stimulation of N_2 fixation was associated with increasing respiration and inhibition of fixation with a further sharp rise in respiration (Bergersen, 1962a). The form of the data in these experiments suggested that the

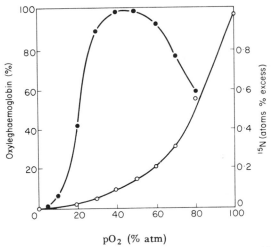

Fig. 5. The effects of external oxygen tension (pO_2) upon N_2-fixation (^{15}N excess, ●) and oxygenation of leghaemoglobin (○) in intact, detached soybean nodules. (Data adapted from Bergersen, 1962a, b).

inhibitory effects of high pO_2 could be overcome by increasing the N_2 concentration. This was not possible at total pressures of 1 atmosphere because of the limited solubility of N_2 in water. Recent work with detached nodules using the acetylene-reduction assay (see 3 below), has shown that O_2-inhibition only occurs at low acetylene concentrations. When 10% (v/v) acetylene is present in the gas phase, the reaction is not inhibited at 90% (v/v) O_2 (Bergersen, 1970). These results confirm the implications of the earlier work, viz. that high substrate concentrations would reverse the O_2-inhibition. The results are one of the consequences of the high solubility of acetylene in water, which permits higher substrate concentrations than can be obtained when N_2 is the substrate. Another important effect of O_2 with practical implications has been noted. At low pO_2, nitrogen fixation by nodules was curtailed and the respiratory quotient (R.Q. = CO_2/O_2) was 1·3. As the pO_2 rose to about 200 mm Hg, the R.Q. fell to 1·0. The net result was that at 60 mm HgO_2, the ratio $CO_2:NH_3$ fixed was 323:1 while at 185 mm Hg of O_2 the ratio was 23:1. That is, as the pO_2 falls below atmospheric concentrations, the efficiency of the fixation of N_2 in terms of carbohydrate consumed, falls steeply. These effects have great importance in practical agriculture, where soil pO_2 values are always lower than atmospheric and are reduced further under wet or compacted soil conditions (Bergersen, 1971b).

3. H_2 Evolution and Other Activities

Hoch et al. (1960) showed that detached soybean nodules evolved H_2. This was the first example of this activity of the N_2-fixing enzyme system. It has now been found that all nitrogenases reduce the H^+ ion to H_2 in the absence of substrate, when reductant and ATP are supplied. The reaction is inhibited by N_2 and other reducible substrates and in the case of N_2, the inhibition is competitive (Bergersen, 1963).

We have already seen that H_2 is a competitive inhibitor of N_2 fixation in nodule systems. Detached nodules have also been used to demonstrate that CO is a potent inhibitor of N_2-fixation but it does not inhibit H_2 evolution by nodules. N_2O was shown to be a competitive inhibitor of N_2-fixation (Hoch et al., 1960) and it is now known that it is reduced to N_2 and H_2O. This was the first example of an alternative substrate for nitrogenase. A number of alternative substrates for nitrogenases are now known and all of them inhibit N_2-fixation by the intact nodule system. The best-known is acetylene. Schöllhorn and Burris (1966) showed that acetylene was a competitive inhibitor of N_2-fixation. Simultaneously Dilworth (1966) showed that is was reduced to ethylene. In the same year, Koch and Evans (1966) used the reduction of acetylene as an assay for the nitrogenase activity of detached soybean nodules. This assay is now established as a convenient, cheap and rapid method for estimating N_2-fixing potential. However, there are limitations to its quantitative use (Bergersen, 1970).

4. Leghaemoglobin

No discussion of legume root nodules would be complete without some reference to the characteristic red pigment which they contain.

Kubo (1939) first identified it as a haemoglobin and it was studied further by Keilin and Wang (1945) and Smith (1949) who showed that the leghaemoglobin was capable of oxygenation and was located in the central, bacteroid-containing tissue of soybean nodules at a concentration of about 0·5 mM. There is some uncertainty about the location of leghaemoglobin within the cells. Dilworth and Kidby (1968) using autoradiography of ^{59}Fe labelled serradella nodules, concluded that the leghaemoglobin was located primarily between the bacteroids and the membrane envelopes which enclosed them. This has also been our opinion based on many different types of observations, some of which have been reported (Bergersen, 1966a). A contrary view has been presented by Dart (1968).

Both host-cell and bacteroid appear to be involved in leghaemoglobin synthesis. Cutting and Schulman (1969) showed that the bacteroids synthesized haem which was incorporated into leghaemoglobin. The same authors (Cutting and Schulman, 1968) stated that the host was responsible for the synthesis of the globin. This was confirmed by Dilworth (1969) who showed that leghaemoglobins from nodules produced by the same strain of R. lupini on lupin and serradella, were different proteins. When different bacterial strains were used, the protein was still characteristic of the host plant.

In intact soybean nodules, the leghaemoglobin is mainly in the ferrous-non-oxy form (Bergersen, 1962b; Appleby, 1969a); up to about 20% may be in the ferrous-oxy form in young nodules. When the pO_2 is increased beyond the level for maximum N_2 fixation, increased oxygenation can be observed (Fig. 5). There is no evidence for any other form of leghaemoglobin in vivo and several other forms which have been suggested, have recently been shown to be artifacts (Appleby, 1969b, c).

Scholander (1960) showed that O_2-diffusion at low concentrations was facilitated by solutions of haemoglobin and myoglobin. Yocum (1961) proposed that a similar role could be assigned to leghaemoglobin in nodules. Work with bacteroid suspensions has shown that leghaemoglobin is not essential for N_2 fixation once the nodule has been disrupted (Bergersen and Turner, 1967). It thus seems likely that leghaemoglobin functions in nodules by facilitating the flux of O_2 to the bacteroids at very low O_2 concentrations. The properties of this haemoprotein and its state in the tissue are consistent with this function (Appleby, personal communication; Tjepkema and Yocum, 1970).

C. Bacteroid Suspensions

In 1964 it was realized that it would be difficult to learn more about the N_2-fixing system of legumes by the study of intact nodules. By that time the study of cell-free preparations from *Clostridium* and *Azotobacter* was well advanced. A number of laboratories attempted to obtain nitrogen fixation by disrupted nodule preparations without success (reported by Delwiche, 1966). Nitrogen fixation by soybean nodule homogenates was reported in 1966 (Bergersen, 1966b) and soon after this it was found that the activity was completely located in the bacteroid fraction of these preparations and that all of the host cell components could be discarded (Table II).

TABLE II

Data which show that all the N_2-fixing activity of soybean nodule homogenates (breis) resides in the bacteroids and is not enhanced by the addition of other nodule components (Data of Bergersen and Turner, 1967)

Treatment	N_2 fixed (μg)
Original brei	12·1
Washed bacteroids	11·9
Bacteroids + membrane fraction	11·7
Bacteroids + membrane fraction + soluble fraction	11·6

Nitrogen-fixing bacteroid suspensions, prepared anaerobically from soybean nodules, produced 95% of the fixed N as NH_4^+ in solution. At the same time, significant evolution of H_2 occurred, even when there was sufficient $^{15}N_2$ present (168 mm Hg) to almost completely saturate the system (K_m (N_2) = 20 mm Hg) (Bergersen and Turner, 1967; 1968). Highest activities achieved with bacteroid suspensions were of the order of 0·5 n mole NH_4^+ min^{-1} mg^{-1} bacteroid protein. N_2 fixation by bacteroids is competitively inhibited by CO with a K_i of 0·4 mm Hg.

1. Effects of O_2

N_2-fixing bacteroid suspensions responded to increased pO_2 in the range 21-84 mm Hg, with increasing initial fixation rates. However, the time for which the reaction proceeded decreased from more than 40 min at 21 mm Hg O_2 to less than 10 min at 84 mm Hg O_2

(Bergersen and Turner, 1968). Although no inhibition of initial fixation rates was observed at the highest pO_2 value used, the effects of pO_2 upon K_m and V_{max} were very similar to those observed for intact nodules (Bergersen, 1962a; Bergersen and Turner, 1968). Bubbling air through a bacteroid suspension removed all N_2-fixing activity after 15 minutes.

2. Energy Yielding Substrates and Respiration

When bacteroid suspensions prepared from soybean were washed in phosphate buffer instead of buffer containing 0·3 M sucrose, N_2-fixing activity was greatly diminished. Restoration of activity, or enhancement above the initial activity, was obtained when pyruvate, fumarate or succinate were supplied. Sucrose was not a substrate. Similar results had been obtained when the respiration of aerobically prepared bacteroids was measured (Bergersen, 1958). It was concluded that the endogenous activity of sucrose-washed bacteroids was due to the presence in the cells of substantial amounts of available substrates which could be leached from the cells in media of low osmotic pressure (Bergersen and Turner, 1967). Succinate was a prominent component of organic acids leached from soybean bacteroids (Sylvester, pers. comm.). Soybean bacteroids contain quite large amounts of poly-β-hydroxybutyrate which would be expected to give rise to aceto-acetate and so feed the tricarboxylic acid cycle which is known to be active in bacteroids (Jordan, 1962). If the rate of depolymerization was slow, this step could be the controlling reaction in determining the rate of endogenous activity. Present evidence does not seem to support this possibility, but no other function for this polymer has been suggested so far. Hexoses are not good substrates for respiration or N_2-fixation by bacteroid suspensions (Burris and Wilson, 1939) but bacteroids do possess the enzyme systems for their utilization. It has been suggested that a permease is inactivated during preparation of the suspensions and that the natural substrate in situ is glucose (Kidby and Parker, pers. comm.).

Bacteroids in legume nodules obtain their energy from aerobic pathways, although in the host cell the amount of free O_2 available to them is limited. The reduction of N_2 to NH_3 also requires that some of the reducing power generated in the bacteroids be diverted from energy-producing pathways. This implies that electron transport pathways in bacteroids branch into two pathways, one terminating with oxidases and yielding ATP and the other terminating in the reduction of N_2. Some observations have confirmed this concept. We found that bacteroid respiration increased after N_2-fixation ceased (Bergersen and Turner, 1968) and also that respiration was depressed during N_2-fixation (Bergersen, 1969). The electron carriers of these pathways have not been characterized with the exception of an NAD reductase (Evans and Russell, 1965) and a non-haem-iron protein involved in the reduction of nitrogenase (Koch *et al.*, 1970) which will be discussed later. A number of haemoproteins are present in soybean bacteroids which are not found in *R. japonicum* grown in air. They have been characterized by Appleby (1967, 1969d, 1969e). Some, or all of these are undoubtedly involved in electron transport, but so far the main terminal oxidases have not been identified. Bacteroids contain no cytochromes a-a_3 and the respiration of aerobically prepared bacteroid suspensions is only erratically affected by carbon monoxide (Appleby, 1969d). However, in our laboratory, carbon monoxide, at concentrations 100 times higher than those necessary to inhibit N_2-fixation, inhibited respiration by about 50% in the dark. This inhibition was partially relieved when the reaction vessels were illuminated, suggesting the involvement of CO-reactive haemoproteins in terminal respiratory pathways. Recently, Daniel and Appleby (pers. comm.) have been able to grow cultures of *R. japonicum* under completely anaerobic conditions, using nitrate as terminal electron acceptor. Under these conditions, cytochrome patterns similar to those of bacteroids have been induced in the cultured cells. This indicates that, at least in part, the characteristic cytochromes of bacteroids result from the near-anaerobic conditions in which they are found within the host cells.

3. Bacteroids and Assimilation of Ammonia

As we have seen, the NH_4^+ produced by N_2 fixation in intact nodules, is rapidly assimilated, firstly to glutamate and then by trans-amination to other amino acids and amides. However, soybean bacteroid suspensions produce only NH_4^+ and no evidence of assimilation has been found during fixation experiments. Bacteroids contain isocitric dehydrogenase and glutamic dehydrogenase (Kennedy *et al.*, 1966; Mooney and Fottrell, 1968) and other enzymes involved in amination and amino-transfer reactions and thus probably play a part in primary assimilation of fixed NH_4^+. The failure to observe these reactions in the N_2 fixation experiments with bacteroid suspensions is probably due to the aerobic conditions used. The equilibrium of reactions such as:

$$\alpha\text{-ketoglutarate} + NH_4^+ \rightleftharpoons \text{glutamate}$$

would be far to the left because of the rapid oxidation of α-ketoglutarate in the citric acid cycle.

D. Preparation, Properties and Purification of Nitrogenase

1. Preparation of N_2-fixing Extracts

Active cell-free extracts are easily prepared from dense bacteroid suspensions (usually 1:1 (v/v) in buffer), by disruption in an inert atmosphere in a French pressure-release cell, followed by centrifugation at 30,000-100,000 × g. Eighty to 90% of the N_2-fixing activity is recovered in the supernatant liquid. A small proportion remains bound to small, dark-brown particles which are sedimented during centrifugation and a little is lost through inactivation. It is essential that careful anaerobic techniques be empolyed at all stages because nitrogenase is inactivated by O_2 although legumes differ in the degree of sensitivity of their N_2-fixing systems.

2. The Assay of Activity of Cell-Free Extracts

Three assays may conveniently be used: (a) the evolution of H_2 in the absence of reducible substrates, (b) the production of NH_4^+ from N_2, and (c) the reduction of C_2H_2 to C_2H_4. All of these reactions require the addition of a low-potential electron donor (conveniently supplied as $Na_2S_2O_4$), ATP and Mg^{2+}. The $N_2 \rightarrow NH_4^+$ assay requires the use of $^{15}N_2$ with crude extracts because of the presence of quite large amounts of endogenous NH_4^+ in the extracts. Direct analysis of NH_4^+ is possible after the removal of low molecular weight compounds from the extracts by anaerobic gelfiltration or ultrafiltration. Details of the procedures used for the preparation and assays of extracts are given by Koch et al. (1967) and Bergersen and Turner (1968).

3. The Properties of Nitrogenase in Extracts

Concentrations of ATP above 2 mM were at first found to be inhibitory (Bergersen, 1969) and, in order to sustain workable periods of activity, an ATP generating system, usually consisting of creatine-phosphate and creatine kinase, is employed. However, when Mg^{2+} and ATP in equimolar proportions are used, higher concentrations of ATP are tolerated and adequate periods of activity are obtained without the use of a generating system.

Bacteroid nitrogenase, like nitrogenases from other bacteria, catalyzes a number of reactions in addition to the reduction of N_2. The reductions of C_2H_2 to C_2H_4, of CN^- to CH_4 and NH_4^+ are also dependent upon ATP and reductant. Hydrolysis of ATP is largely dependent upon reductant. The exchange reaction between D_2 gas and endogenous H-donors, which produces HD when N_2 is present, was first reported in detached nodules by Hoch *et al.* (1960). The reaction has also been studied with partially purified nitrogenase preparations and found to require N_2, ATP and reductant (Turner and Bergersen, 1969). It was also inhibited by very low CO concentrations. The suggestion that this reaction is mediated by bound intermediates of N_2 fixation corresponding to $-N=N-$ and $-N-N-$ was examined in kinetic experiments with inconclusive results. However, this remains the most likely basis for the D_2 exchange reaction.

Kennedy (1970) has studied the activity of nitrogenase preparations from *R. lupini* bacteroids prepared from yellow lupin nodules. It was concluded that one aspect of the role of ATP in nitrogenase activity is the induction of conformational changes in the enzyme. The author (Bergersen, 1971b) has re-examined the role of ATP in nitrogen fixation in thermodynamic terms and given an outline of the practical effects of the ATP requirement upon the productivity of legumes. The energy utilized in legume nitrogen fixation is probably of the order of 5-10% of the total energy yield of a crop relying completely on symbiotic nitrogen and is about half of that dissipated in the industrial fixation of N_2 in modern, efficient processes. In this connection it should also be remembered that plants assimilating their nitrogen from mineral nitrogen in the soil (usually nitrate) also use energy in its reduction to NH_4^+. There is little difference in the relative growth rates of plants assimilating matched amounts of N from N_2 or from NH_4NO_3 (Gibson, 1966), which indicates that there is little difference in the energetics of the two types of assimilation.

4. Purification of Nitrogenase

The centrifuged bacteroid extracts contain most of the soluble components of the cells, including nitrogenase. A number of preliminary partial purifications have been reported. Titration with protamine sulphate, and/or heating at about 55-60°C for 5-10 min removes 60-70% of the proteins and may result in up to 8-fold increases in specific activity because of the removal or inactivation of inhibitory materials and inactive protein (Klucas *et al.,* 1968; Turner and Bergersen, 1969; Bergersen and Turner, 1970). Fractional precipitation of nitrogenase from crude bacteroid extracts with polypropylene glycol has also been used as a preliminary purification and concentration method (Klucas *et al.* 1968). Anaerobic ion exchange chromatography on DEAE cellulose was used to effect further purification of bacteroid nitrogenase and its separation into two components (Klucas *et al.,* 1968). In our laboratory, we are now routinely using a simple gel-filtration chromatography method which separates the nitrogenase into its two components (Bergersen and Turner, 1970).

The first component of bacteroid nitrogenase is eluted from DEAE cellulose with 0·035 M $MgCl_2$, and from Sephadex G200 with elution volumes corresponding to an apparent molecular weight of about 180,000. It contains Fe and Mo. The second component of bacteroid nitrogenase is eluted from DEAE cellulose with 0·1 M $MgCl_2$ and from Sephadex G200 with an elution volume corresponding to an apparent molecular weight of about 50,000. The DEAE cellulose method gives greater purification of component 2 than does the gel-filtration method. This component contains Fe and acid-labile sulphide. Fractions containing each component could be obtained which alone had no activity in nitrogenase assays, but which gave high specific activity when combined. The nodule bacteroid nitrogenase thus resembles nitrogenases which have been purified from *Azotabacter* and *Clostridium* (Burris, 1969).

Component 2 of nitrogenase may be inactivated at 0°C (Dua and Burris, 1963; Moustafa and Mortenson, 1968). We have obtained similar results with bacteroids produced by strain CB1809 of *R. japonicum* on Lincoln soybeans. However, Klucas *et al.* (1968) found no cold inactivation of extracts of bacteroids produced by a U.S. commercial inoculant on Chippewa soybeans. Bacteroid nitrogenase can be easily stored for periods of 1-3 months under liquid N_2.

We have been able to reactivate inactivated component 2 by adding dithiothreitol (5 mM) and Fe^{2+} (5 mM) and standing at room temperature for 1 h. This indicates that inactivation was due to loss of sulphide-bound Fe. Even the activity of crude extracts could be increased by this treatment, indicating that the apparent deficiency of component 2 in bacteroid extracts (Bergersen and Turner, 1970) may have been due to loss of Fe^{2+} during preparation.

5. Reductants of Nitrogenase

Up to 1970, the natural reductant for bacteroid nitrogenase was unknown and $Na_2S_2O_4$ was used as reductant in cell-free systems. Koch *et al.* (1967) identified a non-haem iron protein containing acid-labile sulphur in soybean bacteroids but could find no function for it. Klucas and Evans (1968) reported that a NADH generating system could function as the reducing system for bacteroid extracts containing nitrogenase when a dye, such as benzylviologen was used as a carrier. Recently (Benemann *et al.*, 1969; Yoch *et al.*, 1969) electron carriers which would serve as reductants for Azotobacter nitrogenase were identified, utilizing spinach chloroplast photosystem I as the source of reducing power. The same system has also been used for the identification and purification of a non-haem iron protein from soybean bacteroid extracts, which served as a reductant for nitrogenase (Yoch *et al.*, 1970; Koch *et al.*, 1970). However, as yet none of the other components of the electron transport system linking dehydrogenases with this non-haem iron protein have been identified.

E. Integration of Host and Bacteroid Metabolism

The development of the understanding of nitrogen fixation in legumes has required progressive dissection of the various components of the system and the location of the various activities. To some extent this has now been done, although the task is by no means complete, and it is possible to build up a scheme in which pathways and reactions can be integrated. This has been done in Fig. 6, which attempts to show something of the way in which host functions and bacteroid metabolism interact in the nodule cells. As with all such schemes, it will need to be modified as further work is done, but all which is shown is supported by published work.

The scheme may be summarized as follows: Photosynthetic products supplied by the plant shoots are stored temporarily as starch in the host cells or as polymers such as poly-β-OH-butyric acid in the bacteroids, or they are used directly in the bacteroids for the generation of reducing power and for use as acceptors for the fixed nitrogen in the production of amino acids. The main pathway involved in substrate oxidation appears to be the tricarboxylic acid cycle. Reducing power generated in this way in the bacteroids is transferred by way of a branched electron transport pathway, one branch of which terminates in oxidases which are not yet defined,

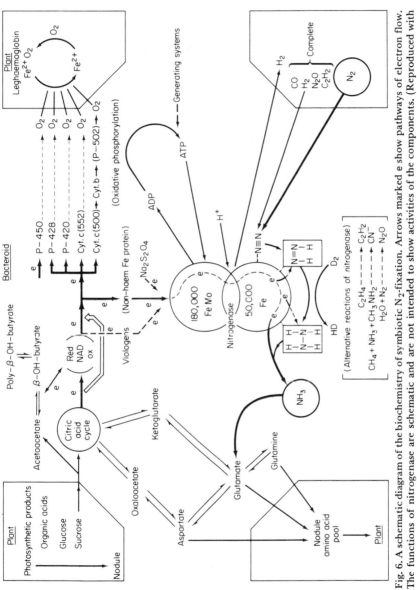

Fig. 6. A schematic diagram of the biochemistry of symbiotic N_2-fixation. Arrows marked e show pathways of electron flow. The functions of nitrogenase are schematic and are not intended to show activities of the components. (Reproduced with permission from Bergersen, 1971a).

but which may involve the haemoproteins which are shown. Terminal oxidation is probably coupled with phosphorylation in the production of ATP. Host structures and leghaemoglobin are involved in the aeration of nodule tissue and in the transport of O_2 at high flux and low concentration to the bacteroid oxidases. In this way inactivation of nitrogenase by O_2 is avoided. The other branch of the electron transport pathway terminates at nitrogenase where the ATP is used, together with the terminal reductant, which appears to be a non-haem iron-protein, for the reduction of N_2 to NH_4^+. In the absence of reducible substrates H^+ ions are reduced to H_2 which is evolved as gas. There are alternative substrates which may also be reduced at this point. Most of these are in some way electronically isosteric with N_2. The fixed NH_4^+ is quickly used up in amination reactions and both the bacteroids and the host tissue may be involved in subsequent transamination reactions. The amino acids produced in these reactions are rapidly translocated to the shoots of the host plant and used in protein synthesis.

REFERENCES

Ahmed, S and Evans, H. J. (1960). *Soil Sci.* **90**, 205-210.
Almon, L. (1933). *Zentbl. Bakt. ParasitKde*, **87**, 289-297.
Amarger, N., Obaton, M. and Blachère, H. (1967). *Can. J. Microbiol.* **13**, 99-106.
Anderson, A. J. and Thomas, M. P. (1946). *Bull. Coun. scient. ind. Res., Melb.* **198**, 7-24.
Appleby, C. A. (1967). *Biochem. biophys. Acta* **147**, 399-402.
Appleby, C. A. (1969a). *Biochim. biophys. Acta* **188**, 222-229.
Appleby, C. A. (1969b).*Biochim. biophys. Acta* **189**, 267-279.
Appleby, C. A. (1969c).*Biochim. biophys. Acta* **180**, 202-203.
Appleby, C. A. (1969d).*Biochim. biophys. Acta* **172**, 71-87.
Appleby, C. A. (1969e). *Biochim. biophys. Acta* **172**, 88-105.
Aprison, M. H. and Burris, R. H. (1952). *Science, N.Y.* **115**, 264-265.
Aprison, M. H., Magee, W. E. and Burris, R. H. (1954). *J. biol. Chem.* **208**, 29-39.
Bach, M. K., Magee, W. E. and Burris, R. H. (1958). *Pl. Physiol., Lancaster* **33**, 118-124.
Bailey, R. W., Greenwood, R. M. and Craig, A. (1971). *J. gen. Microbiol.* **65**, 315-324.
Benemann, J. R., Yoch, D. C., Valentine, R. C. and Arnon, D. I. (1969). *Proc. natn. Acad. Sci. U.S.A.* **64**, 1079-1086.
Bergersen, F. J. (1957). *Aust. J. biol. Sci.* **10**, 233-242.
Bergersen, F. J. (1958). *J. gen. Microbiol.* **19**, 312-323.
Bergersen, F. J. (1960). *J. gen. Microbiol.* **22**, 671-677.
Bergersen, F. J. (1961). *Aust. J. biol. Sci.* **14**, 349-360.
Bergersen, F. J. (1962a). *J. gen. Microbiol.* **29**, 113-125.
Bergersen, F. J. (1962b). *Nature, Lond.* **194**, 1059-1061.
Bergersen, F. J. (1963). *Aust. J. biol. Sci.* **16**, 669-680.

Bergersen, F. J. (1965). *Aust. J. biol. Sci.* 18, 1-9.
Bergersen, F. J. (1966a). *In* IX Int. Cong. Microbiol. Symp. Moscow, 97-101.
Bergersen, F. J. (1966b). *Biochim. biophys. Acta* 130, 304-312.
Bergersen, F. J. (1968). *In* 9th Int. Congr. Soil Sci. Trans. II. 49-63.
Bergersen, F. J. (1969). *Proc. R. Soc. B.* 172, 401-416.
Bergersen, F. J. (1970). *Aust. J. biol. Sci.* 23, 1015-1025.
Bergersen, F. J. (1971a). *A. Rev. Pl. Physiol.* 22, 121-140.
Bergersen, F. J. (1971b). *Pl. Soil.* Special volume, 511-524.
Bergersen, F. J., Kennedy, G. S. and Wittman, W. (1965). L. Johnson. *Aust. J. biol. Sci.* 18, 1135-1142.
Bergersen, F. J. and Turner, G. L. (1967). *Biochim. biophys. Acta* 141, 507-515.
Bergersen, F. J. and Turner, G. L. (1968). *J. gen. Microbiol.* 53, 205-220.
Bergersen, F. J. and Turner, G. L. (1970). *Biochim. biophys. Acta* 214, 28-36.
Bisset, K. A. (1952). *J. gen. Microbiol.* 7, 233-242.
Brockwell, J. (1963). *Fld Stn Rec. Div. Pl. Ind. CSIRO (Aust.)* 2(1), 59-70.
Brockwell, J. and Phillips, L. J. (1965). *Aust. J. Sci.* 27, 332-333.
Brockwell, J., Rea, G. and Asuo, S. K. (1966). *J. Aust. Inst. agric. Sci.* 32, 295-297.
Burris, R. H. (1969). *Proc. R. Soc. B.* 172, 339-354.
Burris, R. H. and Wilson, P. W. (1939). *Cold Spring Harb. Symp. Quant. Biol.* 7, 349-361.
Cowles, J. R., Evans, H. J. and Russell, S. A. (1969). *J. Bact.* 97, 1460-1465.
Cutting, J. A. and Schulman, H. M. (1968). *Fedn Proc. Fedn Am. Socs exp. Biol.* 27, 768.
Cutting, J. A. and Schulman, H. M. (1969). *Biochim. biophys. Acta* 192, 486-493.
Dart, P. J. (1968). Proc. 4th Europ. Reg. Conf. Electron Microsc. 69-70.
Dart, P. J. and Mercer, F. V. (1963). *Arch. Mikrobiol.* 46, 382-401.
Dart, P. J. and Mercer, F. V. (1964). *Arch. Mikrobiol.* 47, 344-378.
Dart, P. J. and Mercer, F. V. (1965a). *Aust. J. agric. Res.* 16, 321-345.
Dart, P. J. and Mercer, F. V. (1965b). *Arch. Mikrobiol.* 51, 233-257.
De Hertogh, A. A., Mayeux, P. A. and Evans, H. J. (1964). *J. biol. Chem.* 239, 2446-2453.
De Ley, J. and Rassel, A. (1965). *J. gen. Microbiol.* 41, 85-91.
Delwiche, C. C. (1966). *Science, N.Y.* 151, 1565.
Dilworth, M. J. (1966). *Biochim. biophys. Acta* 127, 285-294.
Dilworth, M. J. (1969). *Biochim. biophys. Acta* 184, 432-441.
Dilworth, M. J. and Kidby, D. K. (1968). *Expl Cell Res.* 49, 148-159.
Doncaster, C. C., Nutman, P. S. and Bell, F. (1970). Nitrogen fixation in lucerne: a film, British Film Institute, London.
Dua, R. D. and Burris, R. H. (1963). *Proc. natn. Acad. Sci. U.S.A.* 50, 169-174.
Dudman, W. F. (1964). *J. Bact.* 88, 782-794.
Dudman, W. F. (1968). *J. Bact.* 95, 1200-1201.
Dudman, W. F. and Heidelberger, M. (1969). *Science, N.Y.* 164, 954-955.
Evans, H. J. and Russell, S. (1965). *Pl. Physiol., Lancaster* 40, iii-iv.
Foury, A. (1954). Les légumineuses fourragères au Maroc. Service de la Récherche Agronomique. Rabat. 656 pp.
Fred, E. B., Baldwin, I. L. and McCoy, E. (1932). Root nodule bacteria and leguminous plants. Wisconsin University Studies in Science No. 5. Univ. of Wisconsin press, Madison, U.S.A.
Gibson, A. H. (1966). *Aust. J. biol. Sci.* 19, 499-515.
Gibson, A. H. (1967a). *Aust. J. biol. Sci.* 20, 1087-1104.
Gibson, A. H. (1967b). *Aust. J. biol. Sci.* 20, 1105-1117.

Gibson, A. H. (1971). *Pl. Soil* (in press).
Goodchild, D. J. and Bergersen, F. J. (1966). *J. Bact.* 92, 204-213.
Graham, P. H. (1964a). *J. gen. Microbiol.* 35, 511-517.
Graham, P. H. (1964b). *Antonie van Leeuwenhoek* 30, 68-72.
Hallsworth, E. G. (1958). *In* "Nutrition of the legumes" (E. G. Hallsworth, ed.), Butterworth Scientific Publications, London.
Hamdi, Y. A. (1971). *Soil Biol. Biochem.* 3, 121-126.
Harris, J. R. and Rovira, A. D. (1962). *Bact. Proc.* 62, 24.
Hely, F. W. and Ofer, I. (1972). *Aust. J. agric. Res.* 23, 267-284.
Hely, F. W., Bergersen, F. J. and Brockwell, J. (1957). *Aust. J. agric. Res.* 8, 24-44.
Henzell, E. F. and Norris, D. O. (1962). Commonw. Bur. Past. Fld. Crops Bull. No. 46, 1-18.
Hoch, G., Schneider, K. C. and Burris, R. H. (1960). *Biochim. biophys. Acta.* 37, 273-279.
Hubbell, D. H. (1970). *Bot. Gaz.* 131, 337-342.
Humphrey, B. A. and Vincent, J. M. (1959). *J. gen. Microbiol.* 21, 477-484.
Humphrey, B. A. and Vincent, J. M. (1962). *J. gen. Microbiol.* 29, 557-561.
Humphrey, B. A. and Vincent, J. M. (1969). *J. gen. Microbiol.* 59, 411-425.
Jackson, E. K. and Evans, H. J. (1966). *Pl. physiol., Lancaster* 41, 1673-1680.
Jensen, H. L. (1961). *Nature, Lond.* 192, 682.
Jordan, D. C. (1962). *Bact. Rev.* 26, 119-141.
Katznelson, H. and Zagallo, A. C. (1957). *Can. J. Microbiol.* 3, 879-884.
Keele, B. B., Hamilton, P. B. and Elkan, G. H. (1969). *J. Bact.* 97, 1184-1191.
Kefford, N. P., Brockwell, J. and Zwar, J. A. (1960). *Aust. J. biol. Sci.* 13, 456-467.
Keilin, D. and Wang, Y. L. (1945). *Nature, Lond.* 155, 227-229.
Kennedy, I. R. (1966a). *Biochim. biophys. Acta* 130, 285-294.
Kennedy, I. R. (1966b). *Biochim. biophys. Acta* 130, 295-303.
Kennedy, I. R. (1970). *Biochim biophys. Acta* 222, 135-144.
Kennedy, I. R., Parker, C. A. and Kidby, D. K. (1966). *Biochim. biophys. Acta* 130, 517-519.
Kleczkowska, J., Nutman, P. S., Skinner, F. A. and Vincent, J. M. (1968). *In* "Identification methods for Microbiologists" part B (B. M. Gibbs and D. A. Shapton, eds), Academic Press, London and New York, pp. 51-65.
Klucas, R. V. and Evans, H. J. (1968). *Pl. Physiol., Lancaster,* 43, 1458-1460.
Klucas, R. V., Koch, B., Russell, S. and Evans, H. J. (1968). *Pl. Physiol., Lancaster,* 43, 1906-1912.
Koch, B. and Evans, H. J. (1966). *Pl. Physiol., Lancaster* 41, 1748-1750.
Koch, B., Evans, H. J. and Russell, S. (1967). *Proc. natn. Acad. Sci. U.S.A.* 58, 1343-1350.
Koch, B., Wong, P., Russell, S., Howard, R. and Evans, H. J. (1970). *Biochem. J.* 118, 773-781.
Kubo, H. (1939). *Acta Phytochem., Tokyo* 11, 195-200.
Lange, R. T. (1961). *J. gen. Microbiol.* 26, 351-359.
Lawrence, D. B., Schoenicke, R. E., Quispel, A. and Bond, G. (1967). *J. Ecol.,* 55, 793-813.
Lechtova-Trnka, M. (1931). *Botaniste* 23, 301-530.
Li, D. L. and Hubbell, D. H. (1969). *Can. J. Microbiol.* 15, 1133-1136.
Ljunggren, H. and Fahraeus, G. (1961). *J. gen. Microbiol.* 26, 521-528.
McCoy, E. (1932). *Proc. R. Soc. B.* 110, 514-533.
McLachlan, K. D. and Norman, B. W. (1961). *J. Aust. Inst. agric. Sci.* 27, 244-245.

Martinez de Drets, G. and Arias, A. (1970). *J. Bact.* **103**, 97-103.

Mooney, P. and Fottrell, P. F. (1968). *Biochem. J.* **110**, 17-18.

Mortenson, L. E., Morris, J. A. and Jeng, D. Y. (1967). *Biochim. biophys. Acta* **141**, 516-522.

Moustafa, E. and Mortenson, L. E. (1968). *Analyt. Biochem.* **24**, 226-231.

Munns, D. N. (1968a). *Pl. Soil.* **28**, 129-146.

Munns, D. N. (1968b). *Pl. Soil.* **28**, 246-257.

Munns, D. N. (1968c). *Pl. Soil.* **29**, 33-47.

Munns, D. N. (1968d). *Pl. soil.* **29**, 257-262.

Munns, D. N. (1969). *Pl. Soil.* **30**, 117-120.

Norris, D. O. (1956). *Emp. J. Exp. Agric.* **24**, 247-270.

Norris, D. O. (1965). *Pl. Soil.* **22**, 143-166.

Nutman, P. S. (1956). *Biol. Rev.* **31**, 109-151.

Nutman, P. S. (1962). *Proc. R. Soc. B.* **156**, 122-137.

Nutman, P. S. (1963). *In* "Symbiotic associations" 13th Symp. Soc. Gen. Microbiol. Cambridge University Press.

Nutman, P. S. (1969). *Proc. R. Soc. B.* **172**, 417-437.

Pankhurst, C., Schwinghamer, E. A. and Bergersen, F. J. (1972). *J. gen. Microbiol.* **70**, 161-177.

Parker, C. A. (1968). *In* "Festkrift til Hans Laurits Jensen", Gadgaard Nielsens Bogtrykkeri, Lemvig, Denmark.

Phillips, D. A. and Torrey, J. G. (1970). *Physiologia Pl.* **23**, 1057-1063.

Robinson, A. C. (1967). *J. Aust. Inst. agric. Sci.* **33**, 207-209.

Roughley, R. J., Dart, P. J., Nutman, P. S. and Rodriguez-Barrueco, C. (1970). *Proc. 11 Int. Grassld Congr.*, pp. 451-455. University of Queensland Press.

Rovira, A. D. (1961). *Aust. J. agric. Res.* **12**, 77-83.

Sahlman, K. and Fahraeus, G. (1963). *J. gen. Microbiol.* **33**, 425-427.

Scholander, P. F. (1960). *Science, N.Y.* **131**, 585-590.

Schollhorn, R. and Burris, R. H. (1966). *Fedn Proc. Fedn Am. Socs exp. Biol.* **25**, 710.

Schwinghamer, E. A. (1968). *Can. J. Microbiol.* **14**, 355-367.

Schwinghamer, E. A. (1969). *Can. J. Microbiol.* **15**, 611-622.

Schwinghamer, E. A. (1971). *Soil Biol. Biochem.* **3**, 355-363.

Schwinghamer, E. A. and Belkengren, R. P. (1968). *Arch. Mikrobiol.* **64**, 130-145.

Smith, J. D. (1949). *Biochem. J.* **44**, 585-591.

Stanton, W. R. (1966). Grain legumes in Africa, F.A.O. Rome, 183 pp.

Tjepkema, J. D. and Yocum, C. S. (1970). *Pl. Physiol., Lancaster* **46**, suppl., 44.

Torrey, J. G. (1961). *Expl Cell Res.*, **23**, 281-299.

Torrey, J. G. and Barrios, S. (1969). *Caryologia*, **22**, 47-61.

Turner, G. L. and Bergersen, F. J. (1969). *Biochem. J.* **115**, 529-535.

Valera, C. L. and Alexander, M. (1965). *Nature, Lond.* **206**, 326.

Vincent, J. M. (1962). *J. gen. Microbiol.* **28**, 653-663.

Vincent, J. M. (1967). *Aust. J. Sci.* **29**, 192-197.

Vincent, J. M. (1970). A manual for the practical study of root-nodule bacteria. Blackwell Scientific Publications, Oxford. 164 pp.

Virtanen, A. I., Moisio, T. and Burris, R. H. (1955). *Acta chem. scand.*, **9**, 184-186.

Wilson, P. W. (1940). *In* "The biochemistry of symbiotic nitrogen fixation". Madison, The University of Wisconsin press. 302 pp.

Wipf, L. and Cooper, D. C. (1938). *Proc. natn. Acad. Sci. U.S.A.* **24**, 87-91.

Wipf, L. and Cooper, D. C. (1940). *Am. J. Bot.*, **27**, 821-824.

Yao, P. Y. and Vincent, J. M. (1969). *Aust. J. biol. Sci.* **22**, 413-423.

Yoch, D. C., Benemann, J. R., Valentine, R. C. and Arnon, D. I. (1969). *Proc. natn. Acad. Sci. U.S.A.* **64**, 1404-1410.
Yoch, D. C., Benemann, J. R., Arnon, D. I., Valentine, R. C. and Russell, S. A. (1970). *Biochem. biophys. Res. Commun.* **38**, 838-842.
Yocum, C. S. (1961). *Science, N.Y.* **146**, 432.

CHAPTER 22

The Nitrogen Cycle of Pasture Ecosystems

*Division of Tropical Pastures, C.S.I.R.O., Cunningham
Laboratory, Mill Road, St. Lucia, Queensland, 4067,
Australia*

and

Division of Soils, C.S.I.R.O., Cunningham Laboratory

I.	Introduction	228
II.	Nitrogen in Plants	228
	A. Amounts used for Growth	228
	B. Fixation by Legumes	229
	C. Pathways	230
III.	Nitrogen in Animals	232
	A. Amounts in Sheep, Cattle, Wool, and Milk	233
	B. Intake, Retention, and Excretion of Dietary N	233
	C. Faeces and Urine	236
	D. Animals other than Sheep and Cattle	236
IV.	Nitrogen in Soils	237
	A. Amounts and Forms	237
	B. Inputs	238
	C. Transformations	238
	D. Losses of Mineral N	239
V.	Nitrogen Gains and Losses	241
	A. Gains	241
	B. Losses	242
	C. Net Changes in Soil N	242
VI.	Dynamic Aspects of the Nitrogen Cycle	243
	References	244

I. INTRODUCTION

A pasture ecosystem can be defined as consisting of the plants, animals, and soil (to the depth of root penetration) in an area of grazing land supporting herbivorous domestic livestock. The area assumed in this chapter is a field or group of fields within which the grazing livestock are confined. The atmosphere is outside the system.

II. NITROGEN IN PLANTS

A. Amounts used for Growth

The N used by pasture plants comes from uptake of combined N through the roots, or in the case of nodulated legumes, from symbiotic fixation of N_2. Uptake of combined N through the shoots is apparently not important. The forms of N found in pasture plants are described in Chapters 1 and 2.

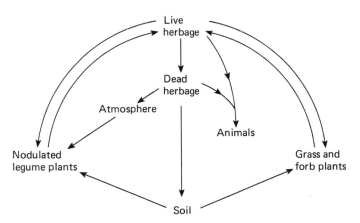

Fig. 1. A flow-diagram for herbage N.

Figure 1 illustrates the flow of N through herbage, i.e. edible plant material above the soil. The flow through roots follows similar pathways. Roots are not normally grazed by domestic livestock, but they are eaten by some other animals (Section II.C). Many unimproved grazing lands contain vegetation, e.g. trees, that is not much eaten by domestic livestock. The role of this vegetation in the N cycle is probably similar to that of the roots of edible species.

Although rates of flow of N through herbage are usually measured on ungrazed swards, they should be similar under grazing. Annual yields of oven-dry herbage from cutting-and-removal experiments vary from less than 1000 kg ha^{-1} for natural semi-desert rangelands to more than 80,000 kg ha^{-1} for infrequently-cut, N-fertilized grass plots in the humid tropics (Vicente-Chandler et al., 1959), but the dry-matter (DM) yields from the majority of pasture ecosystems lie between 2000 and 10,000 kg ha^{-1}. N concentrations in the herbage generally range from about 1·0 to 3·0% in the tropics and 1·5 to 3·5% in temperate regions, giving estimates of 20 to 300 kg N ha^{-1} a year for tropical herbage and 30 to 350 kg ha^{-1} a year for temperate herbage.

Higher values have been obtained in experiments done under favourable conditions. For example, the annual flow of herbage N, in round figures, was 500 kg ha^{-1} for N-fertilized perennial ryegrass in Scotland (Reid, 1970), 600 kg ha^{-1} for clover-ryegrass in New Zealand (Melville and Sears, 1953), and 900 kg ha^{-1} for N-fertilized pangola grass (*Digitaria decumbens*) and Napier grass (*Pennisetum purpureum*) in the Caribbean (Salette, 1970; Vicente-Chandler et al., 1959).

Existing techniques for measuring N used in root growth are particularly unsatisfactory. Some small roots are not harvested (Shamoot et al., 1968), and N used in new root growth cannot be distinguished from that in old roots (Milner and Hughes, 1968). The "roots" usually include an unspecified proportion of stubble, i.e. shoot tissue.

According to the values published for pasture plants in their first year of growth in pot and field experiments, the annual yield of N in herbage needs to be increased by the following proportions to give a figure for the whole plant: for temperate legumes, from 19 to 67% (Fribourg and Johnson, 1955; Greaves and Jones, 1950); for tropical legumes, from 8 to 60% (Henzell et al., 1968; Whitney et al., 1967); and from 38% for grasses with a high N supply to 105% for N-deficient grasses (Henzell et al., 1964; Woldendorp et al., 1966). The variability can be attributed partly to differences in the cutting heights used to separate herbage from "roots".

B. Fixation by Legumes

Fixation of N$_2$ by nodulated legumes plays an essential role in pasture improvement in both temperate and tropical agriculture. Most of the existing information on rates of symbiotic fixation has been obtained under systems of cutting and removal (Henzell, 1970).

Estimates for fixation by clovers range up to about 450 kg ha^{-1} a year in Britain and about 670 kg ha^{-1} in New Zealand (Whitehead, 1970). No doubt the average rates under practical grazing management are a good deal lower than this, especially as soil N approaches equilibrium (Section V). Average annual rates of fixation by tropical legumes in grazed pastures on N-deficient soils in northern Australia have been estimated at 20 to 180 ka ha^{-1} (Henzell, 1968).

Although legumes are able to use available soil N, they appear to be weaker competitors for it than grasses (Henzell, 1970; Walker *et al.*, 1956).

C. Pathways

Part of the N used for growth of herbage is subsequently eaten by animals, removed by man, or recirculated within the plant either for new growth or for storage in reproductive tissues. A significant proportion remains in the same organ until it dies. There is no evidence that significant quantities of N are leached or volatilized from herbage before senescence.

1. N Eaten by Animals

Efficient grazing by domestic livestock during the growing season may remove up to 90% of the herbage N that could be harvested by cutting (Blaser *et al.*, 1961), but in practice the proportion is often much lower, especially when the growing season is short and stock are maintained on the pasture all year round. The proportion eaten varies directly with stocking intensity (biomass of animals relative to herbage yield).

It is much more difficult to estimate the proportion of herbage N eaten by wild animals. They range in size from large mammals (African game animals, kangaroos) through rabbits, rodents, and birds to insects and mites. Their consumption of DM and N is usually ignored unless they compete seriously with domestic animals for the available herbage, e.g. during plagues of grasshoppers and moth larvae.

Even less is known about the grazing of roots, which certainly are eaten by animals such as beetles, beetle larvae, and nematodes.

2. N Removed by Man

Significant quantities of herbage N are cut and removed from pastures, and may not be returned, in some types of intensive grassland farming. For zero grazing the entire annual crop of herbage is cut, but usually only part (often less than half) is harvested for hay or silage.

N is also removed in seed crops. Assuming that the N concentration in legume seeds is 3 to 6% and in grass seeds 1 to 3%, and that average seed yields do not exceed 500 kg ha^{-1} a year, up to 15 to 30 kg N ha^{-1} are removed in legume seed crops and up to 5 to 15 kg in grass seeds. Yields exceeding 2000 kg ha^{-1} have been recorded from some lucerne and perennial ryegrass seed crops in the U.S.A.

3. N Stored in Reproductive Tissues

Between episodes of plant growth N is stored in seeds and/or in surviving vegetative tissues (notably the lower stems and larger roots). Thus, for subterranean clover (*Trifolium subterraneum,* an annual) Lapins and Watson (1970) recorded that, of about 160 kg N ha^{-1} used for a season's growth of whole plants, 30 kg were found in the pods and seeds at maturity. Weinmann (1948) reviewed experiments with perennial grasses indicating that, at maturity, up to 36% of the herbage N had been translocated to the basal internodes and roots. Less than 15% of the herbage N was found in the mature inflorescences of some perennial tropical grasses (Henzell and Oxenham, 1964).

4. N in Plant Residues

Normally, herbage and root residues consist mainly of material that senesced naturally before it died. Most parts of herbage plants live for less than a year, some for only a few weeks.

The dead leaves of herbaceous legumes generally are either deciduous or so fragile that they soon fall onto the soil, whereas the dead culms of grasses may take more than a year to disintegrate. Standing dead herbage is rapidly invaded by saprophytic microorganisms, especially fungi, and is already partly decomposed when the residues reach the soil.

There is strong circumstantial evidence that some N is volatilized, presumably as ammonia, from dead herbage, and that some is leached (Lapins and Watson, 1970). In addition, N is lost to the air if the dead herbage is burned.

Very little detailed work has been done on the fate of N used in root growth. Roots undergo senescence with associated N transformations similar to those occurring in shoots (Habeshaw and Heyes, 1971; cf. Chapter 24). Progressive attack by microorganisms precedes disintegration of the dead residues.

III. NITROGEN IN ANIMALS

The place of domestic and wild animals in the flow of N to other components of pasture ecosystems is illustrated in Fig. 2. In reality, however, the pathways are much more complicated than the figure suggests. For instance, some of the vegetation is invaded by microorganisms (parasitic on live material, saprophytic on dead) before it is eaten by herbivores, and microorganisms (e.g. bacteria and protozoa in ruminants) are always present in the digestive tracts of animals. Also, the distinction between the different groups of animals is blurred in nature, e.g. herbivorous animals commonly eat dead plant material (the normal role of saprophagous animals.

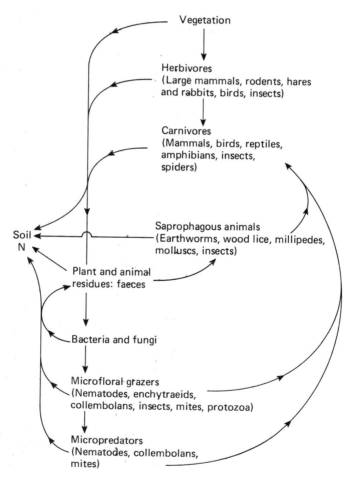

Fig. 2. Pathways for the flow of N through animals in pasture ecosystems. Adapted from Paris (1969).

As a broad generalization the groups of animals at the top of Fig. 2 are larger and live above the ground, and those towards the bottom are smaller and live on or in the soil.

A. Amounts in Sheep, Cattle, Wool, and Milk

The figures for livestock in Table I do not include N in digesta (A.R.C., 1965), but it can be calculated that the weight of this N at any instance comprises less than 0·1% of the total liveweight or 4% of the total N in the animals. Two values for the N concentration in milk are given in Table I: 0·53% is a typical average value for milk of Friesian cows, and 0·61% for milk from Jersey cows (A.R.C., 1965).

TABLE I

Nitrogen contents of livestock and livestock products[a]

Animal and Product	Weight (kg)	N Concentration (%)	N Content (kg)
Sheep:			
Lambs at birth	2—5	2·5	0·05—0·13
Mature ewes and wethers	30—100	2·5, 2·4[b]	0·8—2·4
Annual wool production	1—8	11·4[c]	0·1—0·9
Cattle:			
Calves at birth	15—45	2·6	0·4—1·2
Mature cows and steers	300—550	2·4	7·2—13·2
Annual liveweight gain	90—300	2·4	2·2—7·2
Annual milk production	1000—4000	0·53	5·3—21·2
	1000—4000	0·61	6·1—24·4

[a] Reference: A.R.C. (1965).
[b] 2·5% below and 2·4% above 40 kg liveweight, including the fleece.
[c] Assuming greasy wool consists of 70% of protein containing 16·35% N.

B. Intake, Retention and Excretion of Dietary N

Three examples of the quantities of herbage N that may be eaten by sheep and cattle and the partition of this N between retention and excretion are given in Table II. The first assumes a stocking rate of 8 wethers ha^{-1}, each of 45 kg average liveweight, each producing 5 kg of greasy wool and gaining 3 kg liveweight per annum, and eating a total 2700 kg ha^{-1} of herbage DM a year. The second assumes one Jersey cow ha^{-1}, of average liveweight 360 kg, producing 2700 kg of milk from 3600 kg ha^{-1} of DM consumed annually. The third assumes one beef steer ha^{-1}, having a liveweight of 340 kg when brought into the system and 500 kg when removed a year later, and consuming 3000 kg ha^{-1} of DM a year.

The lower limits of the dietary N concentrations correspond to minimum requirements of apparently-digestible N estimated by the factorial procedure (A.R.C., 1965), converted to total N by the equation of Holter and Reid (1959). A minimum concentration of 0·8 to 1·4% is required to maintain the fermentative activity of rumen microorganisms (Corbett, 1969). The apparent N digestibilities used to partition the dietary N were calculated from the equation of Holter and Reid (1959), assuming that the % N in the diet was constant (unlikely in practice, but the resulting estimates cover most of the relevant experimental values sighted in the literature). The N contents of the livestock and their products were derived from Table I.

The N eaten by sheep and cattle is apparently divided into three fractions (the actual pathways are extremely complicated). Some is excreted in the faeces. The remainder is digested and absorbed, and partitioned between retention and excretion in the urine. There is no evidence of significant gaseous losses of N from live ruminants. Nitrogen fixation by rumen microorganisms has recently been demonstrated (Postgate, 1970), but is quantitatively of little significance.

Only a small proportion of dietary N is retained in animal products. The figure is highest for milk and lowest for meat (Table II), but in practice it also depends on the quality of the animal and its feed supply, especially its intake of available energy. Less productive livestock or feed with a lower energy content than in the examples would reduce the proportion of dietary N retained in body tissues and products.

The proportion of dietary N excreted in the faeces increases as the % N in the diet falls (Table II). For feeds containing 0·4 to 0·6% N the apparent digestibility of dietary N is about zero (Holter and Reid, 1959; Robinson and Stewart, 1968), i.e. as much N is excreted in the faeces as is ingested in the feed. Since some additional N is lost in urine, the animal suffers a net loss.

These examples (Table II) are typical of relatively productive pasture ecosystems. In practice, rates of flow of N through sheep and cattle vary greatly in different parts of the world.

TABLE II

Intake, retention, and excretion of dietary N in three hypothetical systems of livestock production

System	Mean N Content of Animals (kg ha^{-1})	Concentration of N in Diet (%)	Intake of N (kg ha^{-1})	Excreted in Faeces (kg ha^{-1})	Retained in Animals and Animal Products (kg ha^{-1})	Excreted in Urine (kg ha^{-1})
1. Wool	9	1·4	38	18 (47)	5 (13)[a]	15 (39)
		3·5	94	22 (23)	5 (5)	67 (71)
2. Dairying	9	1·7	61	24 (39)	17 (28)	20 (33)
		3·5	126	29 (23)	17 (13)	80 (63)
3. Beef	10	1·3	39	20 (51)	4 (10)	15 (38)
		3·5	105	24 (23)	4 (4)	77 (73)

[a] Figures in parentheses are % of dietary N.

C. Faeces and Urine

Very little has been published on N compounds in the faeces of sheep and cattle, though Mason (1969) found with two forages fed to sheep that undigested plant residue accounted for 11 and 18% of faecal N, and that 61 and 50% was bacterial material (the balance was probably of animal origin); 45 to 65% of the total was α-amino-N. The % N in faeces rises with % N in feed, but they are not closely related.

The chief nitrogenous constituent in the urine of sheep and cattle on high-N diets is usually urea, with some ammonia, allantoin, creatine and creatinine (Church, 1969). On low-N diets hippuric acid or ammonia (Topps and Elliott, 1966) may predominate. The % N in urine varies widely, depending on such factors as the % N in the diet and the animal's intake of water.

Large quantities of excreta are often found where grazing livestock drink or rest. For instance, sheep in Australia deposited about one-third of their faeces on less than 5% of the pasture (Hilder, 1966). Even with random distribution of excreta and a high stocking rate, less than half the area receives dung and urine each year (Petersen et al., 1956; Lotero et al., 1966). This topic is also discussed in Chapter 23. When dairy cattle are milked in barns, or livestock are shut away at night (as in some parts of the tropics), N is lost from the pasture unless the excreta are returned.

D. Animals other than Sheep and Cattle

Information on the N nutrition of wild animals is very sketchy indeed, judging by the comprehensive reviews published in connection with the Grassland Biome, United States International Biological Program (Dix and Beidleman, 1969). The large herbivores, with microbes capable of fermenting fibrous feeds in their digestive tract, can probably digest DM and N as efficiently as sheep and cattle, whereas the smaller animals (Fig. 2) appear to make limited use of plant cell walls. Only a few species of insects (Day and Waterhouse, 1953), snails, protozoa, millipedes, and possibly woodlice are known to possess cellulase (Paris, 1969). Consequently the smaller animals with a simpler digestive system obtain less energy from fibrous plant material than do sheep and cattle, and they may also digest less N (Glover and Duthie, 1958; Mitchell, 1964). Compared with sheep and cattle (Table II), they probably excrete more of the dietary N in the fraction corresponding to faeces and retain less in their bodies. The chief role of small saprophagous animals in pasture ecosystems seems to be in comminuting and redistributing plant and animal residues and excreta, which are decomposed mainly by bacteria, actinomycetes, and fungi.

Whereas mammals and amphibia excrete urinary N mainly as urea, uric acid is the chief end-product in birds, snakes and lizards, insects, and snails, guanine is excreted by spiders, and ammonia by protozoa (Baldwin, 1952; Haggag and Fouad, 1965; Burges and Raw, 1967). Up to one-half the N excreted by earthworms is in the form of protein; much of the remainder is urea and ammonia (Needham, 1957).

IV. NITROGEN IN SOILS

A. Amounts and Forms

Pastures occur on diverse soils, many formerly carrying other vegetation, e.g. forest. The N concentration in the surface 15- to 30-cm layer ranges from less than 0·01% (oven-dry basis) in infertile sands to several percent in peats, with most values falling between 0·04 and 0·40%. The % N in soils usually, but not always, decreases steadily with depth (Stevenson, 1965). The whole profile to the depth of root penetration may contain from 4500 to 24,000 kg N ha^{-1} (Henzell, 1972), which far outweighs the N in plants and animals.

Available mineral N (nitrite, nitrate, and exchangeable ammonium) usually comprises less than 2% of the total in pasture soils. N-deficient soils normally contain only traces of nitrite and nitrate, but they may contain up to about 10 mg kg^{-1} of exchangeable ammonium-N.

As much as 10% of the N in the surface soil and 60% in the subsoil may occur as fixed (non-exchangeable) ammonium-N, which is usually inaccessible to plants and microorganisms (Black, 1968; Bremner, 1968). Over 90% of the N in most surface soils is organically combined. From 20% to about 50% of this N is in bound amino acids and 5 to 10% in hexosamines. Purine and pyrimidine derivatives do not account for more than 1%. The proportion of N in live flora and fauna is also very small. The chemical nature of about half the organic N in soils remains obscure (Bremner, 1968).

B. Inputs

Most of the N passing through plants (Section II) and animals (Section III) in pasture ecosystems subsequently is returned to the soil in dead plant and animal residues and animal excreta.

Synthetic fertilizers supply large quantities of mineral N to grassland in some parts of western Europe (an average of 175 kg N ha^{-1} a year is applied in the Netherlands) and Florida and Georgia. Smaller amounts of combined N may be added to pasture soils by surface flow or underground movement of water from outside the system, and by precipitation. The extensive literature on N in rain- and snow-fall records annual additions of 1 to 30 kg ha^{-1}, but many of the values lie between 2 and 10 kg ha^{-1} (Henzell and Norris, 1962; Henzell, 1972). It has been suggested that useful quantities of N may be absorbed from the air (as ammonia) by soils near industrial areas (Malo and Purvis, 1964), or feedlots.

Considerable uncertainty prevails about rates of N_2 fixation by free-living soil bacteria and blue-green algae. Gains exceeding 100 kg N ha^{-1} a year have been claimed (Moore, 1966), but recent work has indicated that the annual rates are usually less than 10 kg N ha^{-1} in well-aerated pasture soils (Kass et al., 1971; Mishustin, 1970; Steyn and Delwiche, 1970). Higher rates seem more likely when plant residues are decomposed anaerobically (Clark and Paul, 1970; Magdoff and Bouldin, 1970).

C. Transformations

1. Mineralization and Immobilization

Soil organic N is usually mineralized at a rate of less than 3% a year (Bremner, 1968). However, mineralization of even 1% a year from a soil containing 10,000 kg ha^{-1} yields 100 kg N ha^{-1} a year. Up to 40% of the organic N in surface soils can be mineralized in the long term, but attempts to isolate a readily-mineralized fraction of soil N by chemical techniques have met with little success so far (Bremner, 1968).

Fresh organic matter added to soil is rapidly attacked, almost all of its N being accessible to the microflora (Bartholomew, 1965). Microbial growth depends on the supply of energy and N from the decomposing material. If the N concentration is relatively high, ammonium is released into the soil (the process of net mineralization). If it is low, mineral N may be taken up from the soil (net immobilization). ^{15}N studies have revealed that mineralization and immobilization actually occur simultaneously, even during net immobilization. The ecology of the flora and fauna involved in soil N transformations is exceedingly complex (see the excellent review by Clark (1969)).

Microbial N rapidly becomes stabilized against further biological decomposition in soil (Bartholomew, 1965). Various mechanisms have been proposed (Legg et al., 1971), including biosynthesis of compounds resistant to degradation, formation of organo-mineral complexes, and micro-aggregation (i.e. physical inaccessibility), but as yet there is no clear proof that any of them operate in nature. Stabilization can be extremely rapid; 43% of the N added as albumin was mineralized after 8 days of incubation with soil and none over the next 24 days (Vlassak, 1966).

The availability of N from fresh organic matter is a function of its concentration. Plant residues containing >1·8% N usually mineralize N immediately in vitro, and those with <1·2% usually immobilize it (Alexander, 1961). Bartholomew (1965) estimated that about half the N in legume residues (tops) containing 2% N might be mineralized in the first season, whereas residues containing 1·0 to 1·3% N may supply very little mineral N. After several years the availability of N remaining from added organic matter is not much above that of the old soil N.

2. Nitrification

Broadly, nitrification is the reaction or reactions producing nitrite and nitrate (Alexander, 1965). Ammonium is the common substrate. There have been many reports that nitrification is inhibited in soils under grasses (Clark and Paul, 1970), especially N-deficient grasses.

One hypothesis attributes the inhibition to lack of ammonium due to its immobilization by low-N grass root residues; a second claims that living plant roots directly suppress nitrifying bacteria. Evidence has been obtained for both views (Huntjens, 1971; Moore and Waid, 1971).

D. Losses of Mineral N

Mineral N may be lost from the soil in solution or as a gas, thereby reducing the amount available for plant growth.

1. Leaching

The main form of N in groundwater is nitrate. Early lysimeter studies suggested that relatively small quantities of nitrate were leached from grass and grass-legume pastures (Kolenbrander, 1969), but recent work with heavier rates of fertilizer N in western Europe has indicated much larger losses, particularly from sandy soils during winter (Woldendorp et al., 1966; Kofoed and Lindhard, 1968). In the Netherlands, almost no N was lost from spring fertilization, but nearly 40% of the N applied in autumn was leached (Kolenbrander, 1969). However, N was lost in summer if heavy rain fell immediately after fertilization (Woldendorp et al., 1966). Kolenbrander (1969) estimated that about 9 kg N ha^{-1} were leached from an average pasture in the Netherlands receiving 180 kg N ha^{-1} as ammonium nitrate-limestone in three or four dressings.

Few measurements have been made of the amount of mineral N in surface flow on pastures, but even when ammonium nitrate was applied to wet ground the loss in subsequent run-off was only about 15% of a dressing of 224 kg N ha^{-1} (Moe et al., 1967).

2. Volatilization of Ammonia

Ammonia may be lost whenever high concentrations are created by application of ammonium fertilizers or by decomposition of organic residues, animal excreta, or fertilizers such as urea. The loss is largest from alkaline media with low-cation-exchange capacities; up to 90% of the N was lost from urine applied to the surface of columns of dry sandy soil (Stewart, 1970). With pasture, losses of N of up to 60% were recorded from urea fertilizer (Simpson, 1968), more than 50% from urine (Watson and Lapins, 1969), and up to 80% from faeces (Gillard, 1967). It is a reasonable supposition that most of it was lost as ammonia. A small amount of ammonia volatilizes even from acid soils under grass (Martin and Ross, 1968). Ammonia may also be absorbed by soil (Malo and Purvis, 1964), but it is not known what factors determine whether it will be absorbed or volatilized in nature.

3. Denitrification

Two forms of denitrification have been demonstrated in the laboratory. In enzymatic denitrification microorganisms substitute nitrate respiration for oxygen respiration, and in chemical denitrification gaseous N is lost due to instability of nitrite (Clark and Paul,

1970). Both produce N_2, NO_2, and N_2O. While enzymatic denitrification may account for significant losses of nitrate from pasture soils (Woldendorp *et al.*, 1966), there is no proof yet that chemical denitrification occurs in the field (Clark and Paul, 1970).

V. NITROGEN GAINS AND LOSSES

A. Gains

1. In Solid Form

N is imported into pasture ecosystems as a constituent of seeds and vegetative material used for planting; hay, silage, grains, and other feeds; pollen and other plant particles transported by wind or water; domestic and wild animals (including the contents of their digestive tracts); soil particles carried by wind or water; nitrogenous fertilizers and other agricultural chemicals (including urea used as a feed); and animal manures.

2. In Solution

N may be gained by surface or underground flow of water from outside, or in rain and snow.

3. As a Gas

Pasture ecosystems gain N through biological fixation of N_2 or absorption of ammonia.

Use of synthetic fertilizers or symbiotic fixation by nodulated legumes usually provide the main gains of N for intensive pasture systems. In some special situations N may be brought in as feed. For example, milking cows fed on grass may also be allowed 2 to 4 kg of concentrates, containing 2·0 to 3·5% N, per 10 kg of milk. Thus, 16 to 56 kg N may be fed to a cow yielding 4000 kg of milk a year. This N will not be a net gain to the ecosystem, however, unless the excreta are deposited on the pasture. In tropical Australia, urea and molasses are fed as a supplement during the dry season. With 10 to 15 g urea a day given to sheep and 60 to 110 g a day to cattle, 0·8 to 1·2 kg N a sheep and 5·0 to 9·1 kg N a beast are added in six months.

Precipitation and non-symbiotic fixation provide comparatively slow rates of gain but may be significant under extensive conditions, e.g. in unimproved grasslands. It is likely that the other types of gain listed above are of minor importance in most cases.

B. Losses

1. In Solid Form

N is lost in the form of seeds, hay, herbage cut for silage or zero grazing, pollen and other plant material moved by wind and water, soil particles moved by the same agencies, and movement of domestic or wild animals (and their excreta).

2. In Solution

Dissolved N is lost in surface and underground movement of water.

3. As a Gas

N may be lost in the form of ammonia, nitrogen gas, nitrous oxide, and nitrogen dioxide. This is the area of greatest uncertainty in the present knowledge of the N cycle in pastures, chiefly because of the difficulty in measuring fluxes of nitrogenous gases in the field (Burford and Millington, 1968; Ross et al., 1968).

Losses from extensive (low-N) pasture ecosystems are likely to be quite small. The importance of the N removed by sale of domestic livestock and their products has probably been over-emphasized. Much more is lost when the dung and urine are removed (Table II). The same generalization probably applies to wild animals migrating from pasture ecosystems. large quantities of N are removed when herbage is cut and taken away; smaller, but significant, amounts are harvested in seed crops (Section II), but these practices are not used at all in most extensive grazing systems.

Substantial leaching losses are probably restricted to intensive systems of agriculture receiving large inputs of N from fertilizers or nodulated legumes in regions with an excess of precipitation over evaporation.

It seems very likely that some ammonia is volatilized from all types of pasture systems. The importance of denitrification is difficult to assess; it tends to be blamed for any unaccounted-for losses, but is unlikely to be very significant unless surplus mineral N accumulates in the soil. In extensive, low-N systems the chief gaseous loss may be caused by burning the dead herbage.

C. Net Changes in Soil N

So far as the authors are aware there are no published balance sheets giving measured rates for all the pathways of gain and loss of

N in a pasture ecosystem, but net changes in soil N have been measured successfully. Since the soil is the chief store for N in pasture ecosystems, the net change for the whole system is likely to be similar to that for soil alone (which in practice also usually includes fine roots and small animals).

The quantity of N in a soil under pasture tends to increase (or decrease) towards an equilibrium value, presumably because a higher N status in the system exercises a feed-back effect on biological N fixation or because it leads to larger losses, and vice versa. Jenny (1941) found that the equilibrium N concentration under natural grassland in the U.S.A. varied with temperature and rainfall, and Jackman (1964) observed differences between soils; those containing allophane accumulated higher levels of N. In New Zealand it took up to about 50 years under white clover pastures for the rate of change of soil N to become negligibly small; the total soil N (0 to 30 cm) at equilibrium was calculated to lie between 7700 and 16,100 kg ha^{-1}, depending on soil and site (Jackman, 1964).

VI. DYNAMIC ASPECTS OF THE NITROGEN CYCLE

The preceding sections dealt with quantities of N in different components of pasture ecosystems and the rates of flow between them. While N in different components is equally important in the long term, its short-term availability to plants and animals varies widely. There are large quantities of N in the soil, yet, because it is largely unavailable, productivity is often limited by N deficiency.

Less than 3% of the soil N is normally mineralized each year, so in the short term most of this component is outside the active part of the N cycle. Only a portion of the N added in organic residues is mineralized in the following few years, and N fixed by soil microorganisms must likewise undergo mineralization before it is available to plants. In contrast, N in inorganic fertilizers, urine and rain is readily available, while symbiotic fixation in legume nodules supplies N directly to the host plants (but only indirectly to grasses and forbs—Henzell, 1970). Leaching or denitrification remove available N from the system and may therefore be more important in the short term than losses of less available components. Immobilization of mineral N by soil microorganisms is also an important pathway of loss from the most active part of the N cycle.

In general, N flows readily through all components of a pasture ecosystem except the soil in organic N (fixed ammonium-N is in effect outside the cycle). There is thus a large build-up of this component, and until it is large enough to mineralize sufficient N each year, substantial inputs of N from elsewhere are needed to maintain the flow through plants and animals. If it were not for this bottleneck to N flow, highly productive pasture exosystems could function on a small quantity of actively cycling N, inputs being required only to balance losses. In practice, many years must be spent in filling up this organic N sink before enough N is cycling to support high productivity.

REFERENCES

Alexander, M. (1961). "Introduction to Soil Microbiology". Wiley, New York.
Alexander, M. (1965). In "Soil Nitrogen" (W. V. Bartholomew and F. E. Clark, eds), pp. 307-343. American Society of Agronomy, Madison.
A.R.C. (1965). "The Nutrient Requirements of Farm Livestock. No. 2. Ruminants. Technical Reviews and Summaries". Agricultural Research Council, London.
Baldwin, E. (1952). "Dynamic Aspects of Biochemistry". Cambridge University Press.
Bartholomew, W. V. (1965). In "Soil Nitrogen" (W. V. Bartholomew and F. E. Clark, eds), pp. 285-306. American Society of Agronomy, Madison.
Black, C. A. (1968). "Soil-Plant Relationships". Wiley, New York.
Blaser, R. E., Hammes, R. C. Jr., Bryant, H. T., Hardison, W. A., Fontenot, J. P. and Engel, R. W. (1961). Proc. VIIIth Int. Grassl. Congr. Reading, 1960, pp. 601-606.
Bremner, J. M. (1968). In "Study Week on Organic Matter and Soil Fertility, Rome, 1968", pp. 143-185. North-Holland, Amsterdam.
Burford, J. R. and Millington, R. J. (1968). Trans. 9th Int. Congr. Soil Sci. Adelaide, 2, 505-511.
Burges, A. and Raw, F. (Eds) (1967). "Soil Biology". Academic Press, London and New York.
Church, D. C. (1969). "Digestive Physiology and Nutrition of Ruminants", Vol. 1. D. C. Church, Corvallis.
Clark, F. E. (1969). In "Soil Biology: Reviews of Research", pp. 125-161. UNESCO, Paris.
Clark, F. E. and Paul, E. A. (1970). Adv. Argon. 22, 375-435.
Corbett, J. L. (1969). In "Nutrition of Animals of Agricultural Importance" (D. P. Cutherbertson, ed.), Part 2, pp. 593-644. Pergamon, Oxford.
Day, M. F. and Waterhouse, D. F. (1953). In "Insect Physiology" (K. D. Roeder, ed.), pp. 311-330. Wiley, New York.
Dix, R. L. and Beidleman, R. G. (Eds) (1969). "The Grassland Ecosystem: A Preliminary Synthesis", and Supplement (1970). Colorado State University, Fort Collins.
Fribourg, H. A. and Johnson, I. J. (1955). Agron. J. 47, 73-77.
Gillard, P. (1967). J. Aust. Inst. agric. Sci. 33, 30-34.
Clover, J. and Duthie, D. W. (1958). J. agric. Sci., Camb. 51, 289-293.
Greaves, J. E. and Jones, L. W. (1950). Soil Sci. 69, 71-76.
Habeshaw, D. and Heyes, J. K. (1971). New Phytol. 70, 149-162.

Haggag, G. and Fouad, Y. (1965). *Nature, Lond.* 207, 1003-1004.

Henzell, E. F. (1968). *Trop. Grasslds* 2, 1-17.

Henzell, E. F. (1970). Proc. XIth Int. Grassl. Congr. Surfers Paradise, Qd. Aust., pp. A112-A120.

Henzell, E. F. (1972). *In* "Handbook of Tropical Forage Legumes" (P. J. Skerman, ed.) F.A.O., Rome (in press).

Henzell, E. F. and Norris, D. O. (1962). *In* "A review of nitrogen in the Tropics with particular reference to pastures" Symposium, Div. Trop. Past. C.S.I.R.D. pp. 1-18, Commw. Bur. Past. Fd Crops, Bull 46, Hurley, England.

Henzell, E. F. and Oxenham, D. J. (1964). *Aust. J. exp. Agric. Anim. Husb.* 4, 336-344.

Henzell, E. F., Martin, A. E., Ross, P. J. and Haydock, K. P. (1964). *Aust. J. agric. Res.* 15, 876-884.

Henzell, E. F., Martin, A. E., Ross, P. J. and Haydock, K. P. (1968). *Aust. J. agric. Res.* 19, 65-77.

Hilder, E. J. (1966). Proc. Xth Int. Grassld Congr. Helsinki, pp. 977-981.

Holter, J. A. and Reid, J. T. (1959). *J. Anim. Sci.* 18, 1339-1349.

Huntjens, J. L. M. (1971). *Pl. Soil* 34, 393-404.

Jackman, R. H. (1964). *N. Z. Jl agric. Res.* 7, 445-471.

Jenny, H. (1941). "Factors of Soil Formation". McGraw-Hill, New York.

Kass, D. L., Drosdoff, M. and Alexander, M. (1971). *Proc. Soil Sci. Soc. Am.* 35, 286-289.

Kofoed, A. D. and Lindhard, J. (1968). *Tidsskr. PlAvl* 71, 417-437.

Kolenbrander, G. J. (1969). *Neth. J. agric. Sci.* 17, 246-255.

Lapins, P. and Watson, E. R. (1970). *Aust. J. exp. Agric. Anim. Husb.* 10, 599-603.

Legg, J. O., Chichester, F. W., Stanford, G. and De Mar, W. H. (1971). *Proc. Soil Sci. Soc. Am.* 35, 273-276.

Lotero, J., Woodhouse, W. W. Jr. and Petersen, R. G. (1966). *Agron. J.* 58, 262-265.

Magdoff, F. R. and Bouldin, D. R. (1970). *Pl. Soil* 33, 49-61.

Malo, B. A. and Purvis, E. R. (1964). *Soil Sci.* 97, 242-247.

Martin, A. E. and Ross, P. J. (1968). *Pl. Soil* 28, 182-186.

Mason, V. C. (1969). *J. agric. Sci., Camb.* 73, 99-111.

Melville, J. and Sears, P. D. (1953). *N. Z. Jl Sci. Technol.* 35A, Suppl. 1, 30-41.

Milner, C. and Hughes, R. E. (1968). "Methods for the Measurement of the Primary Production of Grassland". (I.B.P. Handbook No. 6). Blackwell, Oxford.

Mishustin, E. N. (1970). *Pl. Soil* 32, 545-554.

Mitchell, H. H. (1964). "Comparative Nutrition of Man and Domestic Animals", Vol. 2. Academic Press, New York.

Moe, P. G., Mannering, J. V. and Johnson, C. B. (1967). *Soil Sci.* 104, 389-394.

Moore, A. W. (1966). *Soils Fertil.* 29, 113-128.

Moore, D. R. E. and Waid, J. S. (1971). *Soil Biol. Biochem.* 3, 69-83.

Needham, A. E. (1957). *J. exp. Biol.* 34, 425-446.

Paris, O. H. (1969). *In* "The Grassland Ecosystem: A Preliminary Synthesis" (R. L. Dix and R. G. Beidleman ed), pp. 331-360, and Supplement (1970), pp. 1-3. Colorado State University, Fort Collins.

Petersen, R. G., Lucas, H. L. and Woodhouse, W. W. Jr. (1956). *Agron. J.* 48, 440-444.

Postgate, J. (1970). *J. gen Microbiol.* 63, 137-139.

Reid, D. (1970). *J. agric. Sci., Camb.* 74, 227-240.

Robinson, D. W. and Stewart, G. A. (1968). *Aust. J. exp. Agric. Anim. Husb.* 8, 419-424.

Ross, P. J., Martin, A. E. and Henzell, E. F. (1968). Trans. 9th Int. Congr. Soil Sci. Adelaide 2, 487-494.

Salette, J. E. (1970). Proc. XIth Int. Grassl. Congr. Surfers Paradise, Qd. Aust., pp. 404-407.

Shamoot, S., McDonald, L. and Bartholomew, W. V. (1968). *Proc. Soil Sci. Soc. Am.* 32, 817-820.

Simpson, J. R. (1968). Trans. 9th Int. Congr. Soil Sci. Adelaide 2, 459-466.

Stevenson, F. J. (1965). *In* "Soil Nitrogen" (W. V. Bartholomew and F. E. Clark, eds), pp. 1-42. American Society of Agronomy, Madison.

Stewart, B. A. (1970). *Environ. Sci. Technol.* 4, 579-582.

Steyn, P. L. and Delwiche, C. C. (1970). *Environ. Sci. Technol.* 4, 1122-1128.

Tops, J. H. and Elliott, R. C. (1966). *Proc. Nutr. Soc.* 25, XIX-XX.

Vicente-Chandler, J., Silva, S. and Figarella, J. (1959). *Agron. J.* 51, 202-206.

Vlassak, K. (1966). *1966). Agricultura Louvain* 14, 49-65.

Walker, T. W., Adams, A. F. R. and Orchiston, H. D. (1956). *Soil Sci.* 81, 339-351.

Watson, E. R. and Lapins, P. (1969). *Aust. J. exp. Agric. Anim. Husb.* 9, 85-91.

Weinmann, H. (1948). *J. Br. Grassld Soc.* 3, 115-140.

Whitehead, D. C. (1970). *In* "The Role of nitrogen in grassland productivity". Commonw. Bur. Past. Fd Crops, Bull. 48, Hurley.

Whitney, A. S., Kanehiro, Y. and Sherman, G. D. (1967). *Agron. J.* 59, 47-50.

Woldendorp, J. W., Dilz, K. and Kolenbrander, G. J. (1966). Proc 1st Gen. Meet. Europ. Grassl. Fed. Wageningen, 1965, pp. 53-68.

CHAPTER 23

Cycling of Mineral Nutrients in Pasture Ecosystems

S. R. WILKINSON

Southern Branch, Soil and Water Conservation Research Division, Agricultural Research Service, USDA, Watkinsville, Ga. 30677, U.S.A.

and

R. W. LOWREY

Department of Animal Science, University of Georgia, Athens, Ga., 30601, U.S.A.

I. INTRODUCTION

A pasture ecosystem may be any pasture system with defined upper, lower and lateral boundaries. These boundaries may be fixed arbitrarily, or may represent natural areas such as a small watershed. Whatever the size of the ecosystem it will be open, at least with respect to one or more properties and will gain or lose energy and matter (Jenny, 1961). A partial list of influxes from outside the defined boundaries includes energy, matter including gases, water, solids dissolved or dispersed in liquid, solids, and immigration of organisms. A partial list of outfluxes includes energy, matter including gases (particularly water as evapotranspiration), water in drainage, solids dissolved or dispersed in water, wind borne particles, and emigration of organisms (Jenny, 1961). The ecosystem is also a community of organisms and their environment treated together as a functional system of complementary relationships (Whitaker, 1970).

The ecosystem represents a grouping of components—soils that support the plants, animals that graze these plants, residues of these plants and animals, microbes which decompose these residues, and atmospheric gases all linked together as functional entities by food webs, flows of mineral nutrients and flows of energy. This definition covers a natural range managed by ecological principles with minor intervention by man as well as sown or derived pastures managed by man through agronomic techniques. The emphasis in this chapter will be on mineral nutrient cycling in those pasture ecosystems where human manipulation is a major factor such as in sown, or intensively managed derived pastures. Such ecosystems are young, with high productivity, and are growth systems (Odum, 1969). High yields of usable product and rapid nutrient exchange rates between organisms are desired. The stability of these systems to external effects is maintained chiefly by man's intercession.

The objectives of this chapter are to describe the factors affecting circulation of mineral nutrients through soil-plant-grazing animal food chains; to discuss the qualitative as well as quantitative aspects of cycles for the mineral nutrients phosphorus, potassium and sulphur as well as brief discussions of cycling of calcium and magnesium; to suggest potential points of control of these mineral cycles by management; and to point out a few areas of needed research. Mineral uptake and function in plants and animals will be presented elsewhere (Chapters 12, Vol. 1; 18, 19) as will nitrogen cycling in pasture ecosystems (Chapter 22). The effect of soil contamination of herbage and soil ingestion by animals is discussed in Chapter 13, Vol. 1.

II. MINERAL CYCLES

Tracing the movement of mineral elements through the complex chain of soils, plants, and animals involves a multiplicity of complex patterns of transfers, transformations, concentration, accumulation and utilization (Whitaker, 1970). In reality the cycle of an element in a pasture is polycyclic, that is, it involves cycles between soil and residues, available soil minerals and plants, plants and animals, etc. These cycles differ in duration: that is, biochemical cycles within organisms may take minutes; those in an ecosystem may take months to thousands of years while geochemical cycles may take millions of years (Pomeroy, 1970). The biochemistry of an element plays a vital role in the structure and function of cells making up the organism in the ecosystem, but the detail required to account for the individual organism's metabolism in the cycle of a nutrient in the ecosystem is prohibitive.

A. General Features of Mineral Cycles

The characteristics of the element undergoing cycling relate significantly to its cycling pattern. Of particular importance is the water solubility of the forms of the elements. Life is essentially aqueous, and the behaviour of an element and compounds in which it functions are critical to reactivity, distribution, and transport in the ecosystem (Deevey, 1970). Biogeochemical cycles and hydrologic cycles of a pasture ecosystem are virtually inseparable (Pomeroy, 1970; Curlin, 1970) (Fig. 1). Minerals are transported through the biological cycle of uptake and utilization by water, although transport in gaseous and particulate forms may occur as well. Sources of mineral input other than chemical fertilizers and irrigation are largely via rainfall and atmospheric fallout, via mineralization of organic matter through the soil-water system and weathering of mineral fractions and consequent release into the soil solution. Losses other than from animal products, emigration of organisms, immobilization, fixation and wind transport are via water transport. Continued leaching of soluble ions and finer textured soil particles may result in significant losses unless balanced by other inputs to the ecosystem (Curlin, 1970). Losses of the element by surface water flow, may also become significant.

Electrochemical, or oxidation-reduction reactions are also significant in the cycling of elements since transfer reactions associated with biological uptake and utilization may involve changes in the oxidation state of the element (Shrift, 1964). Another property particularly important in cycling is the volatility of the element and of the compounds in which it occurs. Cycling is likely to be more spectacular for elements which are both water soluble and volatile. Closed natural cycles for any soluble but non-volatile element are only possible by the action of organisms (Deevey, 1970).

If the element is in so-called "tight" circulation (deficient) the amount present in the plant is related to the amount in the soil, and the addition of the nutrient to the soil will increase the productivity of the system (Whitaker, 1970). According to Pomeroy (1970), it may be less common for a single element to impose a clear cut limit on the growth of an ecosystem than for more complex interactions to occur.

Estimates of the distribution of mineral elements in soils, plants, microbial tissue, and mammals are shown in Table I. These estimates have somewhat limited value since they are not indicative of the amounts and availability of the element in the soil for plant uptake, utilization and cycling in pasture ecosystems. However, they do suggest the relative abundance of elements in the various components of the ecosystem. All of the elements except selenium are essential for plant growth, and all except boron are essential for animal growth. The relative number of atoms in plant tissue illustrates the

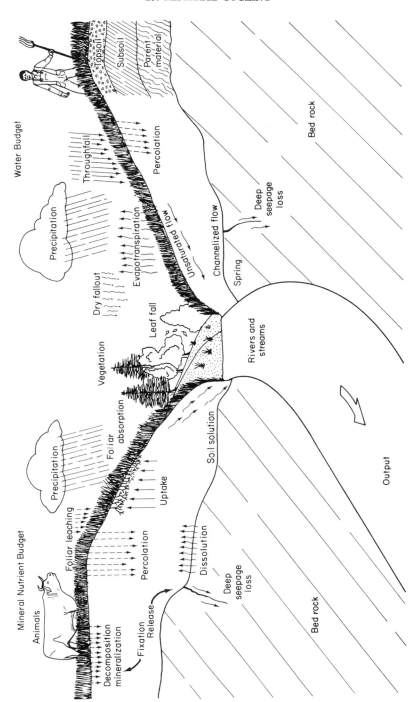

Fig. 1. The complementary relationships between mineral nutrient budgets and water budgets in pasture ecosystems.

TABLE I

The concentration (ppm) of important nutrient elements in soils, plants and mammals

Element	Soil [a] (Total) ppm	Plant matter (oven dry basis) all plants[b] ppm	Bacteria[a] ppm	Fungi[a] ppm	Mammals[a] ppm	Relative number of atoms with respect to molybdenum (plant)[b]
Molybdenum	2 (0·2–5)	0·1		1·5	1	1
Selenium	0·5 (0·1–2·0)	0·2[a]		2	1·7	2·4
Copper	20 (2–100)	6	42	15	2·4	100
Zinc	50 (10–300)	20		150	160	300
Manganese	850 (100–4000)	50	30	25	0·2	1,000
Iron	38,000 (7000–550,000)	100	250	130	160	2,000
Boron	10 (2–100)	20	5·5	5	2	2,000
Chlorine	100	100	2,300	10,000	3,200	3,000
Sulphur	700 (30–900)	1,000	5,300	4,000	5,400	30,000
Phosphorus	650	2,000	30,000	14,000	43,000	60,000
Magnesium	5000 (600–6,000)	2,000	7,000	1,500	1,000	80,000
Calcium	13,700 (7000–500,000)	5,000	5,100	1,700	85,000	125,000
Potassium	14,000 (400–30,000)	10,000	115,000	23,300	7,500	250,000
Nitrogen	1000 (200–2,500)	15,000	96,000	51,000	87,000	1,000,000

[a] Generalized from Bowen (1966). Ranges are listed in parentheses. Mammalia Figures = ppm Dry Mammal = 0·33 ppm Dry Bone + 0·67 ppm Dry Muscle.
[b] Generalized from Epstein (1965).

wide range in elemental requirements by plants. The availability of these elements over the pasture plants' growth phase will be an important factor affecting the productivity of pasture ecosystems. The generally high levels of nutrient concentrations in the bacteria and fungi are significant, since the bulk of energy in an ecosystem appears to be mediated by these decomposers of plant and animal residues. They represent the "detritus" food web, and may serve as sources or sinks for mineral elements (Macfadyen, 1961). The amount of the element in the ecosystem as well as the amount and rate of nutrient actually cycling in the biological phases of the system are important.

B. Mineral Cycling Models

Models are words, diagrams, mathematical programs intended to represent or simulate real systems or processes. They may be static or dynamic. Static models of mineral cycles do not estimate flux between components; dynamic models do. The advantages and requirements of models, the model development process, and systems analysis are discussed in papers by Bourliere and Hadley (1970), Dale (1970), Smith (1970b), Spedding (1970) and Van Dyne (1969). Models are helpful, in organizing data into useful knowledge, in indicating gaps in both understanding and knowledge, and in predicting effects of change in the system on the outputs of the system (Spedding, 1970). Models designed to describe the complexity of dynamic situations will require mathematical programs for either digital or analogue computers.

Van Dyne (1969) presents a generalized macromodel of a grassland ecosystem having 15 ecosystem components, 8 driving forces, two parameters or properties, 9 processes by which matter and/or energy may be transferred from one compartment to another, and four control functions. The four control functions represent plant community, animal community, microbial community and human manipulation effects. These control functions represent potential control points over mineral cycling. Plant community effects include such factors as competition for nutrients, nutrient uptake characteristics, root morphology and distribution as well as longevity, nutrient requirements for optimum plant growth, etc. Animal community effects include grazing patterns, and excreta deposition patterns which will affect the efficiency of cycling of mineral nutrients. Microbial interaction submodels would include antibiotic effects, antagonisms and synergisms which may result from particular species combinations in the soil. The fourth control function, the human, manipulation submodel covers experimental and management practices including fertilization, irrigation, movement of cattle or sheep, seeding, harvesting of crops, physical manipulations such as terracing, etc. Potential control points will be discussed more fully later.

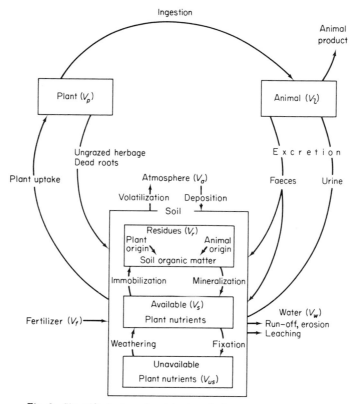

Fig. 2. Simplified mineral nutrient cycle for pasture ecosystems.

A graphical model of the mineral nutrient cycle is shown in Fig. 2. The soil compartment includes the plant and animal residues, the available soil nutrients, and the unavailable soil nutrients. The plant compartment includes the whole plant (tops and live roots). The available plant nutrient compartment involves those nutrients in solution in soil water as well as elements from the soil's labile pool of nutrients. The unavailable soil plant nutrient compartment lumps together all sources of unavailable plant nutrients. Nutrients may be released from this compartment by processes of weathering, by mineralization of organic forms of the nutrient, or by solubilization, and be held by the processes of immobilization and chemical fixation. The flux between these three soil components probably represents a "steady state" condition rather than an equilibrium since the latter condition is often defined as a state of balance or equality between opposing forces in a closed system (Johnson, 1971). Flux may occur in several directions between pools. The fertilizer compartment is shown with mineral flow to the available soil compartment, but minerals from fertilizer also may be absorbed directly by foliage, or may be either immediately unavailable or slowly available for plant uptake.

TABLE II

The amounts (X) of a mineral in each of the compartments of the ecosystem, the rates at which an element is entering (A) and the rate at which it is leaving the ecosystem (Z)

Compartment (V)[a]									Sum
Compartment (V)[a]	V_p	V_s	V_{us}	V_l	V_r	V_f	V_a	V_w	ΣV
Amount (X)	X_p	X_s	X_{us}	X_l	X_r	V_f	X_a	X_w	ΣX
Inflow (A)	A_p	A_s	A_{us}	A_l	A_r	A_f	A_a	A_w	ΣA
Outflow (Z)	Z_p	Z_s	Z_{us}	Z_l	Z_r	Z_f	Z_a	Z_w	ΣZ

[a]Subscript identification:
p = Plant, s = available plant nutrients, us = unavailable plant nutrients, l = animal, r = residues of plant and animal origin as well as soil organic matter, f = fertilizer additions, a = atmospheric compartment, w = water compartment representing nutrient inputs or outputs associated with water flux throughout the system.

Interest in mineral cycling may centre chiefly on whether the ecosystem is losing or gaining a particular plant nutrient. Such studies require a measure, or estimate of the amounts of nutrient in each of the compartments, as well as rates of influx and outflux between compartments (see illustration in Table II). The sum of influxes minus the sum of outfluxes for each of the compartments forms an estimate of change in nutrient content of the ecosystem. The choice of ecosystem boundaries is important for evaluating nutrient flow within the system, as well as estimating net changes in nutrient status of the system. Ecosystems chosen to represent natural drainage areas are advantageous since measurement of water outflow times concentration of nutrient permits estimates of nutrient outflux from the system. Pasture ecosystems are likely to be fields, where grazing animals are confined by artificial barriers, and may not represent natural physiographic boundaries. However, studies of mineral cycling in pasture ecosystems must be conducted in such a way that one may assume that nutrient fluxes across the boundaries are either constant or measurable. The potential number and direction of transfers are illustrated in Table III for the nutrient cycle depicted in Fig. 2. These potential transfer paths illustrate that the nutrient cycle is polycyclic, and is sufficiently complex that differentiation between all cycles in the field may be impossible. For

TABLE III

Potential transfer paths of plant nutrients between compartments in a pasture ecosystem

	Transfer to each compartment[a]							
(l)	V_p	V_s	V_{us}	V_l	V_r	V_f	V_a	V_w
V_p	—	$p(s)$	$p(us)$	pl	pr	—	pa	pw
V_s	$(s)p$	—	$(s)(us)$	$(s)l$	$(s)r$	—	$(s)a$	$(s)w$
V_{us}	—	$(us)(s)$	—	$(us)l$	$(us)r$	—	$(us)a$	$(us)w$
V_l	lp	$l(s)$	$l(us)$	—	lr	—	la	—
V_r	rp	$r(s)$	$r(us)$	rl	—	—	ra	rw
V_f	fp	$f(s)$	$f(us)$	fl	fr	—	fa	fw
V_a	ap	$a(s)$	$a(us)$	al	ar	—	—	aw
V_w	wp	$w(s)$	$w(us)$	wl	wr	—	wa	—

(Transfer from each comprtmnt)

[a] Notations are as described in Table II.

example, a nutrient in the available plant nutrient compartment (V_s) may be transferred directly to plants (V_p) by root absorption, to animals (V_l) by soil ingestion (Healy, 1970), to the unavailable plant nutrient pool (V_{us}) by fixation or chemical precipitation, to the residues compartment (V_r) during microbial decomposition or may be lost with water drained from the soil (V_w). Transfers between compartments may be negligible, non-existent or non-estimatable. The predictive model for the pasture ecosystem will require mathematical relationships which quantitatively describe the most important paths and rates of nutrient transfers between compartments. The art and science of model development and testing has just begun; the reader is referred to Van Dyne (1969), Dale (1970), and Smith (1970b) for more complete discussions of systems analysis, model development and systems simulation.

The compartmental model shown in Fig. 2, and the transfer paths depicted in Table III constitute a model for discussion of nutrient cycling patterns in pasture ecosystems. We shall make no attempt to quantitatively estimate the coefficients for transfer rates between compartments. Hopefully, we will convey a true picture with respect to the infancy of quantitative understanding of mineral cycling in pasture ecosystems.

C. Hypothetical Pasture Ecosystem

For purposes of illustration, a real pasture ecosystem from the Southern Piedmont Conservation Research Center at Watkinsville, Georgia is used as a basis for synthesizing a model. The ecosystem

TABLE IV

Definitions of compartmental pools and relationships between compartments at steady-state conditions (hectare/year basis)

V_l	=	Nutrients in 1121 kg of cow (nutrients contained in 504 kg of calf/ha are shown as losses from the pasture ecosystem).
V_p	=	Nutrients in 11,210 kg herbage + 11,210 kg roots minus nutrient in ungrazed herbage minus nutrient in dead roots.
V_r	=	Nutrients in 2242 kg ungrazed herbage plus nutrients in faeces plus nutrients in 2242 kg dead roots.
V_s	=	Nutrients considered available for plant utilization (extracted from soil with 0·025 NH_2SO_4 and 0·05 $NHCl$) in 0-15 cm layer plus 10% of extractable nutrient in 15-61 cm layers plus nutrients from rain-out, atmospheric deposition plus return of nutrients in urine minus losses of nutrient from surface run-off and leaching. No inputs from fertilizer nutrients were made in the hypothetical system.[a]
V_{us}	=	Total nutrient in the soil minus V_s.

Relationships Between Pools

Cycling pool = $V_p + V_l + V_r + V_s$

Total soil pool = $V_s + V_{us} + V_r$

Total system pool = $V_s + V_{us} + V_r + V_p + V_l$

[a] The size of the soil pool was based on observations that 90% of root mass was contained in the 0-15 cm layer with 95% in the 0-30 cm layer, but that water depletion by a growing fescue crop occurs to a depth of 61 cm (unpublished, Wilkinson).

was selected because values for many of the pool sizes had been measured. This pasture ecosystem consists of a tall fescue *(Festuca arundinacea)* pasture grazed all-year-round by an adult cow with calf on each acre (2·47 cow-calf pairs/ha). Some definitions of compartmental pools and relationships between compartments at steady-state conditions are presented in Table IV. The soils were derived from granitic gneiss, and are classified as clayey, kaolinitic, thermic, of the subgroup typic Hapludults of the order Ultisols. They typically are

TABLE V

Estimated mineral content of Cecil sandy clay loam used in hypothetical model of pasture ecosystem

Mineral element	Top soil (0-15 cm)			Subsoil (15-61 cm)		
	% Total	Unavail. $(V_{us})^a$ kg/ha	Avail. $(V_s)^b$ kg/ha	% Total	Unavail. (V_{us}) kg/ha	Avail. (V_s) kg/ha
Phosphorus	0·048	962	112	0·045	2,939	54
Potassium	0·51	10,650	392	0·51	31,762	243
Sulphur	0·02	444	5	0·02	1,167	178
Calcium	0·26	4,214	1615	0·26	10,457	1201
Magnesium	0·19	3,851	410	0·17	7,134	300

[a] Available soil nutrient pools estimated by extraction with 0·025 NH_2SO_4 and 0·05 $NHCl$, and are considered available for plant utilization. See definitions in Table IV.

[b] Contents of this soil were obtained from unpublished data as well as from literature for this soil type as published by Chapman and Pratt (1961) and Anderson et al. (1971).

low in bases, have subsurface horizons of clay accumulation, and contain appreciable quantities of iron oxides, aluminium oxides and silica. These soils have large capacities to fix or immobilize phosphorus. The pH of the surface soil in this ecosystem was 5·8. The pasture has a history of high fertilization because of previous intensive row crop production. Dilute acid extractions of soil suggest

TABLE VI

Climate associated with the hypothetical pasture ecosystem

Month of Year	Mean rainfall[a] (cm)	Mean potential evapotranspiration[a] (cm)	Mean air temp. 15 cm[b] (C)	Mean soil temp. 5 cm[b] (C)
January	11·35	2·74	7·2	7·2
February	12·27	4·27	10·0	9·4
March	13·00	5·03	13·3	12·8
April	10·16	9·45	17·2	17·2
May	9·19	11·81	22·2	23·9
June	9·22	13·34	25·6	27·2
July	11·18	12·52	26·1	27·2
August	10·13	11·25	26·7	27·8
September	7·34	9·45	23·9	25·0
October	7·37	6·38	18·3	20·6
November	7·11	3·66	13·9	15·0
December	12·52	2·45	7·2	8·3
Annual mean	120·84	92·35	17·8	18·3

[a] From Van Bavel and Carreker (1957).
[b] From Johnstone et al. (1968). Air temperatures recorded at 15 cm height, and soil temperatures 5 cm depth.

that initial nutrient supplies other than nitrogen may be non-limiting to primary productivity (Table V). Estimates of the available plant nutrient compartment were based on the amounts extracted by dilute acid which correlate with nutrients available for plant use. Sufficient nitrogen fertilizer was applied to the ecosystem to maintain a high level of productivity.

In our ecosystem rainfall exceeds potential evapotranspiration during October through till April and potential evapotranspiration losses exceed rainfall during May through till September (Table VI). Simple water balance estimates suggest that after the capacity of the plant root zone to hold water and potential evapotranspiration are taken into account drainage below the root zone will occur periodically during December through till April. Run-off from the pasture ecosystem is a possibility during high intensity rainfall; however, such storms occur infrequently. The slopes of this pasture ecosystem are however such that surface drainage is rapid (2-6% slopes). Losses of water by surface run-off were estimated as 3·1 cm/year. Soil-water storage capacity in the effective root absorbing zone was estimated as 10 cm. Losses of water by percolation below the lower limits of the ecosystem were estimated to be 25·4 cm per year.

The average temperatures of this ecosystem are moderate and low temperatures often limit herbage growth during mid-winter. Droughts of several weeks are not uncommon in summer.

The net primary productivity of the ecosystem was estimated as 11,210 kg dry matter/ha/year (10,000 pounds/acre), of which 8968 kg/ha (8000 pounds/acre) was consumed by the cow-calf units. This latter figure was based on an average consumption of 0·91 kg dry matter/45·4 kg of animal (2 pounds/100 pounds) with the dry matter consumption by the calf assumed to be 0·91 kg/45·4 kg of animal over the last 100 days of the calf growth period of 245 days. The stocking rate was 2·47 cows/ha. The size of the adult cow was estimated as 454 kg and she produced a 204 calf each year or 504 kg of calf produced per ha/year (450 pounds of calf/acre). The cows were mature and their mineral retention was considered negligible over a complete lactation and dry period (no net mineral retention). The calculated removal of minerals in the 504 kg of calf/ha compared to that removed in 11,210 kg of herbage/ha is shown in Table VII. The 2242 kg (2000 pounds/acre) of herbage ungrazed was returned as residues each year. The live root mass was estimated as 8968 kg/ha, with a root turnover of 2242 kg of dead roots each year (based on an estimated 25% death rate of roots in a prairie grass (*Bromus catharticus*) association by Dahlman and Kucera (1969). The growth rate of roots was assumed to replace the dead root mass each year (total root mass of 11,210 kg/ha of which 2242 were dead roots). Definitions of compartments and relationships between compartments are given in Table IV.

The circumstances of this ecosystem poses these very practical questions: Are plant nutrients recycled efficiently enough to preclude the need for maintenance fertilization? Is the nutrient cycle more nearly a closed or open cycle with respect to particular nutrients? These are important questions, from the viewpoints of both economics and the relative fertilizer requirements for grazed pasture ecosystems as contrasted to fertilizer requirements of ungrazed and mechanically defoliated swards. Pastures are fertilized to provide sufficient herbage to sustain animal growth and production. They may also be fertilized to enhance quality, mineral composition, etc. In this chapter, cycling of mineral nutrients is considered from

TABLE VII

Amount of minerals in 504 kg calf, 11,210 kg "Kentucky 31" tall fescue herbage, and 11,210 kg roots per ha

Mineral element	Calf removal %[a]	Calf removal kg/ha	Herbage %[b]	Herbage kg/ha	Roots %[c]	Roots kg/ha[d]
Phosphorus	0·65	3·28	0·34	38·1	0·20	22·4
Potassium	0·22	1·11	2·84	318·4	1·50	168·2
Sulphur	0·15	0·76	0·25	28·0	0·20	22·4
Calcium	1·11	5·59	0·30	33·6	0·30	33·6
Magnesium	0·044	0·22	0·20	22·4	0·20	22·4

[a] Hogan and Nierman (1927).
[b] Unpublished data of Wilkinson, Dawson.
[c] Estimated on basis of few isolated analyses of root tissue.
[d] Based on total root mass of 11,210 kg/ha to 24 inch depth. Assumed 25% turn-over each year, or 2242 kg of dead roots were in 11,210 kg of total root mass.

the standpoint of maintaining high herbage availability, and not from the viewpoint of meeting mineral needs of grazing animals. Reviews on this latter subject may be found in Underwood (1971) as well as others, and the subject is discussed in Chapter 19.

III. NUTRIENT RETURN IN EXCRETA FROM GRAZING ANIMALS

Nutrient return in excreta from animals represents a critical pathway in the nutrient cycle of grazed ecosystems. Since the animal in the system is mobile, cycling of nutrients from animal residues will be a function of that mobility, whereas cycling of mineral nutrients from plant residues is not complicated by mobility factors. In this section, we shall discuss the importance of this mobility on distribution of excreta, and the partitioning of plant nutrients between faecal and urinary excreta.

A. Distribution of Excreta

Factors affecting the time-space distribution of excreta include stocking rate, camping, grazing patterns, type of animal (species, breed and sex), and the amount and frequency of excretion. Management systems will also affect excreta distribution patterns.

The area affected by each excretion, the frequency of excretion, and the distribution of excreta in relation to pasture area are critical to the efficiency of nutrient cycling. Adult cattle defecate about once every 2 h, and urinate once every 3 h. The area affected by each defecation is about 0.09 m^2, and by each urination 0.28 m^2.

TABLE VIII

Number and area affected by faecal and urinary excretions of cattle

Reference	Faecal excretions[a]		Urinary excretions[b]	
	No.	Area (m^2)	No.	Area (m^2)
Petersen *et al.* (1956)	12	0·09	8	0·28
Johnstone-Wallace and Kennedy (1944)	11·75	0·06	9	—
Davies *et al.* (1962)	12	0·07	10	0·19
Bornemissza (1960)	10	0·09	—	—
Weeda (1967)	8·9-13·3	—	—	—
Giobel and Nilsson (1933)	—	0·13	—	—
MacLusky (1960)	11·6	0·07	—	—
Lotero *et al.* (1966)	—	—	—	0·93-1·11[c]
Wilkinson (unpublished)	—	0·05	—	0·52-1·33
Representative values	12	0·09	8	0·28-1·02
(Marsh and Campling, 1970)	11-12	—	—	—

[a] Cattle produce about 16·4 g of faeces dry matter/kg body weight/day, and sheep produce about 8·3 g of faeces/kg body weight/day (Spedding, 1971).

[b] Cattle produce from 10-25 litres of urine per day from a 350 kg animal. Sheep produce from 1-6 litres of urine/day from a 60 kg animal. Sheep and cattle have been estimated to produce 1·4-7·2 g dry matter in urine/kg body weight and cattle 2·8-7·2 g dry matter/kg body weight (Spedding, 1971).

[c] These observations were based on areas where fescue growth was enhanced by single urinations. The affected area chosen to be most representative of adult cattle was 1·02 m^2.

However, the growth of grass may be affected over a much larger area for each urination (1.02 m^2) (see Table VIII). The volume of excreta varies with season, forage condition, time of day, availability of drinking water, as well as size and physiological status of the animals. Urine volumes are most variable (Herriot and Wells, 1963; Lotero *et al.*, 1966; Weeda, 1967 and others). Lotero *et al.* (1966) observed that the greatest growth response occurred in the centre, and decreased toward the periphery of the roughly circular area affected by the urine. Dale (1961) describes the effect of a single sheep urination as occupying a squat inverted cone (at least 7·5 cm deep) surrounded by a broad brim (2·5 cm deep). Other soil and plant properties which may alter the effective area of urine depositions are soil pH, soil texture, soil moisture, percent of herbage cover, and micro- and macro-relief.

Faeces affect the voluntary acceptance of herbage over a larger area than actually covered by faeces. The area of herbage refused appears to depend on intensity of grazing, or the availability of clean herbage, and is considered to be inversely proportional to grazing pressure (Marsh and Campling, 1970). The palatability of grass in urine patches was decreased slightly or not at all (Norman and Green,

TABLE IX

Proportion of pasture ecosystem covered by excreta from a cow-calf pair under several assumptions (one grazing year)

		Urine	
Distribution assumption	Faecal	% of total area	Total
No Overlap[a]	11·8	23·7	35·5
No Overlap[b]	11·8	92·3	104·1
Empirical frequency[b,c]	4·0	51·5	55·5

[a] Area affected daily by faeces and urine of 1·11 m^2 and 2·23 m^2 respectively.

[b] Area affected daily by faeces and urine of 1·11 m^2 and 8·16 m^2 respectively.

[c] The empirical frequency distribution was adapted from Petersen et al. (1956).

1958; McLusky, 1960; Marsh and Campling, 1970). Delay of grazing around dung heaps may result in failure to utilize herbage for periods up to 17 months (Weeda, 1967), and alter the pathway of nutrient circulation.

The proportion of the pasture ecosystem calculated to be covered by excreta is shown in Table IX for three different combinations of excretal density and distribution. With the smaller effective area per urination, and no overlap, about three years is required to cover the entire area of the ecosystem with excreta. In contrast, about one year would be required to complete pasture coverage when the area affected by each urination is assumed to be 1·02 m^2. The application of the empirical frequency distribution of Petersen et al. (1956) to our ecosystem suggests that approximately 44·5% of the pasture area would not be covered by faeces or urine, 14% covered by two excreta, and 12% covered by three excreta or more during one grazing year at 2·47 cow-calf pairs/ha.

Petersen et al. (1956) combined the empirical frequency distribution, a function for loss rates, and amount of nutrient return per excreta to estimate the effect of the freely grazing animal upon the fertility of the pasture. The time for each effect of a new excretion deposited on the pasture to be balanced by the loss of effect from an old excretion was 30 or more months of grazing. At this "steady state" condition, the number of effective excreta remains constant. Loss rates of potassium were assumed to be 10% per month. They concluded that freely grazing animals distributed plant nutrients ineffectively; the greatest benefit could be expected from conditions of high stocking rates and long periods of grazing time.

Although greater faecal deposition frequency was observed near water and fence lines, Petersen et al. (1956) concluded that the "proportion of pasture covered by 0, 1, 2 . . . excretions during a grazing period appears to depend primarily on mean excretal density at the end of the period, and to be practically independent of the size and shape of pasture."

Hilder (1966) suggested that sheep have greater camping tendencies than cattle. The increased number of sheep required to give equivalent grazing pressure to a cow-calf pair should result in an increased number of excretions, and an increased potential for spreading and mixing of faeces with soil by hoof action. Sheep may also graze the pasture more completely and more fully utilize herbage than cattle (Calder, 1970).

Hilder (1966) observed that the rate of faecal deposition and its decomposition were in balance after grazing had been in progress for periods longer than 4 months. Also, in certain paddocks one-third of the faecal output of sheep was returned to less than 5% of the area. The proportion of the area not receiving faecal droppings remained at about 71-73% of the area over stocking rates of 3, 5 and 7 sheep/acre (7·4, 12·4, 17·3/ha, respectively). The percentages of total faeces found on one acre paddocks stocked at rates from 4 to 16 sheep/acre (9·9 to 39·5/ha) indicated over 50% of the faeces was found on 18·9% of the paddock. Similar conclusions were drawn from studies of faecal distribution on 37·2 ha paddocks rotationally grazed (83% of the faeces found in 50% of the area). This difference between cattle and sheep may arise because cattle have different patterns of behaviour with respect to gregariousness, camping, resting, etc. If excretions occur regularly, then the proportion of the time spent grazing and resting may be a good indication of excretion patterns—the assumption being that excretions during grazing contribute to cycling, while excretions in camps do not. Stocking rates may influence the availability of herbage, and herbage availability will influence grazing time. During summer in the Southeast U.S.A., grazing animals spend much of the hot part of the day in woods or under trees, where excretions are effectively lost from the nutrient cycle. Lazenby (1969) emphasized the importance of re-examining the effect of excreta distribution patterns on the fertility and productivity of pastures.

B. Partitioning of Plant Nutrients Between Faeces and Urine

When herbage is consumed by grazing animals, very small proportions of the total mineral intake are retained in animal products. The greatest removal occurs with calcium and phosphorus. The sulphur and potassium concentrations in unwashed wool are high with most

TABLE X

Minerals in animals and animal products[a]

Mineral	Calf	Steer	Lamb	Sheep	Unwashed wool	Cow's milk[d]
	Kilograms per 1000 kg					
Phosphorus	6·71	6·76	4·93	4·53	0·31	1·03
Potassium	1·74	1·49	1·41	1·25	46·6	1·19
Sulphur	1·50[b]	1·50	—	1·50[c]	35·00[c]	0·42
Calcium	11·76	12·76	9·13	8·41	1·28	1·08
Magnesium	1·48	0·36	0·30	0·30	0·24	0·01

[a] From Lawes as cited by Bear (1942) except as noted.
[b] From Hogan and Nierman (1927).
[c] From Williams (1962).
[d] From Forbes as cited by Bear (1942).

of the potassium lost on washing. Losses of plant nutrients in cow's milk may be appreciably greater than those from beef cattle or sheep production. For instance, the production of 13,452 kg of milk/ha/year would remove 13·9 kg of phosphorus whereas 504 kg of calf would remove only 3·3 kg/year.

Mineral retention by animals will be influenced by their age, condition, stage of lactation, etc., and by the level of intake. Mineral retention would be expected mostly in growing classes of livestock. Depending on herbage quality, and physiological state of the animal, mature animals may excrete more mineral than ingested, thus losing body stores. Only a small proportion of the ingested plant nutrient is removed from the pasture by the animal (Donald and Williams, 1954; Williams, 1962; Wilkinson et al., 1971; Rouquette et al., 1971). The remainder of ingested plant nutrients is in general excreted by adult animals, although small amounts of selenium may be volatilized from the animal to the atmosphere (Handreck and Godwin, 1970; Allaway et al., 1967; Rosenfeld and Eppson, 1964).

Potassium and boron are readily absorbed by the animal, and are
excreted in the urine while phosphorus, calcium, magnesium, iron,
manganese, zinc and copper are mainly excreted in the faeces (see
Table XI).

Barrow and Lambourne (1962) found that approximately 0·1 g of
sulphur was excreted in the faeces with each 100 g of feed eaten,

TABLE XI

Partitioning of plant nutrients between urine and faeces
(percent of total excreted)

Plant nutrient	Urine	Faeces %
Potassium	70—90[a]	30—10[a]
Phosphorus	trace	95+
Magnesium	30—10	70—90
Sulphur	6—90[b]	10—94[b]
Calcium	trace	99
Iron	trace	95+
Manganese	trace	95+
Zinc	trace	95+
Copper	trace	95+
Boron	95+	trace

[a] Values adapted from Barrow (1967). Values of Forbes as cited by Bear (1942) are in
good agreement.
[b] From Barrow and Lambourne (1962).

with the remainder of the sulphur being excreted in the urine (after
retention by the animal). Hansard and Mohammed (1968) found
55% of the sulphur excreted in the faeces of sheep. Hansard and
Mohammed (1969) reported that 57% of the sulphur was excreted in
the faeces of cattle and 43% was excreted in the urine. Recalculation
of Hansard and Mohammed's (1969) data based on 0·1 g of sulphur
in faeces per 100 g of feed eaten shows 55·5% of the sulphur excreted
in the faeces, which is in good agreement with the values of Barrow
and Lambourne (1962).

Phosphorus excretion is mainly via the faeces with about 0·06 g
organic phosphorus excreted per 100 g of feed eaten; the remainder
is excreted as inorganic phosphorus (Barrow and Lambourne, 1962).
This suggests that the higher the phosphorus content of the feed the
greater the inorganic phosphorus content of the faeces. Generally,
only trace amounts of phosphorus are excreted in the urine (Barrow
and Lambourne, 1962), although individual sheep have been
observed to excrete up to 32% of dietary phosphorus in the urine
(L'Estrange, 1970).

Barrow (1967) reported that iron, manganese, zinc and copper were excreted in the faeces principally as unabsorbed nutrients from the diet. Iron, manganese and copper are also excreted via the bile, and zinc mainly via the pancreatic juices. Cobalt is mainly excreted in the faeces although that absorbed is excreted in the urine. The excretion of molybdenum is influenced by the sulphate content of the diet with high sulphate promoting urinary excretion and low sulphate enhancing the retention of molybdenum followed by faecal excretion.

Plant nutrients excreted in the urine are considered to be in readily available forms (Barrow, 1967). The availability of plant nutrients in faeces will be discussed in relation to the cycles for the various elements.

IV. PHOSPHORUS CYCLE

A. Characteristics of Phosphorus

Phosphorus cycling is governed by its stability (low solubility), and its low mobility in soils (Bray, 1954). Under normal soil conditions of an oxidizing atmosphere and normal pH ranges it is not subject to large losses by leaching or volatilization (Black, 1968). Losses in gaseous form are apparently confined to unusual circumstances such as anaerobic conditions with an abundance of decomposable organic matter (Alexander, 1964). Phosphorus cycling is, therefore, accomplished through the plant and animal components of the ecosystem.

Phosphorus is a potentially limiting factor in many agricultural ecosystems, perhaps more limiting than calcium or potassium (Black, 1968). The phosphorus in soil is relatively immobile, and since the amounts in soil solution are small, movement by mass flow to plant roots is usually not sufficient to meet plant demands (Barber et al., 1963). Movement of phosphorus by diffusion depends on the diffusion coefficient for phosphorus, the capacity of solid phase phosphorus to renew phosphorus concentrations in solution, and the concentration gradient between the soil solution and the root surface (Olsen and Watanabe, 1970). Plant root extension is an important mechanism by which plant roots absorb sufficient phosphorus to meet growth requirements.

Phosphorus occurs in the system mainly as orthophosphate combined with hydrogen and inorganic or organic cations. Phosphorus in organic forms occurs mainly in ester linkage. In the inorganic forms of phosphorus common in soils, one or more of the hydrogen ions may be replaced by cations such as calcium, iron, and aluminium (Black, 1968).

Although phosphorus exists in different oxidation states, there is usually no change in this state either during assimilation by living organisms or during decomposition of organic phosphorus compounds by microbial populations (Mulder *et al.*, 1969). Organic phosphorus from residues of plants and animals is mineralized by microbes, by hydrolysis and by autolysis. Phosphorus may be immobilized in microbial tissues in organic forms, and fixed as inorganic forms on surfaces of soil particles. Microbes may affect phosphorus supply in different ways; they may immobilize available phosphorus in their tissues, may release orthophosphate by decomposition of organic phosphorus compounds, or may promote mineralization by solubilization of phosphorus from insoluble forms in soils (Mulder *et al.*, 1969). Immobilization of phosphorus has been thought to occur in plant materials containing less than 0·2% phosphorus (Fuller *et al.*, 1956; Mulder *et al.*, 1969) or when C/P ratios are greater than 200 to 1. Microbial tissue then, may be a sink or a reservoir for phosphorus. While agricultural plants contain from 0·1 to 0·5% phosphorus, fungal mycelia contain from 0·5 to 1·0% phosphorus, and soil bacteria contain from 1·5 to 2·5% phosphorus (Mulder *et al.*, 1969).

The average phosphorus content of U.S. soils is 0·062% (Lipman and Conybeare, 1936 as cited by Black, 1968) while the mean content of Australian soils is 0·030% phosphorus, or nearly half that of the U.S. soils (Wild, 1958). Soils of the Atlantic and Gulf Coastal Plain of the U.S. contain less than 0·017% phosphorus, and form an extensive area of low phosphorus in the U.S. Wild (1958) attributed the low phosphorus contents of Australian soils to losses by leaching rather than to a low content of phosphorus in parent rock.

Phosphorus released by weathering of primary phosphorus-bearing minerals and additions of plant residues and fertilizers combines primarily with the surfaces of the clay fraction (Black, 1968). Phosphorus contents of finer textured soils are generally greater than those of the coarse textured soils. Organic phosphorus has been found to represent from as little as 0·3% to as much as 95% of the total phosphorus in soils (Black, 1968). Organic phosphorus is considered essentially unavailable for plant uptake (Pierre and Parker, 1927).

Inorganic phosphorus compounds exist in acid soils mainly as iron and aluminium compounds, and in calcareous soils as predominantly calcium phosphates. Soil pHs of 6·5-7 have been considered to be most desirable because this pH is above the range of minimum solubility of iron and aluminium phosphates, and below the pH of minimum solubility of calcium phosphates. The reader is referred to several excellent discussions of the factors affecting availability of soil phosphorus to plants (Black, 1968; Fried and Broeshart, 1967; and Larson, 1967).

Leaching losses from the pasture ecosystem are probably of significance on a short-term basis in sands and peats that have little tendency to react with phosphorus (Black, 1968; Ozanne, 1962). Ozanne *et al.* (1961) found that 50% of labelled phosphorus leached past 76 cm with 23 cm of rain in 38 days in a sandy soil having 5·3% organic matter in the top 10 cm. The leaching losses reported by Ozanne *et al.* (1961) from the top 10 cm of soil exceeded the losses due to uptake of phosphorus by a subterranean clover crop. Leaching losses of phosphorus from soils over geologic time can be quite significant (Wild, 1961). Since phosphorus is not volatilized, it moves continuously seaward except for the action of plants in recycling phosphorus.

B. *Cycling of Phosphorus in the Hypothetical Pasture Ecosystem*

Table XII contains our estimates of the amounts of phosphorus in the various parts of the pasture ecosystem. These amounts of phosphorus were used in subsequent calculations of cycling approximated on an annual basis using the assumptions described earlier.

If 25·4 cm of drainage water remains in equilibrium with an assumed soil solution phosphorus content of 0·1 ppm, the potential loss of phosphorus by leaching each year represents about 0·3 kg phosphorus/ha. On a short-term basis, losses by leaching would appear to be insignificant as far as phosphorus effects on system productivity are concerned.

Losses of soil particles, and phosphorus losses associated with erosion would be negligible from such an ecosystem. Annual phosphorus losses with 12·29 cm of annual runoff from blue grass sods (*Poa pratensis*) in Missouri averaged 0·18 kg/ha/year (Miller and Krusekopf, 1932). Losses of phosphorus by foliar leaching would also be negligible as well as losses associated with organic residues. However, direct losses of plant residues in water were shown to be an

important loss of plant nutrients from a Pennine Moorland ecosystem (Crisp, 1966).

An important exception may be leaching losses from frozen or
dormant vegetation. Studies by Timmons *et al.* (1970), Jones and

TABLE XII

Amounts of the elements phosphorus, potassium, sulphur, calcium and
magnesium used in calculating sizes of compartments and pools as defined
in Table IV

Compartmental factors	Plant Nutrient				
	P	K	S	Ca	Mg
	kilograms/hectare				
Available soil (0–61 cm)	166	635	183	2,816	710
Unavailable soil (0–61 cm)	3,901	45,102	1,611	14,671	14,984
Total, soil (0–61 cm)	4,067	45,737	1,794	17,487	15,694
Plant (live)					
Above ground	38·1	318·4	28·0	33·6	22·4
Below ground	17·9	134·5	17·9	26·9	17·9
Total	56·0	452·9	45·9	-60·5	40·3
Livestock-mature cow	7·3	2·5	1·7	14·3	0·5
Calf-removal	3·3·	1·1	0·8	5·6	0·3
Total herbage intake by animals	30·5	254·7	22·4	26·9	17·9ˉ
Residues					
Animal origin					
Faeces	27·2	25·4	9·0	21·1	15·8
Urine	Nil	228·2	12·6	0·2	1·8
Plant origin					
Ungrazed herbage	7·6	63·7	5·6	6·7	4·5
Dead roots	4·5	33·6	4·5	6·7	4·5
Total residues	39·3	350·9	31·8	34·7	26·6
Water					
Leaching losses	0·3	139·2	40·6	172·7	76·2
Surface runoff losses	0·2	3·4	0·2	1·2	0·2
Atmosphere					
Deposition rain-out	0·2	1·1	5·7	3·9	1·3
Volatilization	Nil	Nil	?	Nil	Nil
Fertilizer	Nil	Nil	Nil	Nil	Nil

Bromfield (1969), Harley *et al.* (1951), and Kline (1969) suggest that
leaching of phosphorus from frozen or dormant vegetation may be
more important than generally recognized in estimates of phosphorus
flow in humid pasture ecosystems. From 69-80% of the total
phosphorus from frozen or non-green vegetation may be readily
leached with water. Microbial activity in such plant tissue substantially reduces the phosphorus soluble in water. The intensity and

duration of the first precipitation after the vegetation is frozen or becomes dormant will have a significant effect on the amount of phosphorus lost to drainage water or to the soil.

Accessions of phosphorus from the atmosphere are small and probably occur as dustfall and rain-out. Accessions of 0·2 to 0·6 kg phosphorus/ha/year have been cited by Duvigneaud and Denaeyer-DeSmet (1970). Atmospheric accessions were included in the available soil pool of nutrients. However, the effect of 0·2 kg phosphorus/ha on the cycling pool of phosphorus will be nil.

The proportion of phosphorus in herbage recycled by the animal depends on the degree of herbage utilization (grazing pressure or stocking rate). Carter and Day (1970) found that stocking rates of 15 sheep/ha resulted in forage utilization of about 53% while stocking rates of 25 sheep/ha resulted in herbage utilization of 77%. The assumed forage utilization was 80% in our cow-calf ecosystem.

Removal of phosphorus in the weaned calf (504 kg/ha) is low (3·3 kg/ha/year) and retention by the mature cow over a yearly period is nil. Slightly lower losses might be expected from removal of 504 kg of sheep (see Table X). Losses from wool gathering will be negligible.

Total phosphorus in faecal excretion was estimated to be 27·2 kg/year of which 5·29 kg was considered to be organic. Phosphorus excretion in urine was ignored. Consumption of feed high in phosphorus by ruminants increases the proportion of inorganic phosphorus in the faeces (Barrow and Lambourne, 1962).

Of the 80% of herbage utilized by the cow-calf pairs, 71% of the yield of phosphorus will travel through the rumen and GI tract to 12% of the land surface each year. Since phosphorus is an immobile nutrient, the effective area fertilized is not likely to be significantly greater than the actual area of deposition. Twenty percent of potential phosphorus return comes from ungrazed plant residues. Phosphorus return from ungrazed residues would be favourably distributed except where rejection of herbage was influenced by dung deposition. The phosphorus leached from ungrazed herbage is highly available for recycling. Low rates of utilization reduce the proportion of total phosphorus in forage which passes through the cow's digestive system.

The efficiency of phosphorus cycling will not be critical to primary productivity since phosphorus is not limiting tall fescue growth (critical phosphorus percentage of 0·25%). However, the rate and extent of phosphorus immobilization will be important in determining when phosphorus reaches a tight circulation pattern.

When ungrazed herbage from perennial plants matures, phosphorus is re-distributed to new shoots or to roots for growth and metabolism (Biddulph *et al.,* 1958; Greenway and Gunn, 1966; Bouma, 1967). This constitutes an efficient cycling mechanism. Ozanne and Howes (1971) considered intra-plant recycling to be the cause for a lower phosphorus requirement for ungrazed annual pastures in Australia compared to grazed annual pastures (subterranean clover). Monthly cut *Festuca-Agrostis* and *Nardus* pastures resulted in 1·78 and 1·54 times as much phosphorus removal as annually cut *Festuca-Agrostis* and *Nardus* pastures, respectively (Floate, 1970d). The species involved were presumably *Festuca rubra, Agrostis tenuis* and *Nardus stricta.* Since the yield of the annually cut herbage was greater, part of this difference in phosphorus yield represents intra-plant re-distribution, and decreased plant requirement. In perennial pastures with low intensity of herbage utilization, intra-plant recycling may be an important quantitative factor in the phosphorus cycle.

Increases in proportions of inorganic phosphorus in faeces were found when both annually and monthly cut *Nardus* and *Festuca-Agrostis* herbage were converted to faeces (Floate, 1970a). Floate (1970d) found an increase from 62·3 to 78·4% of total phosphorus that was inorganic upon conversion of herbage to faeces. Bromfield and Jones (1970) found up to 80% of the organic phosphorus in green feed was mineralized via the animal with the amount mineralized decreasing with increasing maturity of forage.

Organic matter in faeces is more resistant to further decomposition than fresh plant material (Barrow, 1961; Floate, 1970a). The concentrating effect during digestion by the animal partially accounts for the increases in all forms of phosphorus in faeces.

Inorganic phosphorus in sheep dung is soluble in dilute acid but not very water soluble, whereas inorganic phosphorus from plant materials is highly water soluble (Bromfield and Jones, 1970; Bromfield, 1961). The solubility relationships with cattle faeces are not known although 25% of the phosphorus in stockyard manure after 20 years storage was organic (Peperzak *et al.,* 1959). Mechanical damage or crushing of faeces has been shown to increase leaching of phosphorus. Sheep faeces retained 40% of their initial total phosphorus after two years of exposure to the weathering and leaching action of 109·2 cm of rain. Ninety per cent of this residual phosphorus was organic (Bromfield and Jones, 1970).

The availability of phosphorus for plant growth from sheep faeces mixed with soil is high (Bromfield, 1961; McAuliffe *et al.*, 1949; Bornemissza and Williams, 1970), but the inaccessibility and low mobility of phosphorus in faeces deposited on the soil surface results in low availability for root absorption and cycling.

Dung beetles, earthworms and other soil fauna increase the rate of mineralization of faeces by burying and by increasing mixing of soil and faeces. Phosphorus uptake and yield of Japanese millet were increased by *Onthophagus australis* Guer. beetles in a greenhouse study. Phosphorus recoveries by millet from faeces applied to the soil surface was 3% without dung beetles, 17% with confined dung beetles and 26% for faeces mixed with the soil (Bornemissza and Williams, 1970). Gillard (1967) reported that, during the wet summer growing season when beetles were active, 12·3 of 13·6 kg of nitrogen excreted by a steer were incorporated into the soil, whereas only 2·7 kg of nitrogen was incorporated at other times. Dung beetles reduced the time for decomposition of faeces from 12 months to 3 months.

A similar action might be expected from earthworms, but little information as to their concentrations in faecal pads and their capacity to mix dung with soil is available. Earthworm activity has been reported to be very high under productive pastures (Walker, 1962). Earthworm numbers and weights per ha of pasture soils as high as 7·41 million/ha with a weight of 2354 kg/ha have been reported (Russell, 1963). Russell (1963) stated that earthworms are probably the principal agents in the incorporation of plant and animal residues into the soil.

Earthworms, dung beetles and other soil fauna enhance nutrient cycling by providing foci of high nutrient concentrations possibly enhancing root growth and nutrient uptake. Soil fauna may also serve to inoculate soil with microflora which can hasten decomposition of residues (Macfadyen, 1961).

Whether or not net mineralization of phosphorus occurs from residues of plant or animal origin depends not only on spatial relationship to the soil, but also on such factors as the initial organic phosphorus content, total phosphorus content, temperature, water content, and time.

From 3 to 30% of the original organic phosphorus in sheep faeces was mineralized while from 2 to 15% of the phosphorus in plant materials was mineralized (Floate, 1970). Net amounts mineralized were positively correlated with the initial levels of organic phosphorus, and Floate suggested that positive net mineralization is

CHEMISTRY AND BIOCHEMISTRY OF HERBAGE

unlikely to occur when organic phosphorus contents are less than 0·09% during the first few weeks of decomposition.

Decomposition of residues which contain less than 0·2% total phosphorus probably results in little net mineralization of phosphorus, Fuller et al. (1956); Swaby (1962). Organic phosphorus in both plant and animal residues appears to be more of a "sink" than a source of phosphorus for cycling (Floate, 1970a, b, c, d; Bromfield and Jones, 1970; Donald and Williams, 1954).

The percentage of organic phosphorus mineralized from faeces was 10% at 30°C, 2·5% at 10°C, and −12% at 5°C (net immobilization of phosphorus) while phosphorus mineralization from Nardus and Agrostis-Festuca herbage was 0·2% of the original phosphorus at 30°C to −41% at 5°C (Floate, 1970b). Increased mineralization at temperatures greater than 30°C indicates that the thermophillic range is more favorable than the mesophillic range for mineralization of phosphorus (Alexander, 1964).

Rumen bacteria operate at temperatures of 39-41°C, and have phosphorus contents of the order of 5%. Rumen fluids normally contain about 10 m moles of phosphorus per liter (several orders of magnitude greater than the concentration of phosphorus in soil solution) (Hungate, 1966 as cited by Pomeroy, 1970). The higher temperature of rumen bacteria, plus their higher phosphorus content compared to soil bacteria (1·5-2·5% phosphorus) suggests that when bacteria move from the rumen ecosystem to the lower intestinal tract considerable mineralization will occur. Ruminants thus enhance dissolution and mineralization of phosphorus. The process of digestion will also result in narrowing carbon-phosphorus ratios because of the greater loss of carbon than phosphorus during herbage digestion by the ruminant animal.

Floate (1970c) suggested that water content effects on mineralization of phosphorus were of less significance than the effects of temperature. Heating or drying the soil, and then re-wetting often results in a flush of decomposition of organic matter several times greater than the rate associated with continuously moist conditions (reviewed by Russell, 1963; Alexander, 1964). This may be important in the release of phosphorus from surface deposited residues, particularly in areas of erratic rainfall, or in areas with regular dry or rainy periods. Under our environment mineralization of organic phosphorus in surface residues would of greatest importance during the spring and summer months (supported by the observations of Saunders and Metson, 1971).

The phosphorus in the root part of the plant component in our model represents a total of 22·4 kg/ha (dead and live roots) and is a reservoir of phosphorus, some of which may be recycled to plant tops for assimilation and growth (Table XII). We have assumed that little of the phosphorus was translocated out of the root prior to

TABLE XIII

The amounts of phosphorus in the compartments of the hypothetical cow-calf pasture ecosystem, the sizes of the pools considered cycling on an annual basis and the amounts in the soil and total ecosystem [a]

	kg/ha	%[b]	%[c]
Total in various pools			
Total ecosystem pool	4161·7	100·0	—
Total soil pool	4106·0	98·7	—
Cycling pool	260·7	6·3	100·0
Total in compartments			
V_l Cow	7·3		2·80
V_p Plant	48·4		18·57
V_r Residues	39·3		15·07
V_s Soil Available	165·7		63·56
V_{us} Soil Unavailable	3901·0		
Losses from annual cycle			
Leaching	0·3		0·12
Surface Run-off	0·2		0·08
Animal Product	3·3		1·27
Total Losses	0·8		1·46

[a] See Table IV for definition of pools and relationships between pools. Calculations based on information from Table X.
[b] Phosphorus in pool or compartment as a percentage of total ecosystem pool.
[c] Phosphorus in each compartment as a percentage of cycling pool.

death, and that mineralization of phosphorus and release to the available pool of phosphorus was required to return the phosphorus to the cycling pool. Whitehead (1970) suggested that legume roots would neither mineralize or immobilize inorganic phosphorus while grass roots would be likely to immobilize phosphorus. If the annual fractional loss rate (0·004) suggested for resistant humus fractions of a prairie grass ecosystem parallel mineralization, then the release of phosphorus from organic matter would offset losses from leaching and run-off.

The proportion of total system phosphorus cycling on an annual basis was 6·3% (Table XIII), thereby emphasizing the large, relatively inert pool of unavailable phosphorus in the soil system. The phosphorus cycle presented in Table XIII suggests that the cycle is nearly closed, with losses amounting to about 1·5% of the cycling pool each year. This conclusion is misleading since losses of phosphorus from the cycling pool by phosphorus fixation, microbial immobiliaztion, and by inefficient distribution of phosphorus return in animal excreta are not estimated. Such losses probably exceed those shown for leaching, and animal product removal. For example, if one assumes that 0·2% phosphorus is a level in residues below which no net mineralization occurs then phosphorus estimates for transfer to the unavailable soil pool because of microbial immobilization become 4·48 kg for ungrazed herbage, 2·24 kg for dead roots and about 8.97 kg for faeces, or a total of 15.69 kg/ha/year. If the herbage were ungrazed, but clipped to prevent nutrient redistribution within the plant, and the residues returned to the land, the estimated loss from the cycling pool under the same assumptions would be 26·9 kg/ha/year. Complete removal of herbage from the land thereby preventing its phosphorus from re-entry into the cycle would represent a loss of 46·2 kg/ha/year (assuming microbial immobilization losses in dead roots to be the same under all harvest methods). Such over simplifications suggest that the cycle for phosphorus under grazed pastures more nearly approaches a closed cycle than under mechanical harvest and removal conditions. However, if the maximum pasture area affected by faeces is 12% then phosphorus cycles on the remaining 88% of the pasture in a manner similar to that where herbage is clipped and removed.

Nitrification, with its associated decreases in pH, represents an important potential loss of phosphorus by fixation. Increases of organic phosphorus in pasture soils (Jackman, 1964) and the accumulation of partially decomposed mats of organic materials at the surface of pasture also represent important potential losses from the pool of cycling phosphorus (Floate and Torrance, 1970).

C. *Response to Excreta Return in Grazing Studies*

Numerous trials have been conducted with small plot experimental designs where sheep excreta have been returned mechanically or naturally, or not returned. Most of these studies suggest that the return of phosphorus in the dung was of little immediate value to the growth of grass or clover because of uneven return, and low availability of phosphorus for plant use (Brockman *et al.,* 1970;

Wolton *et al.*, 1970; Herriot and Wells, 1963; Watkin, 1957; Wolton, 1955, 1963) and that only during the long-term would there be a measurable effect of sheep dung on the level of available phosphorus in the soil. Brockman *et al.* (1970) concluded that after 5 years of sheep dung return in individually grazed plots there was an increase in resin-extractable phosphorus in the top 2·5 cm of soil, but that it was ineffcctive in supplying the increased demand for phosphorus associated with increased nitrogen levels. Phosphorus transfer in communally grazed small plot experiments has been suggested to be negligible (Herriot and Wells, 1963; Brockman *et al.*, 1970). Carter and Day (1970) indicated that phosphorus fertilization was as necessary for increased herbage availability at their highest stocking rates of 25 merino wethers/ha as at lower stocking rates. Ozanne and Howes (1971) reported an increased phosphorus fertilizer requirement for grazed annual pastures in contrast to ungrazed annual pastures. However, McLachlan and Norman (1966) suggested that the nutrient cycling process may result in lower phosphorus demand, or that greater utilization of pastures at higher stocking rates would place no greater strain on phosphorus requirements than standing unused herbage. Further studies at sites having previous 3-fold difference in stocking rates suggested lower phosphorus requirements under the higher stocking rate. McLachlan (1968) suggested that the highcr stocking rate promoted more frequent cycling of phosphorus by the greater frequency of defoliation, and enhanced accumulation of cycled phosphorus in the surface 4 inches of soil. Increascd frequency of defoliation as occurs under higher stocking rates results in reduced plant growth (Carter and Day, 1970), reduced root growth (Crider, 1955; Oswalt *et al.*, 1959), reduced root respiration and nutrient uptake (Davidson and Milthorpe, 1966), and lowered redistribution of phosphorus within the plant. Ozanne and Howes (1971) suggested that the effects of previous differential grazing history as measured by McLachlan (1968) were less than the effects caused by current grazing.

Blue and Gammon (1963) found that phosphorus recovery was much greater for phosphorus applied to grazed areas than for plots where the forage was clipped and removed. Sears *et al.*, (1953) also found that return of dung and urine increased pasture yield partly through an effect of nitrogen in enhancing grass growth. Available phosphorus accumulated three-fold over a period of two years at a stocking rate of 4·7-5·0 animal units/ha on coastal bermudagrass (*Cynodon dactylon*) pastures on loamy fine sandy soil in East Texas,

U.S.A. (Rouquette *et al.*, 1971). This accumulation was probably enhanced by yearly applications of 48 kg phosphorus/ha, and chain-harrowing to spread the dung after each 180 day grazing season. Coastal bermudagrass with its extensive stoloniferous and rhizomatomus growth habit may also have more favorable dung-root contact. These authors estimated 52 and 78% of herbage phosphorus was cycled through Brahman-Hereford F_1 cross animals at stocking rates of 3 and 2·7 animal units/ha, and 78 and 83% of the phosphorus cycled through the animal at stocking rates of 5 and 4·7 animal units/ha, respectively.

Phosphorus accumulated to 7·5 cm depth and was linearly related to the amounts of grain fed up to 58 metric tons/ha over a 4 year period (Benacchio *et al.*, 1970). Pasture growth was also enhanced by increased levels of grain feeding in this study (Benacchio *et al.*, 1969). Stocking rates were 0·28 ha/steer with no grain fed to 0·1 ha/steer at the highest grain feeding levels on uniform mixtures of *Lotus corniculatus, Trifolium repens* in combination with *Phleum pratense,* and *Festuca elatior.* Phosphorus accumulations at the highest stocking rate and level of grain feeding were about 1·4 times the level of the no-grain fed treatment. This suggested that animals effectively recycled the phosphorus contained in 58 metric tons of corn grain.

Andries and Van Slijcken (1969) recommended that the phosphorus requirements of grazed herbage of mixtures of *Lolium perenne, F. pratensis, Poa trivialis* and *T. repens, T. pratense,* and *T. hybridium* be reduced from 300 kg P_2O_5/ha for mowing to 40 to 80 kg P_2O_5/ha for grazing based on soil analysis and export data from a five year trial comparing grazing, mowing then grazing, and mowing methods of harvest. Pasture mixtures containing both grasses and legumes may present an additional complication since continued inputs of phosphorus may be required to maintain clovers in competition with grass (Sears *et al.*, 1953; Cullen, 1971).

D. Summary

Phosphorus moves in the pasture ecosystem from soil to plant roots by diffusion, is absorbed, translocated to leaves which are grazed, or not grazed depending on degree of herbage utilization, and then returned to the soil in either faeces, or plant residues. The negligible return through urinary excretions, the small potential area covered directly by faeces each grazing season, and the low mobility and spatial unavailability of phosphorus in faeces suggests that phosphorus cycling through grazing animals is inefficient in the short-term. However, phosphorus may cycle more effectively under high stocking rates associated with intensive grassland utilization.

V. POTASSIUM CYCLE

A. Characteristics of Potassium

Potassium occurs in soil and plant tissue as a monovalent cation and forms compounds which, in general, are nonvolatile. Assimilation and uptake in the biosphere are not accompanied by valence change.

Potassium occurs in soils in amounts ranging from less than 0·05% to greater than 2·5% (Chapman and Pratt, 1961). Lipman and Conybeare (1936) estimated the average potassium content of U.S. soils as 0·83%. The potassium content of soils is related to the parent material, and to the degree of weathering. Potassium contents usually increase with increasing depth of the soil profile. However, in some older soils the surface horizons may have higher potassium contents than those immediately below because plants have absorbed potassium from lower horizons, and released potassium at the soil surface. The bulk of soil potassium resides in non-exchangeable form, with much smaller amounts occurring in exchangeable and dissolved forms (Black, 1968, and others). The exchangeable and dissolved forms of potassium are considered available for root uptake (Black, 1968; Fried and Broeshart, 1967; Russell, 1963). The non-exchangeable forms include mostly potassium-bearing minerals and insoluble salts formed from complexes with iron and aluminum.

The amount of potassium dissolved in soil solutions is very small in relation to the amounts of potassium absorbed by herbage and consequently continued renewal is needed to supply plant potassium requirements. However, potassium ions in the exchange phase are readily replaceable by other ions. Consequently, the zones of availability around plant roots will be larger for potassium than phosphorus (Barber et al., 1963). Concentrations of potassium in soil solutions will vary from near zero to over 200 ppm (Reisenaur, 1966). Concentrations of potassium in soil solution would be expected to increase with decreasing soil moisture content. The transport between non-exchangeable, exchangeable and solution potassium will be shifted toward the more available forms as potassium is removed from soil solution by plant uptake or leaching. Similarly the addition of potassium from urine or fertilizer will tend to shift this equilibrium toward the exchangeable and non-exchangeable forms.

Additions of potassium to soils or finer texture and higher exchange capacities usually results in increases of exchangeable potassium, and may also increase the non-exchangeable soil potassium. The fixation of potassium is influenced by the clay content as well as type of clay, soil pH, and the percentage of the exchange capacity saturated with potassium (Munson and Nelson, 1963). Soils of higher clay content have greater capacity to fix potassium than do coarser textured soils. Liming of acid soils may increase potassium retention through a "hydroxylating" effect on the exchangeable aluminum at the surfaces of the aluminosilicate layers thereby removing the trivalent ions from competition with potassium, and freeing blocked sites so that potassium can compete with other cations such as calcium for adsorption (Thomas and Hipp, 1968). The degree of fixation depends partly upon the type of clay mineral present; for example, soils where kaolinitic clays predominate fix relatively little potassium, whereas those soils having 2:1 type clay minerals fix appreciable amounts of potassium (Buckman and Brady, 1968). Beneficial effects of potassium fixation may be the prevention of leaching losses, and luxury consumption by plants. Chief detrimental effects of fixation are the increases in potassium required to overcome potassium deficiency. The importance of potassium fixation to potassium cycling will be dependent on the level of potassium sufficiency in the soil.

The absorption of potassium by plants increases with increasing water content of the soil probably because the proportion of the soil effective for diffusion of potassium and that volume of water that can carry potassium to the roots by mass movement increases with the proportion of soil volume occupied by water (theory proposed by Black, 1968, p. 716). An important aspect of water balance is that as soil dries from the top, the supply of available potassium and effective root absorbing surface may decrease drastically (Black, 1968). Barber (1968) has recently reviewed the factors affecting the mechanisms of potassium absorption by plants.

Intra-plant redistribution of potassium may be important in the potassium economy of the pasture ecosystem. Greenway and Pitman (1965) reported that 51 of 129 μg of potassium which entered the youngest leaf of barley at the 3 leaf stage came from older leaves. Potassium deficiency symptoms in plants normally first appear on older leaves which is an indication of the high mobility of potassium within the plant. Potassium sufficiency in herbage plants usually ranges from 1 to 2% potassium. Potassium requirements for legumes grown in competition with grasses usually exceed those of legumes grown in monoculture (Blaser and Kimbrough, 1968). The occurrence of potassium in plant tissue is mostly in ionic form, and is not strongly bound, with approximately two-thirds of the total amount in plants immediately soluble in water and with the remaining one-third requiring some microbial action for release (Alexander, 1964). Bacteria contain up to 2·0% of their dry weight as potassium while mycelium of fungi contain as little potassium as 0·1% (Alexander, 1964) whereas the range in higher plant tissue ranges from 0·2% to over 6·8% (Ulrich and Ohki, 1966). Stevenson (1967) and Barber (1968) concluded that organic acid production by soil microbes may be important factors in solubilization, mobilization and transport of minerals including potassium in soil.

Potassium uptake by plants is normally in proportion to the available potassium in the soil. Luxury consumption of potassium occurs in herbage grasses under conditions of high potassium availability (all potassium above the critical level, or amount required for optimum yields may be considered "luxury" consumption) (Buckman and Brady, 1968). Luxury consumption of potassium in pasture ecosystems may be undesirable because of the wasteful use of potassium, as well as its potential involvement in the metabolic disorders of grass tetany (see review by Grunes et al., 1970). However, the potassium accumulated in herbage also represents a relatively labile pool capable of rapid release and recycling.

B. Cycling of Potassium in the Hypothetical Pasture Ecosystem

The mobility of potassium suggests that rapid turnover rates of potassium will occur between compartments of the ecosystem. The large pool of available potassium suggests that leaching losses will be high (see Table XII and Table XIV). The available pool was 58·4% of the cycling pool. Leaching losses were estimated as 139·2 kg potassium/ha/year or 11·3% of the cycling pool. This estimate was derived on the basis of drainage of 25·4 cm of water having an equilibrium solution concentration of 54·8 ppm potassium. Losses of potassium on this soil with its finer textured subsoils would be expected to be relatively less than those occurring on sandier soils. Atmospheric accessions of potassium were estimated as 1·2 kg/ha/year (based on 120·84 cm of rain containing about 0·1 ppm potassium) (based on composition of rain water from Carroll, 1962). Potassium contents of rain water were about double those reported by Likens *et al.* (1970) for a forested watershed in New England, U.S.A. (0·05 ppm potassium). Although dustfall may contribute to input of potassium, Likens *et al.*, (1970) suggested that the bulk of the chemical input to their ecosystem came with precipitation. Accessions reported in Sweden (cited by Tamm, 1958) indicate from 0·6

TABLE XIV

The amounts of potassium in the compartments of the hypothetical cow-calf pasture ecosystem, the sizes of pools considered cycling on an annual basis and the amounts in the soil and total ecosystem[a]

	kg/ha	%[b]	%[c]
Total in various pools			
Total ecosystem pool	46,338·2	100·0	—
Total soil pool	45,859·8	98·97	—
Cycling pool	1,236·2	2·67	100·0
Total in compartments			
V_l Cow-Calf	2·5		0·20
V_p Plant	389·2		31·49
V_r Residues	122·8		9·93
V_s Soil Available	721·7		58·38
V_{us} Soil Unavailable	45,102·0		—
Losses from annual cycle			
Leaching	139·2		11·26
Surface Run-off	3·4		0·28
Animal Product	1·1		10·09
Total Losses	143·7		11·63

[a] See Table IV for definition of pools and relationships between pools. Calculations based on information from Table X.
[b] Potassium in pool or compartment as a percentage of total ecosystem pool.
[c] Potassium in each compartment as a percentage of cycling pool.

to 3·7 kg/ha of potassium input from the atmosphere. Potassium from the atmosphere in relation to the amount of potassium cycling in an improved pasture ecosystem will have minor effects on the amount and flux between various compartments of the ecosystem.

The animals consume 254·7 kg of potassium with a loss of about 1·1 kg of potassium in the calves. About 228·2 kg of readily available potassium are returned to the pasture (90%). Although a relatively small proportion of potassium intake is excreted in faeces (10% of intake) the availability of this potassium is likely to be high because of its high degree of solubility in water (97% for sheep and cow manure, Bear, 1942). Davies *et al.* (1962) reported all of the potassium in dairy cattle dung was water soluble. If no overlap is assumed this suggests that the entire pasture area in the hypothetical pasture will be re-fertilized each year with potassium (see Section III). If, however, the empirical frequency distribution determined by Petersen *et al.* (1956) is applied, approximately 44·5% of the pasture area would not be covered by faeces or urine, 29·5% covered with one excreta, 14% covered by two excreta, and 12·0% covered by three excreta or more during one grazing year.

Potassium return to the soil pool from ungrazed herbage residues will be substantial (63·7 kg/ha), as well as return from roots during death and decay processes (33·6 kg/ha).

The net effect of cycling of potassium through the animal will be to enhance its availability because of the high proportion of ingested potassium excreted in the urine. Crucial to the effectiveness of potassium cycled through the animal will be enrichment of camp areas and the enhanced loss rate from such enriched areas. Loss rates of 139·2 kg of potassium/ha from leaching suggest a rather open cycle for potassium under this climatic condition. Ecosystems having less percolating water would be expected to have more nearly closed potassium cycles.

Losses of potassium because of fixation in the pasture ecosystem were assumed small since the predominant clay mineral is kaolinite. Release of unavailable potassium (potassium in non-exchangeable forms plus that from potassium bearing minerals) will be related to weathering, crop removal, and soil pH. The large pool of available potassium extracted by dilute acid extraction undoubtedly contains some non-exchangeable forms of potassium. Of the cycling pool, 31·5% is in the plant component. The percentage of the total soil pool of potassium cycling through the plant compartment is extremely small (0·84%). If forage were mechanically harvested and removed from the ecosystem, the losses of cycling potassium would be 318·4 kg or 25·8% of the cycling pool. Losses from the cycling pool by removal of animal products would be negligible. Losses because of leaching and inefficient return of excreta may be expected to deplete the overall pasture potassium supply to levels where potassium response in terms of herbage yield might be expected by additions of fertilizers after several years of grazing (139·2 kg/ha plus 44·5% unfertilized with excreta per year).

C. Response to Excreta Return in Grazing Trials

Brockman *et al.* (1970) compared three methods of experimental management—individually grazed plots, communally grazed plots, and cutting with removal of herbage—as to their effect on the growth and yield of grass/clover swards and found that excreta return markedly increased clover yields and raised herbage potassium content. Appreciable transfer of potassium from plot to plot from animal returns occurred when small plots were communally grazed. Their studies suggest that re-circulation of potassium by the grazing animal is significant. Studies by Wolton *et al.* (1970) of natural return of excreta, return of excreta from caged sheep, and simulated return of 75%, 125% N-P-K equivalent in excreta as inorganic fertilizers suggested that herbage yields were comparable between grazing and the 75% equivalent nutrient return level. Grazing reduced the clover content of the sward over the 3 year period of the study and consequently reduced yields in the grazed treatment. The relative potassium yields in herbage relative to a "no return of excreta" treatment (100%) were 122% with the "grazed with natural return", treatment while return of excreta from animals fed grass indoors yielded 142% of the "no return clipping" treatment. The relative potassium yield in the "75% simulated urine return" treatment was 139% of the cut, "no return" treatment, which suggested an effectiveness of urine return alone equivalent to about 70% of full return of excreta. These authors concluded that although fertilizer mixtures simulating excreta may replace the nutrient effect of the grazing animal, they will not replace their physical effects on the pasture.

Earlier studies by Wolton (1955, 1963) established the significance of urine return in increasing the level of available soil potassium. Wolton (1955) also suggested that urine return may result in increased availability of potassium since the presence of ammonia from urine might depress the amount of potassium fixed by the clay colloids in the soil. This same mechanism could enhance leaching losses as well.

Cuykendall and Martin (1968) concluded that sheep excreta increased potassium levels of soil, and plant tissues in communally grazed grass, or alfalfa plots. Grazing caused higher yields of all herbage until N for grasses and K for grasses and alfalfa were added at rates of 112 and 224 kg/ha, respectively. They suggest that clipping to simulate grazing will require high levels of fertility to compensate for the effects of animal excreta.

Watkin (1957) evaluated the return of dung and urine to small plot treatments and observed that urine resulted in marked increases in the potassium content of the pasture when returned in quantity to a grass-dominant sward. Urine potassium was superior to fertilizer potassium (K_2SO_4) per pound of potassium returned to the pasture.

Other studies involving communal grazing with return of dung and urine, dung, urine or no return to small plots suggest that potassium return in urine contributes significantly to the potassium status of the soil, the plant, and to cycling of potassium (Wheeler, 1958a; 1958b; Metson and Hurst, 1953; Wolton, 1963; Herriot and Wells, 1963; Andries and Van Slijiken, 1969; Blue and Gammon, 1963). However, Herriot and Wells (1963) suggested that the experimental techniques used in such studies involves intense stocking, intensive utilization of herbage, close-fencing, and a highly favorable pattern of excreta return. These factors enhance the probability of an almost "closed" nutrient cycle. These conditions are not typical of the many pastures where herbage utilization is low and extensive pasture, in relation to animal numbers, is available for grazing. Small scale animal grazing trials may not reflect excretal patterns effected in the field by continuous grazing, but may more nearly reflect patterns of excretal distribution obtained from intensive rotational or strip grazing systems. Other potential limitations of small scale animal grazing experiments were that over short periods of grazing sheep may not return excreta related in quality or quantity to that consumed from the plot (Herriot and Wells, 1963).

Soil potassium levels increased from about 200 ppm to 281 ppm at the 0-2·5 cm depth, and from 138 to 189 ppm at the 2·5-7·5 cm depth as grain fed to steers on pasture was increased from none to 58 metric tons/ha over a 4 year period (Benacchio et al., 1970). Potassium was considered a good indicator of the fertility added to the soil from the excreta of grazing animals, and the animal was considered effective in recycling potassium. This trial also involved intensive utilization of small pasture areas and because of the large grain inputs may not represent typical pasture ecosystems.

Rouquette et al. (1971) reported that the percent of potassium yield in coastal bermudagrass herbage cycled through cattle was 77·8 and 82·7% at a stocking rate of 5·0 yearling heifers and 4·7 animal units per ha in 1969, 1970, respectively and 44·2 and 70·7% at a lower stocking rate of 3·0 yearling heifers and 2·7 animal units in 1969, 1970, respectively for a study conducted in East Texas, U.S.A. Potassium extractable with N ammonium acetate increased five-fold at the high stocking rate and about four-fold at the lower stocking rate over the level of potassium initially extractable from the soil. The bermudagrass pasture was fertilized annually with 224, 48, 93 kg/ha of N, P, K, respectively. These striking potassium accumulations were obtained over two 180 day grazing seasons. Companion studies involving clipping coastal bermudagrass in small plots indicated little accumulation of potassium in the 0-15 cm layer of loamy fine sand over a three year period from an application of 623 kg/ha of potassium, in contrast to the 5-fold accumulation of potassium on highly stocked pastures fertilized with 186 kg/ha of potassium over a two year period (Matocha *et al.,* 1971). Such accumulations suggest a role of plants in re-distributing potassium from lower layers in the soil profile. Coastal bermudagrass is noted for its deep and extensive root distribution in sandy soils of the Coastal Plain of Georgia (Burton *et al.,* 1954).

D. Summary

Evidence from experimental trials involving soil-plant and grazing animals suggest that potassium is effectively cycled. However, most of these trials represent ideal conditions for the establishment of a closed cycle for potassium, i.e. intensive utilization of herbage (high stocking rates), close-fencing with limited opportunity for the livestock to establish camping patterns, and a system of animal management that resembles intensive rotational grazing. The observations of Hilder (1966); Petersen *et al.* (1956); Petersen *et al.* (1956) suggest that under a commercial grazing ecosystem where different camping patterns may be established, where utilization of herbage may be less efficient, and loss rates because of leaching and fixation from camp sites higher, the potassium cycle will be more nearly "open". It is doubtful that the cycle in grazed pasture ecosystems will ever be as "open" as that represented by clipping and removal of herbage.

VI. SULPHUR CYCLE

A. Characteristics of Sulphur

Sulphur is essential for the growth of all organisms and is required in amounts slightly less than phosphorus for higher plants. Sulphur, unlike phosphorus, may undergo valence changes during cycling which are similar to those for carbon and nitrogen. Sulphate, the principal form of sulphur absorbed by roots, and utilized by plants may be reduced from the +6 valence state to the −2 valence state when incorporated into the sulphur-containing amino acids, cystine, methionine, etc. Sulphur also exists in the valence state of +4 as sulphur dioxide. Sulphur dioxide is the source of much of atmospheric sulphur inputs which occur by rain-out, dry deposition, and direct absorption of sulphur dioxide by vegetation and soils. Sulphur dioxide is readily converted to sulphate, which is the form in which the majority of sulphur is found in the ambient troposphere (Robinson and Robbins, 1970).

Up to 90% of the sulphur content of tobacco plants grown in sulphur deficient media was found to be derived from sulphur dioxide labeled with S-35 (Faller, 1970). Foliar absorption both through stomata and by dissolution of SO_2 into films of water on leaf and soil surfaces may be significant entry points into the sulphur cycle of pasture ecosystems near heavily industrialized areas. When the sulphur content of the atmosphere is such that $11 \cdot 2$ kg/ha of sulphur are carried down with precipitation, the occurrence of sulphur deficiency is unlikely (Whitehead, 1964). Although this amount is not adequate for most crops, the direct absorption of sulphur by soil and vegetation may be adequate for vegetation requirements. Soil has been estimated to absorb as much as 30 kg/ha in Indiana (Bertramson et al., 1950). At the lower levels which might be expected over most pasture ecosystems ($1 \cdot 0$-$2 \cdot 0$ $\mu g/m^3$) sulphur uptake through leaf surfaces would represent a minor route of sulphur assimilation by herbage. Hydrogen sulphide represents a biologically important volatile sulphur compound. In pasture soils under reducing conditions, volatilization of sulphur as H_2S may represent a significant loss route. The source of much of the SO_2 in the air may be H_2S from terrestrial sources which is later oxidized to SO_2 and to sulphate (Erikson, 1960 and Jensen and Nakai, 1962 as cited by Deevy, 1970). The average concentrations of H_2S and SO_2 in the troposphere were estimated as $0 \cdot 2$ ppb H_2S and $0 \cdot 2$ ppb SO_2 (Robinson and Robbins, 1970). The significance of microbial reduction of soil sulphur in the nutrient cycle of pasture ecosystems is difficult to assess partly because of the presence of soluble zinc and iron compounds which may result in the formation of insoluble iron and zinc sulphides. The soil may be a sink or a source of sulphur for losses or gains from the atmospheric compartment. Robinson and Robbins (1970) estimate that natural emissions of sulphur in the form of H_2S are about 30% greater than industrial emissions of H_2S and SO_2 on a world-wide basis.

Transformations of sulphur resemble those for nitrogen in that sulphur is also a constituent of protoplasm and that the conditions necessary for reduction of sulphates are similar to those for nitrate (Alexander, 1964). The principal organisms associated with reduction of sulphate in normal soils are bacteria of the genus *Desulfovibrio,* which are obligate anaerobes. These organisms seem to be quite sensitive to acid soils and experience little growth at pHs below 5·5. They also require electron donors or energy sources which may include a number of carbohydrates, organic acids, alcohols and also molecular hydrogen. Sulphides may be oxidized to sulphate by *Thiobacilli.* Although most oxidation-reduction reactions of sulphur in soil are attributed mainly to soil microorganisms, oxidation-reduction reactions may occur without microbial action (Freney, 1967; Alexander, 1964).

Freney (1967) reported the average percentage distribution of sulphur in Australian soils as 93% organic and 7% inorganic. Fifty-one percent of the sulphur was of the sulphate ester type and 41% was carbon bonded and not hydrolysed to sulphate (e.g., cystine and methionine). The majority of the inorganic sulphur occurred as sulphates. The transformations of sulphur in soil have been divided into the following four distinct phases by Alexander (1964):

1. Mineralization or the release of sulphur from organic to inorganic combination.
2. Immobilization or the assimilation of inorganic sulphate into microbial tissue.
3. Oxidations—the conversion of sulphides, thiosulphates, polythionates and elemental sulphur to sulphate.
4. Reductions—conversion of sulphates and other oxidized forms of sulphates to hydrogen sulphide.

The mineralization rate of organic sulphur in soil depends on those factors which affect microbial activity such as soil temperature, aeration, water content, pH, type of organic residue and the effect of plants. Plants affect the amounts of available sulphur both by competing for sulphate with microbes and also by the improvement of aeration as a result of increased utilization of water. This reduces the chances for volatilization losses of sulphur as a result of microbial action (Nicolson, 1970). The reader is referred to reviews by Whitehead (1964), Freney (1967), and Alexander (1964) for more detail with respect to sulphur and its transformations in the soil.

Total sulphur contents of soils vary from less than 100 ppm to over 2000 ppm (Chapman and Pratt, 1961; Whitehead, 1964). Sulphur may react with soil in much the same way as phosphate although the degree of adsorption of sulphate is less marked than for phosphate in acid soils. Adsorption of sulphates occurs in soils having large amounts of extractable aluminum and iron (Barrow, 1969). Barrow (1970) found that adsorbed sulphate increases with increased rainfall, and was greatest on soils derived from basic rocks, and least on soils derived from siliceous rocks such as granites and sediments. Adsorbed sulphates tend to be greatest in sub-surface layers high in clay content (Freney *et al.*, 1962; Jordan, 1964; and Ensminger, 1954). Leaching losses of sulphate tend to be high on sandy soils (Barrow, 1970; McKell and Williams, 1960; Whitehead, 1964). Deep rooted perennial crops may effectively absorb this sulphate and the return of these residues to the soil surface increases the proportion of sulphur in the organic fraction (Freney *et al.*, 1962). Where drainage occurs below the lower limits of the ecosystem, leaching losses constitute a major loss pathway of sulphur from the cycle (Fried and Broeshart, 1967; Cooke, 1967; and others).

The importance of sulphur as a needed plant nutrient in agriculture may be increasing because of (1) increased usage of high analysis, low sulphur content fertilizer materials, (2) increased crop yields associated with increased use of heavy rates of N, P, K fertilizers, more intensive cropping and utilization of herbage, (3) greater emphasis on control of air pollution and reduction of sulphur emissions into the air, and (4) declining use of sulphur in fungicides and pesticides (Bixby and Beaton, 1970).

Intra-plant re-distribution of sulphur after initial absorption and deposition is not as rapid as redistribution of phosphorus. The rapid synthesis of sulphur into protein and other plant constituents is part of the reason for its low mobility (Biddulph, 1959). The inorganic sulphate ion has been shown, however, to move from older leaves to newly developed leaves of cotton plants (Eaton, 1966). Thus the mechanism of intra-plant circulation may be of less importance than for phosphorus and potassium, and the rate of sulphur circulation may be more dependent on mineralization from plant residues. This may represent an important quantitative variation in cycling between phosphorus and sulphur.

B. Cycling of Sulphur in the Cow-Calf Pasture Ecosystem

The yearly flux of sulphur through the system is presented in Table XV. Since the soil available pool size is so large (72·3% of the cycling pool), sulphur can be considered to be in "loose" circulation.

Sulphur in the above ground herbage was estimated as 28 kg/ha/year (Table XII), of which 80% was consumed by the animal (22·4 kg/ha of total intake by the livestock). Calf removal was only 0·8 kg/ha, or 3·6% of total intake. Applying the results reported by Barrow and Lambourne (1962) to herbage containing 0·25% S, about 9 kg of

TABLE XV

The amounts of sulphur in the compartments of the hypothetical cow-calf ecosystem, the size of the pool considered cycling on an annual basis and the amounts in the soil and total ecosystem[a]

	kg/ha	%[b]	%[c]
Total in various pools			
Total ecosystem pool	1832·6	100·0	—
Total soil pool	1790·6	97·7	—
Cycling pool	221·6	12·1	100·0
Total in compartments			
V_l Cow-Calf	1·7		0·77
V_p Plant	40·3		18·19
V_r Residues	19·1		8·62
V_s Soil Available	160·5		72·43
V_{us} Soil Unavailable	1611·00		—
Losses from animal cycle			
Leaching	40·6		18·32
Surface Run-off	0·2		0·09
Animal Product	0·8		0·36
Total Losses	41·6		18·77

[a] See Table IV for definitions of pools and relationships between pools. Calculations based on information from Table X.
[b] Sulphur in pool or compartment as a percentage of total ecosystem pool.
[c] Sulphur in each compartment as a percentage of the cycling pool.

sulphur/ha will be excreted in the faeces and about 12·5 kg of sulphur/ha excreted in urine. The effect of increasing intake of sulphur is to increase the levels excreted in urine. Therefore, consumption of herbage having high sulphur contents enhances cycling. Sulphur appears to be intermediate between phosphorus which is primarily excreted in the faeces, and potassium which is primarily excreted in urine. Turn-over rates in the cycle would suggest more effective cycling of sulphur than phosphorus and perhaps less effective cycling than potassium. Sulphur from urine spots is readily leached under conditions of rapid rates of water percolation through the profile. Volatilization losses from urine are not known. Concentration of sulphur in animal camping sites through urine return also enhances leaching losses. Losses of sulphur in the 25·4 cm of drainage in the ecosystem were estimated as 40·6 kg/ha, equivalent to that in the plant compartment and fifty times that lost from animal products (Table XV).

Very little of the cycling sulphur is in the livestock pool (2·5 kg/ha) while 19·1 kg/ha is in the residues pool. Sulphur in urine was considered to enter the soil available pool. The availability of sulphur from the ungrazed herbage (5·6 kg), dead roots (4·5 kg), and faeces (9·0 kg) will depend on the proportion of each in organic form, as well as the form and solubility of the sulphur in each residue. With sulphur contents of ungrazed herbage of 0·25%, roots of 0·20%, and faeces estimated as 0·30%, and the critical level of sulphur below which no net mineralization occurs of about 0·2% (Whitehead, 1970; Barrow, 1960) or 0·15% S (Stewart et al., 1966), little net mineralization would be expected from roots with somewhat larger amounts expected from ungrazed herbage, and faeces. Barrow (1961) reported that higher sulphur contents of faeces were required before net mineralization of sulphur occurred than with plant materials (an effect of prior decomposition during digestion by the animal).

The importance of soil-faeces contact and mixing for availability of sulphur for plant uptake has been demonstrated by Bornemissza and Williams (1970). Sulphur recovery from cow dung by Japanese millet in pot culture was 16·5% with dung beetle infestation, 24·0% when dung was mixed with the soil, and 6·3% with dung placed upon the soil surface without mixing or beetles. The slightly greater recoveries by millet of sulphur than phosphorus from dung placed on the soil surface suggests that sulphur mobility from dung piles is greater than that for phosphorus (Bornemissza and Williams, 1970). If all sulphur contents greater than 0·2% are released to the available pool of nutrient sulphur, about 1·1 kg sulphur will be released from the ungrazed herbage, 3·0 kg sulphur from faeces, 12·6 kg of sulphur via urine return, and none from decaying roots, giving a potential return of 16·7 kg/ha. With ungrazed herbage, we might expect a potential return of 5·6 kg/ha, and no return where herbage was clipped and removed. Such calculations suggest a highly favourable effect of the grazing animal on cycling of sulphur.

Direct leaching of sulphates from plant and animal residues to the soil pool would be expected to be relatively low and not represent a very large contribution of sulphur to the available pool. Intra-plant redistribution may be significant in grazed and ungrazed ecosystems, but unimportant when herbage is removed from the pasture.

Releases by mineralization and weathering were not included in the available pool calculations (Table XV). The annual flux of sulphur through the plant compartment was about 18·2% of the cycling pool. The largest single compartment of the cycling pool was the soil available pool, while the animal compartment was the smallest with less than 1% of the sulphur in the cycling pool. Leaching losses were similar to the amounts of sulphur in the plant compartment. Leaching losses were nearly one-fourth of the available pool and continued leaching losses without additions from fertilizers or other sources would soon "tighten-up" the circulation of sulphur. The cycling pool was estimated to contain 12% of the total system sulphur. The assumption of a 61 cm deep soil profile as a part of the ecoystem results in large total amounts of the element in the system. This choice of the lower limit of the ecosystem was based on observed soil water depletion to 61 cm. As with cycles for phosphorus and potassium, most of the element in the system is in the soil.

Losses of sulphur from the cycle by volatilization were not estimated, but there is evidence for some loss of volatile sulphur from herbage plants (Asher and Grundon, 1970). Similarly, losses of H_2S from normal pasture soils may occur during periods of reducing soil conditions and microbial decomposition of organic matter. Such losses are probably small in most moderate to well-drained pasture ecosystems.

C. Field Studies of the Sulphur Cycle

The use of radioisotope techniques appears to offer possibilities in evaluating the soil-plant-animal cycle of mineral elements. Two conditions that an isotope should meet before being considered a satisfactory tracer are long half-life to permit long term studies and an energy emission strong enough to be measured easily but not strong enough to be hazardous when used in large quantities to treat large areas. Unfortunately, most of the mineral elements do not have an isotope which meets the above criteria.

The most extensive use of a "tracer" in studying the soil-plant-animal cycling of minerals has been made by Till, May and coworkers at the Pastoral Research Laboratory in Armidale who used ^{35}S to study nutrient availability to plants, estimate absorption, leaching and recycling.

The first study by this group (May *et al.*, 1968) established that labelling only $\frac{1}{32}$ of the grazed area gave results similar to labelling the entire area. The ^{35}S was applied in the form of CaS^{35}O$_4$2H$_2$O) (gypsum) to the small areas and the study was continued for 760 days. Most of the applied sulphur remained in the top 10 cm of the

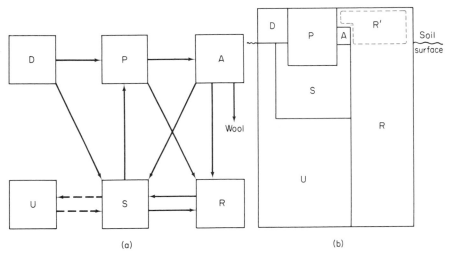

Fig. 3. The rectangles represent compartments within the system. Each compartment contains a pool of sulphur as defined in Table XIV. Known transports are shown by continuous lines and possible transports by broken lines. At equilibrium the initial sulphur dressing (D) has become distributed among other compartments. The approximate relative sizes of the compartments and their position relative to the soil surface are shown in Fig. 3b (reproduced with permission of Till and May, 1970).

soil. ^{35}S was redistributed and incorporated into plant material over the entire grazing area, with deposition of labelled material being 5-7 times greater in the resting area than in the overall paddock. By comparing specific activities of the soil and plants, it was concluded that a large percentage of the soil sulphur did not exchange with the ^{35}S and that a large portion of the ^{35}S continued to cycle.

In a second study, Till and May (1970a) used a series of ^{35}S applications and several groups of different sheep to estimate the size of the cycling sulphur pool (by isotope dilution). The study was conducted for a 1874-day period. The sulphur cycle proposed is summarized in Fig. 3 and Table XVI, which gives definitions and assumptions used. Figure 3 indicates that a large portion of the total sulphur is not entering the sulphur cycle (pool U) and that a large part of the sulphur in the cycling pool is in the residue (R and R^1) compartment.

TABLE XVI

Definitions of pools and relationships among them at equilibrium
(from Till and May, 1970a)

(a) Definitions

D Fertilizer sulphur,
P Plant sulphur,
A Animal sulphur (in sheep),
R Organic sulphur (in dead plant matter, organisms, excreta, etc.),
R' That portion of R not included in soil samples (e.g. faeces and plant litter),
S Soil sulphur (in forms suitable for immediate utilization by plants),
C Cycling sulphur (total cycling pool),
B Total soil sulphur,
G Total sulphur in the system,
U Sulphur that apparently does not enter the cycle (apparently inert).

(b) Relationships between pools at equilibrium

$$D = O,$$
$$C = P + A + R + S,$$
$$B = S + U + R - R',$$
$$G = S + R + P + A,$$
$$ = B + R' + P + A,$$
$$U = G + C.$$

(c) Estimation of U

The following values have been assumed:

1. P_{max} (c. 15 kg S) taken as sulphur content of 1 year's plant production,
2. A_{max} (c. 0·5 kg S) taken as sulphur content of six sheep,
3. R'_{max} (c. 4·5 kg S) taken as half annual above-ground plant production with the observed average value of B (c. 100 kg S) from Table II, in the relationships in Table Ib to obtain the relationship $U = 1·2B - C$ which was used to derive the estimates of U given in Table II.

The following assumptions are made in calculating the size of the cycling sulphur pool:

1. The specific activity of the sulphur in wool grown during a particular period is the same as the average specific activity of the sulphur in the ingested plant material providing that there are no large changes in the plant specific activity during that period.
2. There is no large loss or long-term hold-up of the applied sulphur before it enters the sulphur cycle.
3. The applied sulphur that enters a pool becomes mixed with the sulphur already in that pool.

Using formulae outlined in the above table, Till and May made four different estimates of the cycling sulphur pool and the inert sulphur pool during the study. The cycling pool averaged 69 kg and the inert pool 53 kg. The amount of sulphur in the cycle was about 12 times as great as the amount applied and 5 times as great as the amount in the plant material for one year. Since most of the cycling sulphur is in the residue compartment, this would be the limiting step in the cycle and if it were not being made available fast enough could result in a deficiency for plant growth. Only about 1 kg/ha/year was removed by the wool of the sheep. In this study little sulphur was lost from the cycle but in other ecosystems losses from the cycle by leaching and/or run-off might be expected. The soil type in this study was a yellow podzolic type derived from alluvial and colluvial material.

In a third study, Till and May (1970b) found that strip application of ^{35}S gypsum gave similar results to covering the entire area. The same was true of applying the ^{35}S in solid form rather than in aqueous solution. The cycling sulphur pool was estimated as about 100 kg/ha which was about 55% of the total sulphur in the top 10 cm of soil.

In the three experiments above, levels of ^{35}S used were not high enough to measure specific activity of the soil and plant fractions over a long period. A fourth experiment (Till and May, 1971) was conducted using high specific activity ^{35}S-gypsum applied to small areas (0·04% total area) at random. The pasture plots were stocked at two rates, 10 and 20 sheep/ha. At 264 days after applying the ^{35}S, gypsum "hot spots" were fenced off and one-half of the sheep replaced.

It was found that total soil sulphur was fairly constant with time of sampling, varying from 350 ppm in the top 2·5 cm to about 25 ppm at a depth of 25 cm. There was little change in the radioactivity in the various fractions of the soil after 150 days with about 60% of the activity still in the first 2·5 cm. About 43% of the applied ^{35}S had been re-distributed by day 264 when the "hot spots" were fenced off. The workers concluded that the extractable soil sulphur was a precursor of plant sulphur, that the organic fractions in the soil were the source of replenishment of the extractable pool and that an "unavailable" pool was turning over very slowly.

Using the data of the above four experiments, May et al. (1971) used an analogue computer to conduct a systems analysis of ^{35}S kinetics in pastures grazed by sheep. They concluded that (1) the observed transient behaviour of the specific activity of wool is consistent with a closed-system recycling process involving ca. 60% of the total sulphur in the system; (2) the major part, up to 90% of the cycling sulphur, may be held in organic residues in forms not directly available to plants; (3) the rate of release of organic sulphur from these residues in forms suitable for plant uptake is a major determinant of the sulphur supply to plant and animal compartments of the system; (4) the productive potential of the system is in general determined by the rate of sulphur transport between compartments rather than the amount present in any compartment at any time; and (5) the presence of the animal may have an appreciable effect on maintaining the level of recycling of sulphur within the pasture system.

VII. CALCIUM AND MAGNESIUM CYCLES

A. Characteristics of Calcium and Magnesium

These elements are considered together because of their important effects on soil properties such as soil structure, percentage base saturation, soil pH, biological activity of the soil, and availability of other nutrients for plant uptake, growth and cycling.

Calcium is usually the dominant cation on the exchange complex of most soils. Calcium and magnesium make up 60-80% of the exchangeable cations in most soils, and together with the hydrogen ion determine the pH of the soil. As soils become acid by leaching of exchangeable bases of calcium and magnesium, or as a result of nitrification, calcium and magnesium on the exchange complex are replaced by hydrogen ions, or aluminum ions. When soil pH decreases many important changes may occur in the soil environment, with effects on the cycling of mineral nutrients through plants. Some of the more important of these effects are the following:

1. Increased solubility of manganese, aluminum, copper, and other trace elements which may result in toxicity to herbage species.
2. Phosphorus compounds may be converted to insoluble iron and aluminum phosphates having very low availability for plant uptake.
3. The cation exchange capacity of soils may decrease and result in reduced capacity of soil to hold cations from leaching.
4. The availability of molybdenum for plant absorption and growth may be decreased (Parker and Harris, 1962).
5. Rate of nitrification falls off greatly below a soil pH of 6·0 and becomes negligible below pH 5·0 (Alexander, 1964).

6. Root growth of many improved pasture species is reduced at low soil pH (less than 5·0). Legume root growth is often more affected than that of grasses. Although the reduction of root growth by extremely acid soil conditions is fairly non-specific, different plant species vary in their sensitivity and tolerance to soil acidity, as well as different varieties within species (Vose, 1963; Foy et al., 1965a, b).

Plants absorb calcium and magnesium from soil solution in the ionic form, Ca^{2+}, Mg^{2+}. Calcium and magnesium occur in soil as non-exchangeable, exchangeable, and dissolved or freely-diffusible forms. Total calcium contents of some U.S. soils ranged from 0·06 to 1·8% and magnesium from 0·05 to 1·09% (Chapman and Pratt, 1961). During absorption, translocation, and synthesis of biologically important compounds, calcium and magnesium do not undergo valence change. Compounds of importance to biological systems containing calcium and magnesium are non-volatile, and cycling occurs by the action of organisms in the ecosystem. Calcium plays a very important role in maintaining root surfaces in condition for cation absorption (Black, 1968), and thus may be especially significant in the absorption and cycling of other nutrients.

In most soils, the concentrations of calcium and magnesium in soil solution are sufficiently large, and the volume of soil explored by most root systems sufficiently large that transport of soil solution to root surfaces is sufficient to meet most of the plant's requirements (Fried and Shapiro, 1961; Barber et al., 1963). Concentrations of calcium in soil solutions range from near zero to over 1000 ppm, while magnesium concentrations range from near zero to over 1000 ppm (Reisenaur, 1966). Obviously, concentrations of calcium are lowest in acid soils. However, in alkaline or sodic soils, calcium may be of low availability because of extreme insolubility of calcium compounds formed, because of a low supply of exchangeable calcium, and because of an excess of sodium over calcium.

B. Cycling of Calcium and Magnesium in the Hypothetical Pasture Ecosystem

The amounts of calcium in various segments of the hypothetical pasture ecosystem are given in Table XII. Atmospheric input was estimated as 3·9 kg/ha/year, with erosion losses of 1·2 kg/ha/year (Table XVII). Miller and Krusekopf (1932) found losses of calcium from bluegrass sod on a 7% slope in Missouri to be 1·2 kg/ha with

12·3 cm of run-off. Inputs from the atmosphere were based on an average rainfall composition of 0·32 ppm calcium (from Carroll, 1962). Losses of calcium by leaching were estimated as 172·7 kg/ha based on 25·4 cm of leachate containing 68 ppm calcium (about 6% of cycling pool). Total calcium in the plant compartment was estimated

TABLE XVII

The amounts of calcium in the compartments of the hypothetical cow-calf pasture ecosystem, the size of the pools considered cycling on an annual basis and the amounts in the soil and total ecosystem[a]

	kg/ha	%[b]	%[c]
Total in various pools			
Total ecosystem pool	17,419·8	100·0	—
Total soil pool	17,351·7	99·6	—
Cycling pool	2,748·8	15·8	100·0
Total in compartments			
V_l Cow-Calf	14·3		0·52
V_p Plant	53·8		1·96
V_r Residues	34·5		1·26
V_s Soil Available	2,646·2		96·27
V_{us} Soil Unavailable	14,671·0		—
Losses from animal cycle			
Leaching	172·7		6·28
Surface Run-off	1·2		0·04
Animal Product	5·6		0·20
Total Losses	179·5		6·52

[a] See Table IV for definition of pools and relationships between pools. Calculations based on information from Table X.
[b] Calcium in pool or compartment as a percentage of total ecosystem pool.
[c] Calcium in each compartment as a percentage of cycling pool.

as 67·2 kg/ha of which 13·2 kg was estimated to be returned to the soil as residues.

Losses by leaching constitute the greatest loss route from the calcium cycle, as might be expected in a perennial grass pasture whose primary productivity is maintained by nitrogen fertilization. Of the 22·4 kg of calcium ingested by the livestock component each year, about 5·6 kg was estimated to be removed each year. This constitutes a higher rate of removal of herbage mineral ingested than with phosphorus, potassium and sulphur, but constitutes less than 1% of the cycling pool of calcium.

Calcium excreted by the animal is primarily in the faeces. Faeces distribution will therefore determine cycling of calcium in soil-plant-grazing systems. The availability of calcium in faeces for plant growth has not been extensively studied, although Davies *et al.* (1962) found that exchangeable calcium of soil was raised by 9·8, 2·4, and 0·9 me % at the 0-4, 4-8, and 8-15 cm depths under dung patches when sampled six to eight months after deposition. Since

TABLE XVIII

The amounts of magnesium in the compartments of the hypothetical cow-calf pasture ecosystem, the size of the pools considered cycling on an annual basis and the amounts in the soil and total ecosystem[a]

	kg/ha	%[b]	%[c]
Total in various pools			
Total ecosystem pool	15,681·8	100·0	—
Total soil pool	15,645·5	99·8	—
Cycling pool	697·8	4·4	100·0
Total in compartments			
V_l Cow-Calf	0·5		0·07
V_p Plant	35·8		5·13
V_r Residues	24·8		3·55
V_s Soil Available	636·7		91·24
V_{us} Soil Unavailable	14,984·0		—
Losses from annual cycle			
Leaching	76·2		10·92
Surface Run-off	0·2		0·029
Animal Product	0·3		0·043
Total Losses	76·7		10·99

[a] See Table IV for definition of pools and relationships between pools. Calculations based on information from Table X.

[b] Magnesium in pool as a percentage of total ecosystem pool.

[c] Magnesium in each compartment as a percentage of cycling pool.

much of the calcium is in structural components of cells and in compounds of low solubility, low immediate availability of calcium in dung and plant residues deposited on the soil surface would be expected. Immobilization of calcium by microbial populations would not appear to be as important as with sulphur and phosphorus since bacterial and fungal tissue requirements are small (Nicholas, 1963). Bacteria contain about the same calcium content as higher plants and fungi contain much less (Table I). Direct competition between soil microbial populations and higher plants for calcium would appear to be a relatively minor factor in the calcium cycle. For many microorganisms calcium requirements are only a few ppm.

Magnesium losses by leaching were estimated as 76·2 kg/ha based on a soil solution content of 30 ppm and drainage losses of 25·4 cm/year (Tables XII and XVIII). Accessions to the ecosystem of 1·3 kg/ha/year were based on estimated average rainwater composition of 0·11 ppm. Atmospheric inputs of both calcium and magnesium would be expected to have negligible effects on both the amount and rate of cycling. The losses of about 0·3 kg of magnesium from the cycle in the removal of 504 kg of calf would have negligible effects on the cycling pool. As with calcium, the major losses from cycling will be by leaching, and from positional unavailabilities of residues of animal origin to the majority of the pasture. Davies *et al.* (1962) found 62% of the magnesium in two dung samples to be water soluble; 6-8 months after deposition exchangeable magnesium in soil under dung patches was raised by 5·5, 2·4, and 0·6 me % at the 0-4, 4-8, and 8-15 cm depths.

Magnesium requirements for microorganism growth are larger than for calcium (Nicholas, 1963). As with calcium, immobilization of magnesium in microbial tissue may not be as critical a link in the cycle of magnesium in pasture ecosystems as microbial immobilization is for phosphorus and sulphur.

C. *Response to Excreta Return*

Herriot and Wells (1963) determined the influence of full return and no return of sheep excreta on calcium and magnesium in the top soil and suggested that nitrogen return from grazing animals depressed clover content of the mixed clover/grass sward and resulted in lower calcium and magnesium removal. Watkin (1957) found that dung return had little effect on calcium content of grass/clover pastures. Neither of these reports suggest very positive evidence for effective calcium cycling from grazing. Magnesium and calcium contents of harvested herbage were depressed in urine return treatments of Watkin (1957) probably because of the high potassium content of the urine. The effect of more efficient potassium return than calcium and magnesium was pointed out by Watkin (1957). This enhancing of potassium with less effective cycling of calcium and magnesium may result in an upward spiral effect on the ratio of potassium to calcium plus magnesium in soil and plant components of the ecosystem (Voisin, 1963).

D. Summary

The cycling of calcium and magnesium through soil-plant-animal systems, or pasture systems has not been directly studied to our knowledge. Major losses from the cycle by leaching are likely when drainage occurs below the lower limits of the ecosystem and by uneven return of calcium and magnesium in faecal material. The use of high levels of N fertilizers also enhances calcium losses. Losses to animal products are relatively minor except for milk. The importance of calcium and magnesium not only lies in their essentiality for plant and animal growth, but in their effects on physical and chemical properties of soil conducive to high productivity of pasture eco-systems. Magnesium cycling is likely to be more efficient than calcium in terms of turn-over times, but may be more critical to maintenance of productivity in the ecosystem because of smaller pools of magnesium in many soils. Immobilization of calcium and magnesium in microbial tissue may not be a critical factor in availability of these elements for cycling in most improved pasture ecosystems.

VIII. POTENTIAL CONTROL POINTS

Control points for mineral cycling involve ways of (1) increasing available nutrient pool size (gains in the cycling pool), (2) increasing the transport rate between compartment pools and (3) decreasing the losses of nutrients from the cycling pool of nutrients. Potential management means for controlling nutrient cycling in pasture eco-systems involve the following: soil selection, soil and pasture fertiliz-ation, soil management, pasture crop selection and management systems, and animal management systems.

A. Soil Selection

The concentrations of nutrient elements in any one plant species generally reflect the character of soil in which it is grown since the soil solution is the primary source of the element to enter the biological cycle, and the soil pool is normally the largest pool in the cycle. This results in the soil compartment exerting a specific buffering action on the rate and amount of element cycling.

Physical characteristics of soil such as its infiltration rate, rate and depth of drainage, its capacity to hold water and nutrients all play significant roles in the observed loss rates of mineral elements from leaching and from erosion. The proportions of sand, silt and clay along with organic matter content as well as parent material, relief and time largely determine these properties. Increasing concentrations of organic matter increases the capacity of soils to retain cations from leaching, and to supply cations for plant uptake. Organic matter, and organic compounds therein may significantly affect nutrient uptake processes by their ability to form complexes with mineral ions (mechanism of chelation).

The relative age of soils or maturity of soils is also important since gains from weathering and consequent mobilization may be higher on young, immature soils relatively rich in unweathered minerals than in older more weathered soils. Mature soils are considered to be those in steady-state with soil-forming factors in their environment. Soil properties are a function of the climate, organisms, topography, parent material, age of soil, etc. (Jenny, 1941; Jenny, 1961). Soil represents the historical record of effects of biotic and abiotic factors on its properties during formation. Soil selection along with the climate and relief, play major roles in determining the structure of the pasture ecosystem. Soil types interact with the processes of microbial immobilization, chemical fixation, chemical and physical weathering, diffusion of ions to root and adsorbing surfaces, mineralization of organic matter and residues, and water flow, and consequently affect cycling of mineral nutrients. Although soil type selection may not always be possible, soil type matching with pasture crop characteristics is practical and important from the standpoint of mineral cycling.

B. Soil and Pasture Fertilization

Direct application of plant nutrients to pasture ecosystems is highly effective in increasing the amount of element in the system, and in the cycling pool. However, economic restraints including failure to utilize the increased herbage production and potentially important social restraints arising from nutrient leakage contributing to undesirable eutrophication of other ecosystems may restrict the use of fertilizer to increase the productivity of the ecosystem. The potential for quantitatively modifying nutrient cycles by fertilization is great because of the technology available to vary the form, rate, and time of application of fertilizer materials. Timing of plant nutrient applications may be an effective management method for maximizing effectiveness of mineral cycling.

Large single doses of fertilizer may drastically change the "steady-state" conditions of an ecosystem, but the size of the cycling pool may return rapidly to near its initial level because of increased loss rates from the cycle. On the other hand, if plant nutrients are introduced to the cycle on a continuing basis then fertilization becomes one of the major factors determining the "steady-state" of the system. Wise pasture ecosystem management always chooses to optimize the size of the mineral cycling pool consistent with high net primary productivity, and minimum nutrient leakage from the ecosystem.

C. Soil Management

Irrigation, terracing, land forming are useful in improving soil-plant-water relationships, and consequently encourage effective cycling of minerals. Control of soil acidity to near neutral pHs by chemical amendments such as lime for acid soils and sulphur for alkaline soils may create more favourable conditions for root growth and function, enhance nutrient availability, as well as stimulating decomposer and transformer organism activity in the soil. Fungi are considered to be dominant in acid soils whereas bacteria are dominant in neutral or alkaline soils (Alexander, 1964). The importance of this association to mineral cycling has not been established.

Chain-harrowing the pasture or other physical means for re-distributing faeces offers possibilities to reduce the potential ill effects from poor distribution of faeces, but may damage the herbage (Weeda, 1967). One powerful, but not always practical technique for increasing turnover of nutrients immobilized in residues at the surface of the soil is to plough-up pastures and plant arable crops. The use of small grains for grazing has become an important part of forage systems for livestock in the Southeast U.S.A. Tillage and/or sod-fallow of pastures has increased phosphorus supply to subsequent small grain crops (Dawley, 1965; Wheeler, 1958b).

Soil management practices which establish soil physical conditions at or near optimum for effective operation of the "detritus" food web are likely to influence favourably both the size of the available soil nutrient pool, and the transport rate between compartments of the mineral cycle. Malone (1970) states that grassland fungi are the most sensitive of soil microflora and that their densities reflect soil moisture content. Control of biotic and abiotic factors is of great importance in maintaining continuity in the microbial turnover of mineral elements (Witkamp, 1969a, b). Witkamp *et al.* (1966) observed that stagnation of mineral element turnover resulted in mineral accumulation in litter, and eventually slowed down the rate of cycling of minerals in all compartments of the ecosystem.

Burning of pastures would appear to enhance cycling by rapid release of non-volatile nutrients in the residues. Evans and Allen (1971) found that burning heather (*Calluna vulgaris*) at low temperatures (310-580°C) resulted in losses of 5-10% of the potassium, calcium, magnesium and phosphorus in the heather, while burning at higher temperatures (590-750°C) resulted in 20% losses of potassium, and 10-15% losses for phosphorus, calcium and magnesium. Sulphur losses were 18% at the lower temperature burn, and 36% at the higher temperatures. Severity of heather burn was affected by age, density and moisture content of the heather as well as wind speed at the time of burn. A common practice in the Southeast U.S.A. is to burn coastal bermudagrass pastures during late winter to provide disease, insect and weed control. However, residues from burns of coastal bermudagrass have been observed to be washed from the land during intense spring rain and as such may constitute a potential loss of mineral nutrient from the cycling pool. Morris (1968) has shown that at high levels of fertilization burning has increased forage yields because of improved weed control. Burning did not increase yields at medium levels of fertilization. Daubenmire (1968) suggests that regular burning of grassland was not of great significance to mineral nutrient availability although there was an indication of accumulation of exchangeable cations at the surface of the soil. Under circumstances where there is a large standing stock of minerals in dead herbage, burning may result in rapid mineralization with minor losses to the atmosphere under controlled low temperature burns. The cations released upon burning were mainly water soluble, and the phosphorus was dilute-acid extractable, in a study by Smith (1970a).

D. Pasture Crop Selection and Management

Selection of pasture crop or crops represents an important means for controlling nutrient cycling for most improved ecosystems. In practice, choice of herbage plant is based on its net primary productivity under the conditions of the ecosystem, and its ability to produce useful animal products. Crop varieties and species are known to vary in their mineral requirements and tolerance to toxic levels of aluminum, etc. (Vose, 1963; Foy *et al.*, 1965a, b). Crop characteristics can be matched to soil and climatic characteristics. For example, the use of deep rooted perennial plants in deep sandy soils will effectively extend the lower limits of the pasture eco-system. Where climatic conditions permit, double-cropping or mix-tures of species will extend the growing season and reduce the potential loss rates of nutrients because of leaching, etc. Inter-seeding cool season annuals or perennials into perennial warm season sods just prior to or during their dormant seasons, permits an extended grazing season as well as fuller utilization of soil, land and water resources, and a probable reduction of mobile nutrient loss with percolating water. (Welch *et al.*, 1967; Wilkinson *et al.*, 1968). Anatomical and morphological characteristics of herbage plants also represent potentially useful points for modification of mineral cycling. For example, herbage plants with stoloniferous, rhizomatous growth habits, such as *Cynodon* species have been observed to recover more quickly from the effects of faecal droppings than have bunch type grasses.

When rainfall is less than potential evapotranspiration, crop selec-tion for those grasses tolerant of plant-water stress, and/or those which more effectively utilize both water and nutrients will be important to maintenance of efficient mineral cycles.

Control of the proportion of the total plant population in unpalatable and undesirable plants (weeds) represents an important means of maximizing the effectivenesss of nutrient supplies in the ecosystem. During any grazing season, unharvested plants (weeds, ungrazed herbage) represent nutrients not contributing to the economic gain from the pasture.

Other management control points include the design and layout of the pasture in relation to the presence of both natural and planned physical features. Water, minerals, shade, etc. should be located to encourage the most effective excreta patterns, and in such ways discourage permanent camp sites. The exclusion of large areas of forest, or other non-productive areas of pasture ecosystems from the grazing or resting area should also be helpful.

E. Animal Management

Management of animals to obtain optimum herbage utilization consistent with maintenance of plant productivity represents an important method of modifying the mineral cycle, since animal excretion becomes the dominant nutrient re-cycle pathway. The animal short-circuits slowly mineralized nutrients in plant residues for those nutrients principally excreted in the urine (May *et al.*, 1971; Floate, 1970d; Barrow, 1967). Grazing practices may be controlled by man through (1) the season and stage of growth at which grazing takes place, (2) the frequency of grazing, and (3) the severity of grazing (Alcock, 1964). Grazing management practices such as high stocking rates intended to rapidly and completely utilize herbage promote rapid circulation of nutrients between soil-plant-animal compartments, and facilitate more effective excreta patterns. Such systems of grazing may be referred to as strip and rotational grazing. Continuous grazing, at the same stocking rate over an entire grazing season, often results in less intense rates of defoliation over larger areas of the pasture and may promote camp development and inefficient distribution of nutrients in excreta. All grazing classifications interact with climatic, soil and plant parts of the system, giving many potential combinations having mineral nutrient cycles qualitatively similar but different quantitatively. The main impact of these grazing systems and animal management on mineral cycling will be expressed as a function of the rate of herbage consumption per unit area, the distribution of excreta, and the effects of defoliation on the re-growth of herbage, and growth of root systems.

Zero grazing or mechanical harvesting of herbage, fed to animals in confinement, with return of excreta to land surfaces appears to offer the most potential for approaching a closed mineral-nutrient cycle. In practice, the losses of liquid excrement during confinement may be substantial, particularly when animals are confined to lots having permeable surfaces, or where little effort is expended to save excreta.

The species or breed of animal and whether it is growing, reproducing, producing meat, milk, or wool influences the flux of mineral nutrients between compartments. For example, cattle breeds of zebu origin have more heat tolerance than the so-called British breeds whereas the British breeds have greater cold tolerance. Grazing habit response to the same climatic stress will be different for these cattle of different origin, and it would be reasonable to expect different patterns of excreta return in response to climatic and environmental conditions.

IX. NEEDED RESEARCH

The study of improved pasture systems as a whole has been much neglected. Loomis (1969) has expressed the apparent need for this research as follows: "During the past 100 years, much of our progress in improving agricultural efficiency has come through the rational application of the limiting factor concept. In a cyclic, iterative manner, the factor currently most limiting to production is identified and its impact ameliorated through research. The successes of such successive approximations are numerous and obvious, but as Jensen (3) has pointed out, the technique may not be adequate for the future. The time constants of present cycles are too long, and because of the subcultures present in our science, conflicting solutions are frequently arrived at by ecologists, physiologists and geneticists." Two subsidiary principles must be added to the concept of the "Law of minimum" in order for it to be useful in practice. The first of these principles is that the "Law of minimum" applies to conditions where inflows of energy, minerals, etc. balance outflows of energy, minerals, etc. (steady-state conditions), and secondly, factors interact and modify the effects of the individual factors on the system (Odum, 1971, p. 106). When these subsidiary principles are recognized, the importance and need of mathematical models of mineral cycling in pasture ecosystems which take into account factor interaction and the effects of changes in system operation becomes apparent. It is appropriate to combine ecological knowledge and technological resourcefulness to the solution of production and environmental problems of pasture ecosystems. The development of models of pasture ecosystems will demand interdisciplinary team research with systems analysts as members of the team.

Some problems in need of solution are the following:

1. The development of techniques for accurately estimating the nutrient requirements of pastures without involving the entire pasture system. Soil and crop fertility requirements are still determined predominantly on an empirical basis, and there is a need to move from an empirical basis to more comprehensive solutions to soil fertility requirements.

2. Although it is generally agreed that the "detritus" food web is far more complex than the grazing food web (Lewis, 1969), and that its operation is an essential part of the nutrient cycle, can it be shortened or simplified and so increase productivity without hazard to the ecosystem or to the environment? Further elucidation of the role microbial populations play in mineral nutrient cycling under pasture conditions is needed. Of particular interest is the effect mycorrhizal associations have on nutrient availability and cycling. Most plants in the field have mycorrhizal associations (Gerdemann, 1968) while those in many nutrient uptake studies conducted in the

laboratory do not. Thus a logical question is, what is the role and importance of these mycorrhizal fungi in the nutrient cycle of pasture ecosystems? Is there a possibility of plant to plant transfer of mineral nutrients by symbiotic fungi? (See Odum, 1971, p. 104). Jackson (1965) reported that the commonest hyphae in pasture soils of New Zealand were those of vesicular arbuscular fungi. Similarly, the role of soil fauna such as dung beetles and earthworms in the nutrient cycle of pasture ecosystems needs further examination and investigation. The importance of flies in decomposition of dung pads in pastures has been reported by Papp (1971) and Anderson (1966). Papp believed that fly larvae were more effective in consuming dung than scarabeid beetles. Of what significance are flies in the mineral cycle of pastures? In general, much more needs to be learned about the microflora and -fauna interface in pasture soils, and their importance to the mineral cycle in pasture ecosystems.

3. Distribution of excreta functions are needed which estimate excreta distribution in time and space under a variety of grazing systems with various classes of grazing animals. The availability of nutrients in excreta and the quantitative relationships for predicting their contribution to the mineral cycle are also needed.

4. Although increased stocking rate appears to enhance circulation of most nutrients, evidence summarized by Gardiner (1969) suggests that deficiency of selenium may be aggravated by improved pasture techniques, and by intensive husbandry. The selenium cycle in soil-plant-animal systems has been described by Allaway et al., (1967), and by Olson (1967). Allaway (1968) has reviewed the cycling of trace elements in the environment as well as the potential for agronomic control of the cycles of trace elements. However, there is a need for more information on the cycling of these elements in soil-plant-grazing animal systems. This need is prompted more by the requirement of the grazing animal for the element, or because of potential toxicities, than by the effect of the element on primary productivity of the ecosystem.

5. Although instantaneous estimates of the amounts of particular forms of most elements in components of the ecosystem are all that can be expected with existing technology, the results of a research group in Australia (May et al., 1968; Till and May, 1970a, b; May et al., 1971) show that the radioisotope ^{35}S may be used to estimate the rate of circulation of sulphur through various compartments of the pasture ecosystem. Further research with this radioisotope in other pasture ecosystems having different water

balance situations, and different soil, plant, and animal components is desirable. Other isotopes having potential use for short-term studies are ^{32}P, ^{45}Ca, and ^{134}Cs with its potential relationship to potassium. Dodd and Van Amburg (1970) successfully used ^{134}Cs in a recent study of methods, rates and transfer from one part of a little bluestem (*Andropogon scoparius*) grassland ecosystem to another. Certainly techniques are needed whereby flux of minerals between compartments can be accurately measured.

6. Long-term, well designed pasture fertilization experiments on soils deficient in specific nutrients, are needed to evaluate the contribution of the grazing animal to the cycling patterns of plant nutrients under a variety of pasture ecosystems. These studies should be coordinated on a national or regional basis (perhaps international) whereby objectives and procedures as well as data collection are standardized to permit exchange of data, systems analysis, model testing and improvement. Coordinated research efforts of this type are essential for the efficient use of support funds, and the results of such studies have practical as well as academic interest. Such studies should also include the hydrologic cycle as well as energy flow patterns so that interrelationships between mineral, water, and energy flow may be established.

REFERENCES

Alcock, M. B. (1964). *In* "Symposium on Grazing" (D. J. Crisp, ed.), pp. 25-42. Blackwell, Oxford.

Alexander, M. (1964). *In* "Introduction to Soil Microbiology." Second Printing. 1st Edition. John Wiley & Sons, Inc., New York and London.

Allaway, W. H. (1968). *Adv. Agron.* 20, 235-274.

Allaway, W. H., Cary, E. E. and Ehlig, C. F. (1967). *In* "Symposium: Selenium in Biomedicine" (O. H. Muth, ed.), pp. 273-296. AVI Publishing Co., Westport, Connecticut.

Anderson, J. (1966). *In* "Management of Farm Animal Wastes. Proceedings National Symposium on Animal Waste Management" (E. P. Taiganides, ed.), ASAE Publication No. SP-0366, pp. 20-23. Amer. Soc. Agr. Eng., St. Joseph, Michigan.

Anderson, O. E., Carter, R. L., Perkins, H. F. and Jones, J. B. Jr. (1971). U. of Ga. Col. Expt. Sta. Res. Report 102.

Andrics, A. and Van Slijcken, A. (1969). *Revue Agric., Brux.* 22, (9), 1197-1229.

Asher, C. J. and Grundon, N. J. (1970). *Proc. XI Int. Grassld Congr.* 329-332.

Barber, D. A. (1968). *A. Rev. Pl. Physiol.* 19, 71-88.

Barber, S. A., Walker, J. M. and Vasey, E. H. (1963). *J. agric. Fd Chem.* 11, 204-207.

Barber, S. A. (1968). *In* "The Role of Potassium in Agriculture" (V. J. Kilmer, S. E. Younts, N. C. Brady, eds), pp. 293-310. Amer. Soc. Agron., Crop Sci. Soc. Amer., Soil Sci. Soc. Amer., Madison, Wisconsin.

Barrow, N. J. (1960). *Aust. J. agric. Res.* 11, 960-969.

Barrow, N. J. (1961). *Aust. J. agric. Res.* 12, 644-650.

Barrow, N. J. (1967). *J. Aust. Inst. agric. Sci.* 33, 254-262.

Barrow, N. J. (1969). *Soil Sci.* 108, 193-201.

Barrow, N. J. (1970). *Proc. XI Int. Grassld Congr.* 370-373.

Barrow, N. J. and Lambourne, L. J. (1962). *Aust. J. agric. Res.* 13, 461-471.

Bear, F. E. (1942). *In* "Soils and Fertilizers" Third Edition, pp. 208-221. John Wiley & Sons, Inc., New York.

Benacchio, S. S., Mott, G. O., Huber, D. A. and Baumgardner, M. F. (1969). *Agron. J.* 61, 271-274.

Benacchio, S. S., Baumgardner, M. F. and Mott, G. O. (1970). *Proc. Soil Sci. Soc. Am.* 34, 621-624.

Bertramson, B. R., Fried, M. and Tisdale, S. L. (1950). *Soil Sci.* 70, 27-41.

Biddulph, O. (1959). *In* "Plant Physiology: A Treatise. Vol. 11: Plants in Relation to Water and Solutes" (F. C. Steward, ed.), pp. 553-603. Academic Press, New York and London.

Biddulph, O., Biddulph, S., Cory, R. and Keontz, H. (1958). *Pl. Physiol., Lancaster*, 33, 293-300.

Bixby, D. W. and Beaton, J. D. (1970). "Sulphur Containing Fertilizers, Properties and Applications". The Sulphur Inst. Tech. Bull. No. 17. Washington, D.C. 30 pp.

Black, C. A. (1968). Phosphorus. *In* "Soil Plant Relationship", pp. 558-653. John Wiley & Sons, Inc., N.Y., London, and Sidney.

Blaser, R. E. and Kimbrough, E. L. (1968). *In* "The Role of Potassium in Agriculture" (V. J. Kilmer, S. E. Younts and N. C. Brady, eds), pp. 423-445. The Amer. Soc. Agron., Crop Sci. Soc., and Soil Sci. Soc. Amer., Madison, Wisconsin.

Blue, W. G. and Gammon, N., Jr. (1963). *Proc. Soil Crop Sci. Soc. Fla.* 23, 152-161.

Bornemissza, G. F. (1960). *J. Aust. Inst. agric. Sci.* 26, 54-56.

Bornemissza, G. F. and Williams, C. H. (1970). *Pedobiologia*, 10, 1-7.

Bouma, D. (1967). *Aust. J. biol. Sci.* 20, 601-612.

Bourliere, F. and Hadley, M. (1970). *In* "Ecological Studies. I. Temperate Forest Ecosystems, Vol. 1" (David E. Reichle, ed.), pp. 1-6. Springer-Verlag, Berlin, Heidelberg, and N.Y.

Bowen, H. J. M. (1966). "Trace Elements in Biochemistry". Academic Press, N.Y.

Bray, R. H. (1954). *Soil Sci.* 78, 9-22.

Brockman, J. S., Shaw, P. G. and Wolton, K. M. (1970). *J. agric. Sci., Camb.* 74, 397-407.

Bromfield, S. M. (1961). *Aust. J. agric. Res.* 12, 111-123.

Bromfield, S. M. and Jones, O. L. (1970). *Aust. J. agric. Res.* 21, 699-711.

Buckman, H. O. and Brady, N. C. (1968). "The Nature and Properties of Soils". 6th Edition, 11th Printing. Macmillan Co., N.Y.

Burton, G. W., DeVane, E. H. and Carter, R. L. (1954). *Agron. J.* 46, 229-233.

Calder, F. W. (1970). *J. Br. Grassld Soc.* 25, 144-153.

Carroll, D. (1962). Geological Survey Water Supply Paper 1535-G. U.S. Government Printing Office, Washington. 18 pp.

Carter, E. D. and Day, H. R. (1970). *Aust. J. agric. Res.* 21, 473-491.

Chapman, H. D. and Pratt, P. F. (1961). "Methods of Analysis for Soils, Plants, and Waters". U. of Calif. Press, Div. Agr. Sci.

Cooke, G. W. (1967). "The Control of Soil Fertility" Crosby Lockwood & Son, Ltd. 526 pp.

Crider, F. J. (1955). Tech. Bull. 1102, U.S. Dept. of Agric. 23 pp.

Crisp, D. T. (1966). *J. appl. Ecol.* 3, 327-348.

Cullen, N. A. (1971). *N.Z. Jl agric. Res.* 14, 10-17.

Curlin, J. E. (1970). *In* "Ecological Studies. I. Temperate Forest Ecosystems. Vol. I" (David E. Reichle, ed.), pp. 268-285. Springer-Verlag, Berlin.

Cuykendall, C. H. and Martin, G. C. (1968). *Agron. J.* 60, 404-408.

Dahlman, R. C. and Kucera, C. L. (1969). *In* "Proceedings of the Second National Symposium in Radioecology," Ann Arbor, Mich., May 15-17, 1967 (D. J. Nelson and F. E. Evans, eds), 652-660.

Dale, M. B. (1970). *Ecology* 51, 1-16.

Dale, W. R. (1961). *Proc. N.Z. Grassld Ass.* 23rd Conf., 118-124.

Daubenmire, R. (1968). *Adv. Ecol. Res.* 5, 209-266.

Davidson, J. L. and Milthorpe, F. L. (1966). *Ann. Bot.* 30, 185-198.

Davies, E. B., Hogg, D. E. and Hopewell, H. G. (1962). *In* "Transactions of the Joint Meeting Commissions IV and V of the International Society of Soil Science," pp. 159-172.

Dawley, W. K. (1965). *Can. J. Pl. Sci.* 45(2), 139-144.

Deevey, E. S., Jr. (1970). *Scient. Am.* 223, 148-159.

Dodd, J. D. and Van Amburg, G. L. (1970). *Can. J. Soil Sci.* 50, 121-129.

Donald, C. M. and Williams, C. H. (1954). *Aust. J. agric. Res.* 5, 664-687.

Duvigneaud, P. and Denaeyer-De Smet, S. (1970). *In* "Ecological Studies. I. Temperate Forest Ecosystems. Vol. I" (David E. Reichle, ed.), pp. 199-225. Springer-Verlag, Berlin.

Eaton, F. M. (1966). *In* "Diagnostic Criteria for Plants and Soil" (H. G. Chapman, ed.), pp. 444-475. Univ. of Calif., Div. of Agr. Sci., Riverside, Calif.

Ensminger, L. E. (1954). *Proc. Soil Sci. Soc. Am.* 18, 259-264.

Epstein, E. (1965). *In* "Plant Biochemistry" (J. Bonner and J. E. Varner, eds), pp. 438-461. Academic Press, N.Y.

Evans, C. C. and Allen, S. E. (1971). *Oikos* 22, 149-154.

Faller, N. (1970). *Sulphur Inst. J.* 6, 5-7.

Floate, M. J. S. (1970a). *Soil Biol. Biochem.* 2, 173-185.

Floate, M. J. S. (1970b). *Soil Biol. Biochem.* 2, 187-196.

Floate, M. J. S. (1970c). *Soil Biol. Biochem.* 2, 275-283.

Floate, M. J. S. (1970d). *J. Br. Grassld Soc.* 25, 205-302.

Floate, M. J. S. and Torrance, C. J. W. (1970). *J. Sci. Fd Agric.* 21, 116-120.

Foy, C. D., Arminger, W. H., Briggle, L. W. and Reid, D. A. (1965a). *Agron. J.* 57, 413-417.

Foy, C. D., Burns, G. R., Brown, J. C. and Fleming, A. L. (1965b). *Proc. Soil Sci. Soc. Am.* 29, 64-67.

Freney, J. R. (1967). *In* "Soil Biochemistry" (A. D. McLaren and G. H. Peterson, eds), pp. 229-259. Marcel Dekker, N.Y.

Freney, J. R., Barrow, N. J. and Spencer, K. (1962). *Pl. Soil* 27, 295-308.

Fried, M. and Broeshart, H. (1967). "The Soil-Plant System in Relation to Inorganic Nutrition". Academic Press, New York and London.

Fried, M. and Shapiro, R. E. (1961). *A. Rev. Pl. Physiol.* 12, 91-112.

Fuller, W. H., Neilson, D. R. and Miller, R. W. (1956). *Proc. Soil Sci. Soc. Am.* 20, 218-224.

Gardiner, M. R. (1969). *Outl. Agric.* 6, 19-28.

Gerdemann, J. W. (1968). *A. Rev. Phytopathology*, 6, 397-418.

Gillard, P. (1967). *J. Aust. Inst. agric. Sci.* 33, 30-34.

Glöbel, G. and Nilsson, N. (1933). *Svenska Betes-o. Vallför Årsskr.* 15, 159-173.

Greenway, H. and Gunn, A. (1966). *Planta*, 71, 43-67.

Greenway, H. and Pitman, M. G. (1965). *Aust. J. biol. Sci.* 18, 135-147.

Grunes, D. L., Stout, P. R. and Brownell, J. R. (1970). *Adv. Agron.* 22, 331-374.

Handreck, K. A. and Godwin, K. O. (1970). *Aust. J. agric. Res.* 21, 71-84.

Hansard, S. L. and Mohammed, A. S. (1968). *J. Nutr.* 96, 247-254.

Hansard, S. L. and Mohammed, A. S. (1969). *J. Anim. Sci.* 28, 283-287.

Harley, C. P., Moon, H. H. and Regeimbal, L. O. (1951). *Proc. Am. Soc. hort. Sci.* 57, 17-23.

Healy, W. B. (1970). *In* "Transactions of 9th Int. Congr. of Soil Science, Vol. III," pp. 437-445.

Herriot, J. B. D. and Wells, D. A. (1963). *J. agric. Sci., Camb.* 61, 89-99.

Hilder, E. J. (1966). *Proc. X Int. Grassld Congr.* 977-981.

Hogan, A. G. and Nierman, J. L. (1927). U. of Mo. Agr. Exp. Sta. Res. Bull. 107.

Hungate, R. E. (1966). *In* "The Rumen and Its Microbes." Academic Press, New York and London. 533 pp.

Jackman, R. H. (1964). *N. Z. Jl agric. Res.* 7, 445-471.

Jackson, R. M. (1965). *N. Z. Jl agric. Res.* 8, 865-877.

Jenny, H. (1941). *In* "Factors of Soil Formation." McGraw-Hill Book Co., N.Y.

Jenny, H. (1961). *Proc. Soil Sci. Soc. Am.* 25, 385-388.

Johnson, N. M. (1971). *Ecology*, 52, 529-531.

Johnstone, F. E., Jr., Cobb, C. and Carter, H. S. (1968). *Georgia Agr. Exp. Stn. Res. Bull.* 31, 26 pp.

Johnstone-Wallace, D. B. and Kennedy, K. (1944). *J. agric. Sci., Camb.* 34, 190-197.

Jones, O. L. and Bromfield, S. M. (1969). *Aust. J. agric. Res.* 20, 653-663.

Jordan, H. V. (1964). U.S. Dept. Agric. Tech. Bull. 1297, U.S. Gov. Printing Office, Washington, D.C., pp. 45.

Kline, J. R. (1969). *In* "The Grassland Ecosystem. A Preliminary Synthesis" (R. L. Dix and R. E. Beidleman, eds), pp. 71-88. Range Sci. Dept. Sci. Series No. 2, Colorado State Univ., Ft. Collins, Col.

Larson, S. (1967). *Adv. Agron.* 19, 151-210.

Lazenby, A. (1969). *In* "Intensive Utilization of Pastures" (B. F. James, ed.), pp. 105-124. Angus and Robertson, Ltd., Sydney.

L'Estrange, J. L. (1970). *Irish J. agric. Res.* 9, 161-178.

Lewis, J. K. (1969). *In* "The Ecosystem Concept in Natural Resource Management" (G. M. Van Dyne, ed.), pp. 91-187. Academic Press, New York and London.

Likens, G. E., Bormann, H. F., Johnson, N. M., Fisher, D. W. and Pierce, R. S. (1970). *Ecol. Monogr.* 40, 23-47.

Lipman, J. G. and Conybeare, A. B. (1936). *New Jersey Agric. Exp. Stn. Bull.* 607.

Loomis, R. S. (1969). *Hortscience* 4, 14-16.

Lotero, J., Woodhouse, W. W., Jr. and Petersen, R. G. (1966). *Agron. J.* 58, 262-265.

Macfadyen, A. (1961). *Ann. appl. Biol.* 49, 215-218.

MacLusky, D. S. (1960). *J. Br. Grassld Soc.* 15, 181-188.

McAuliffe, C., Peech, M. and Bradfield, R. (1949). *Soil Sci.* 68, 185-195.

McKell, C. M. and Williams, W. A. (1960). *J. Range Mgmt* 13, 113-117.

McLachlan, K. D. (1968). *Aust. J. exp. Agric. Anim. Husb.* 8, 32-39.

McLachlan, K. D. and Norman, B. W. (1966). *Aust. J. exp. Agric. Anim. Husb.* 6, 22-24.
Malone, C. R. (1970). *J. appl. Ecol.* 7, 591-501.
Marsh, R. and Campling, R. C. (1970). *Herb. Abstr.* 40, 123-130.
Matocha, J. E., Rouquette, F. M., Jr. and Duble, R. L. (1971). *Agron. J.* (In press).
May, P. F., Till, A. R. and Cumming, M. J. (1971). *J. appl. Ecol.* (in press).
May, P. F., Till, A. R. and Downes, A. M. (1968). *Aust. J. agric. Res.* 19, 531-543.
Metson, A. J. and Hurst, F. B. (1953). *N.Z. Jl Sci. Technol.* 35A, 327-359.
Miller, M. F. and Krusekopf, H. H. (1932). Mo. Agr. Expt. Sta. Res. Bull. 177, 32 pp.
Morris, H. D. (1968). *Agron. J.* 60, 518-521.
Mott, G. O., Eddleman, B. R. and Timm. D. H. (1969). *Proc. Soil Crop Sci. Soc. Fla.* 29, 238-253.
Mulder, E. G., Lie, T. A. and Woldendrup, J. W. (1969). In "Soil Biology Reviews of Research" pp. 188-192. UN Educational, Scientific, and Cultural Organization, Paris.
Munson, R. D. and Nelson, W. L. (1963). *J. agric. Fd Chem.* 11, 193-201.
Nicholas, D. J. D. (1963). In "Plant Physiology, A Treatise, Vol. III. Inorganic Nutrition of Plants" pp. 363-447. Academic Press, New York and London.
Nicolson, A. J. (1970). *Soil Sci.* 110, 345-350.
Norman, M. J. T. and Green, J. O. (1958). *J. Br. Grassld Soc.* 13, 39-45.
Odum, E. P. (1969). *Science, N.Y.* 164, 262-270.
Odum, E. P. (1971). In "Fundamentals of Ecology, 3rd Edition." W. B. Saunders Co., Philadelphia.
Olsen, S. R. and Watanabe, F. S. (1970). *Soil Sci.* 110, 318-327.
Olson, O. E. (1967). In "Symposium: Selenium in Biomedicine" (O. H. Muth, ed.), The AVI Publishing Co., Inc. Westport, Conn., 297-312.
Oswalt, D. L., Bertrand, A. R. and Teel, M. R. (1959). *Proc. Soil Sci. Soc. Am.* 23, 228-230.
Ozanne, P. G. (1962). In "Transactions Joint Meeting Comm. II and IV of Intl. Soil Sci. (New Zealand)." 139-143.
Ozanne, P. G. and Howes, K. M. W. (1971). *Aust. J. agric. Res.* 22, 81-92.
Ozanne, P. G., Kirton, D. J. and Shaw, T. C. (1961). *Aust. J. agric. Res.* 12, 409-423.
Papp, L. (1971). *Acta zool. hung.* 17, 91-105.
Parker, M. B. and Harris, H. B. (1962). *Agron. J.* 54, 480-483.
Peperzak, P., Caldwell, A. G., Hunziker, R. R. and Black, C. A. (1959). *Soil Sci.* 87, 293-302.
Petersen, R. G., Lucas, H. L. and Woodhouse, W. W., Jr. (1956). *Agron. J.* 48, 440-444.
Petersen, R. G., Woodhouse, W. W., Jr. and Lucas, H. L. (1956). *Agron. J.* 48, 444-449.
Pierre, W. H. and Parker, F. W. (1927). *Soil Sci.* 24, 119-128.
Pomeroy, L. R. (1970). *A. Rev. Ecol. Systematics* 1, 171-190.
Reisenaur, H. M. (1966). In "Environmental Biology" (P. L. Altman and D. S. Dittmer, eds), pp. 507-508. Fed. Amer. Soc. Exptl Biol., Bethesda, Md.
Robinson, E. and Robbins, R. C. (1970). In "Global Effects of Environmental Pollution" (S. F. Singer, ed.), pp. 50-64. Springer-Verlag, N.Y.
Rosenfeld, I. and Eppson, H. F. (1964). *Wyoming Agri. Exp. Stn Bull.* 414.
Rouquette, F. M., Jr., Matocha, J. E. and Duble, R. L. (1971). *Agron. J.* (In press).

Russell, E. W. (1963). *In* "Soil Conditions and Plant Growth," 9th ed. John Wiley, N.Y.

Saunders. W. H. M. and Metson, A. J. (1971). *N. Z. Jl agric. Res.* 14, 307-328.

Sears, P. D., Melville, J. and Evans, L. T. (1953). *N. Z. Jl Sci. Technol.* 35, sec. A., suppl. 1, 77 pp.

Shrift, A. (1964). *Nature, Lond.* 201, 1304-1305.

Smith, D. W. (1970a). *Can. J. Soil Sci.* 50, 17-29.

Smith, F. E. (1970b). *In* "Ecological Studies 1. Temperate Forest Ecosystems. Vol. 1" (D. E. Reichle, ed.), pp. 7-18. Springer-Verlag, Berlin.

Spedding, C. R. W. (1970). *Proc. XI Int. Grassld Congr.* A126-A131.

Spedding, C. R. W. (1971). The Nutrient Cycle. *In* "Grassland Ecology" p. 103. Oxford U. Press, London.

Stevenson, F. J. (1967). *In* "Soil Biochemistry" (A. Douglas McLaren and G. H. Peterson, eds), pp. 119-146. Marcel Dekker, N.Y.

Stewart, B. A., Porter, L. K. and Viets, F. G. (1966). *Proc. Soil Sci. Soc. Am.* 30, 355-358.

Swaby, R. J. (1962). *In* "Transaction of the Joint Meeting Commissions IV and V of the Int. Soc. of Soil Sci.," pp. 159-172.

Tamm, C. D. (1958). *In* "Encyclopedia of Plant Physiology Vol. IV, The Mineral Nutrition of Plants" (W. Ruhland, ed.), pp. 233-242. Springer-Verlag, Berlin-Göttingen and Heidelberg.

Thomas, G. W. and Hipp, B. W. (1968). *In* "The Role of Potassium in Agriculture" (V. J. Kilmer, S. E. Younts and N. C. Brady, eds), pp. 269-291. Amer. Soc. Agron., Crop Sci. Soc. Amer., Soil Sci. Soc. Amer., Madison, Wisc.

Till, A. R. and May, P. F. (1970a). *Aust. J. agric. Res.* 21, 253-260.

Till, A. R. and May, P. F. (1970b). *Aust. J. agric. Res.* 21, 455-463.

Till, A. R. and May, P. F. (1971). *Aust. J. agric. Res.* 22, 391-400.

Timmons, D. R., Holt, R. F. and Latterell, J. J. (1970). *Water Resources Res.* 6, 1367-1375.

Ulrich, A. and Ohki, K. (1966). *In* "Diagnostic Criteria for Plants and Soils" (H. D. Chapman, ed.), pp. 362-393. Univ. of Calif., Div. of Agr. Sci.

Underwood, E. J. (1971). *In* "Trace Elements in Human and Animal Nutrition." Third Edition. Academic Press, New York and London.

Van Bavel, C. H. M. and Carreker, J. R. (1957). *Georgia. Agr. Exp. Stn. Bull.* N.S. 15.

Van Dyne, G. M. (1969). *In* "Grasslands Management, Research, and Training Viewed in a Systems Context." Col. State Univ. Range Sci. Dept. Sci. Series No. 3, 31 pp.

Voisin, A. (1963). "Grass Tetany". Thomas, Springfield, Ill.

Vose, P. B. (1963). *Herb. Abstr.* 33, 1-12.

Walker, T. W. (1962). *In* "Transactions of Joint Meetings of Comm. IV and V of the Int. Soc. Soil Sci." pp. 704-714.

Watkin, B. R. (1957). *J. Br. Grassld Soc.* 12, 264-278.

Weeda, W. C. (1967). *N. Z. Jl agric. Res.* 10, 150-159.

Welch, L. F., Wilkinson, S. R. and Hillsman, G. A. (1967). *Agron. J.* 59, 467-472.

Wheeler, J. L. (1958a). *J. Br. Grassld Soc.* 13, 196-202.

Wheeler, J. L. (1958b). *J. Br. Grassld Soc.* 13, 262-269.

Whitaker, R. H. (1970). *In* "Communities and Ecosystems. Current Concepts in Biology Series." The MacMillan Co., Collier-MacMillan Ltd., London.

Whitehead, D. C. (1964). *Soils Fertil.*, 27, 1-8.

Whitehead, D. C. (1970). *J. Br. Grassld Soc.* 25, 236-241.

Wild, A. (1958). *Aust. J. agric. Res.* 9, 193-204.

Wild, A. (1961). *Aust. J. agric. Res.* 12, 286-299.

Wilkinson, S. R., Welch, L. F., Hillsman, G. A. and Jackson, W. A. (1968). *Agron. J.* 60, 359-362.

Wilkinson, S. R., Stuedemann, J. A., Williams, D. J., Jones, J. B., Jr., Dawson, R. N. and Jackson, W. A. (1971). *In* "Livestock Waste Management and Pollution Abatement. Proceedings International Symposium on Livestock Wastes." ASAE Publ. No. 271, pp. 321-324. Amer. Soc. Ag. Eng., St Joseph, Mich.

Williams, C. H. (1962). *J. Aust. Inst. agric. Sci.* 28, 196-205.

Witkamp, M. (1969a). *Soil Biol. Biochem.* 1, 167-176.

Witkamp, M. (1969b). *Soil Biol. Biochem.* 1, 177-184.

Witkamp, M., Frank, M. L. and Shoopman, J. L. (1966). *J. appl. Ecol.* 3, 383-391.

Wolton, K. M. (1955). *J. Br. Grassld Soc.* 10, 240-253.

Wolton, K. M. (1963). *J. Br. Grassld Soc.* 18, 213-219.

Wolton, K. M., Brockman, J. S. and Shaw, P. G. (1970). *J. Br. Grassld Soc.* 25, 255-260.

CHAPTER 24

Changes Accompanying Growth and Senescence and Effect of Physiological Stress

C. J. BRADY

Plant Physiology Unit, C.S.I.R.O. Division of Food Research,
School of Biological Sciences,
Macquarie University,
North Ryde, Sydney, Australia

I. INTRODUCTION

Growth and development of a plant may conveniently be divided into phases of cell division and cell extension. In some, but by no means all cases, these phases are segregated into distinct zones within tissues and so may be studied separately. The concern of this chapter rests centrally with matters which influence plant composition. While such changes may be determined in the dividing cells they are mostly expressed during or following cell expansion. This being so, there is here little concern for those early phases of growth where the mass of tissue involved is small, and there is no consideration of growth by cell division.

Differentiation is an important part of growth and development, but there is little treatment in this chapter of the processes of cell and tissue differentiation. Rather is it the intention to describe those growth and age associated changes which modify the chemistry of the plant in a substantial manner, and to relate these to the environment. Changes within cells and tissues are a function of the development of the plant as a whole. Such changes are influenced by the content and distribution of plant-growth substances. The nature and some of the influences of these growth regulators are outlined in Section V.

Influences of age on plant development feature in a number of other chapters. Effects on protein components are considered in Chapter 2, Vol. 1, on structural carbohydrates in Chapter 4, Vol. 1 and on organic acids in Chapter 20. Changes in chlorophyll and in carotenoids during leaf senescence are described in Chapter 10, Vol. 1. Chapter 22 includes a consideration of the redistribution of nitrogen within the plant as older tissues senesce, and Chapter 27, Vol. 3 considers nitrogen redistribution during ageing of detached leaves. Effects due to water stress are recorded in Chapters 15 and 17 and the influence of mineral components on plant function at different growth stages is discussed in Chapters 12, Vol. 1, 18 and 19.

II. CHANGES ASSOCIATED WITH GROWTH

A. Growth by Cell Extension

The major increases in volume and surface area of a plant are a result of increases in the size of cells. For this increase to occur, the cells must be able to take up large amounts of water. Not only is the increase in cell volume a result of a vast increase in the content of water per cell, but the increases in length and breadth of the cells are dependent upon the wall or turgor pressure exerted by the protoplast (Lockhart, 1965; Cleland, 1971). It is not surprising then that extension growth is limited when the tissue water potential is lowered during periods of water stress (Boyer, 1970). During extension growth the osmotic pressure of the cell contents is maintained as inorganic ions and small organic molecules are accumulated. The amount of protoplasm increases and within it, there are increases in the numbers of mitochondria, plastids and other membranous organelles (Juniper and Clowes, 1965). Consequently, co-incident with the increases in cell size and fresh weight, large increases in the contents of dry matter and nitrogen per cell occur.

If the cell is to expand changes must occur in the cell wall, and a great deal of study has been made of these changes. The processes involved, however, are not yet well defined. A number of more or less distinct phases of wall extension are recognized, though there is no agreement on just how many.

The driving force for extension growth is the turgor pressure exerted by the protoplast. There is evidence that a certain critical pressure is needed before wall extension occurs, and in some circumstances at least, the rate of extension is linearly related to the turgor pressure in excess of the critical pressure. However, wall extension is not only a physical extension in response to an applied force; the temperature response of the process and a sensitivity to anoxia and to potassium cyanide indicate that biochemical processes—often described as wall loosening—are involved. Breaking and reformation of some covalent linkages are thought to be involved in cell wall extension, but the nature and number of the bonds involved in wall 'loosening" have not been determined (Lamport, 1970). Cell walls contain a number of enzymes which hydrolyse particular carbohydrates of the wall. The activity of some of these hydrolases have been shown to be increased by treatments which result in cell extension. However, these increased activities have not been causally related to wall extension. There is some evidence that the bonds involved in wall loosening are acid labile and alkaline stable, and linkages between arabinose and hydroxyproline residues in cell wall protein have been suggested in this regard. However, there is no conclusive evidence that such linkages are involved in the rigidity or in the loosening of the cell wall.

As the cell extends neither its thickness nor its density decreases. New material must be added to the wall. These additions may be to the inner surface of the wall (apposition) or within the existing wall (intussusception). They involve increases in both carbohydrate and protein components. The assembly of the carbohydrate and proteins occurs in the cytoplasm and the wall polymers are then secreted from the cell (Mühlethaler, 1967; Villemez *et al.*, 1968) (see also Chapter 4, Vol. 1). How the wall is finally assembled and made rigid is not understood. For growth of the cell and the cell wall to proceed there is a need for the continued synthesis of ribonucleic acid (RNA) and of protein; that is, when RNA and protein synthesis are inhibited growth ceases. It is a common observation that RNA and protein synthesis must continue if differentiation is to proceed. While this need is obvious where the products are accumulated, in other circumstances it is often not clear whether the need relates to the

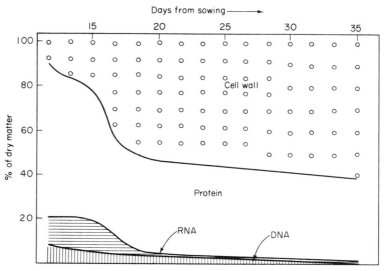

Fig. 1. Changing composition of the ethanol-insoluble portion of wheat leaves (from Williams and Rijven, 1965).

synthesis of proteins specifically required for the change, or reflects a more basic and less specific need for an integrated metabolism.

Apical tissues generally have a high protein content and a relatively low carbon to nitrogen ratio. Although in general chemical terms, differences between species are less pronounced in apical than in more mature tissues, specific differences are apparent as in those species with high concentrations of alkaloids in the apices. Growth brings with it a large increase in vacuolation, and the development of the structural system of the plant. Vacuolation contributes to the development of the biochemical differences which distinguish tissues and species in that it provides one mechanism by which large amounts of low molecular weight, biochemically active molecules may be present within the cells without distributing metabolism. The extension of the structural system broadens the carbon/nitrogen ratio considerably and is the sink into which a variety of carbo-hydrates is deposited. The increasing content of carbon-rich wall material relative to protein and nucleic acid in the developing wheat (*Triticum vulgare*) leaf is shown in Fig. 1. Associated with exten-sion growth of the leaves, and, to a lesser extent of the stem, is chloroplast development. This development contributes largely to the content of lipid and protein within these tissues, and gives the potential for starch accumulation to the leaves.

B. Chloroplast Development

Plastids have some autonomy within the cell. They contain DNA and this contains information for the synthesis of some plastid components. They have the capacity to replicate their DNA, to synthesize some RNA and to make some proteins (Kirk, 1970) (see also Chapter 2, Vol. 1). Included in the information in their DNA is that for the RNA of the plastid ribosomes. Included in the proteins whose assembly depends upon functional plastid ribosomes are a number of the enzymes concerned in carbon fixation (see Chapter 16).

Definitive evidence as to the developmental functions the chloro-plasts perform independently of the nucleus is currently lacking. That the nucleus contains information for chloroplast components is known, and there is evidence that certain portions of chloroplast assembly are dependent on functional cytoplasmic ribosomes. In-tegrated metabolism by nuclear cytoplasmic and chloroplast com-ponents appear to be involved in chloroplast assembly.

Most studies of chloroplast development have attempted to segre-gate chloroplast from cell growth. Most frequently this has involved growth of seedlings in the dark followed by measurements of chloroplast components or functions after the etiolated leaves were irradiated. In the dark-grown tissues proplastids are present and these complete their development only when adequate irradiation is provided.

These studies have shown that the extent of chloroplast develop-ment in the dark varies considerably between species, but that there is a general correlation between the degree of cell expansion, and the extent of proplastid development that occurs in the dark. Thus in pea (*Pisum arvense*) apices there is little expansion growth in the dark, and the proplastids are undeveloped. In the bean (*Phaseolus vulgaris*) and in monocotyledons, leaves expand in the dark and the plastids in these show a development of plastid membranes and an accumulation of stroma enzymes. In peas (Graham *et al.*, 1968), in beans (Mego and Jagendorf, 1961) and in monocotyledons (Feiererabend and Pirson, 1966; Graham *et al.*, 1970) further development of the chloroplast occurs if the tissues are exposed for a brief period to low intensity red (660 nm) light.

Red light activates phytochrome—a protein with a tetrapyrrole chromophore—in the plant cells. This pigment can exist in two

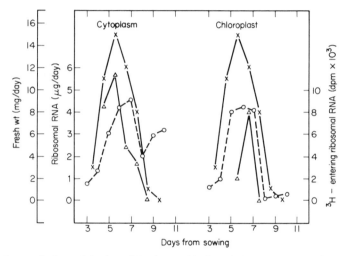

Fig. 2. Accumulation of fresh weight (x, mg/day) and ribosomal RNA (△, μg/day), and incorporation of ^{3}H-uracil into ribosomal RNA (o) of the cytoplasm and chloroplasts of wheat leaves growing under a 12 hours light, 12 hours dark daily regime (adapted from Patterson and Smillie, 1971).

forms. After exposure to red light the pigment has an absorption maximum about 730 nm. Exposure to far-red light (730 nm) changes the absorption spectrum so that there is a peak at 660 nm, but no absorption at 730 nm (Siegelman and Butler, 1965). The far-red absorbing form of the molecule induces various morphological and biochemical effects in plants and chloroplast development is among these. There is not a great deal of information on the effect of different daylight conditions in phytochrome inductions, but it seems likely that phytochrome activation of chloroplast development does occur within the apices of plants developing in daylight.

Consequently, plastids may develop with other cell functions in apical leaves, so that emerging leaves contain highly differentiated plastids of low chlorophyll content. Thus wheat leaves develop their full content of ribosomal-RNA in the dark, and a near normal content of soluble protein after phytochrome activation. Nonetheless, during normal growth chloroplast development does seem to lag behind cytoplasmic development. During normal development of the primary wheat leaf, the peak rate of increase in chloroplast ribosomal-RNA occurs after the period of maximum increases in cytoplasmic ribosomal RNA and in leaf area (Fig. 2). Thus the younger leaves have a higher ratio of cytoplasmic to chloroplast ribosomal-RNA.

Maturation of the chloroplast occurs in white light, and the synthesis of chlorophyll requires continuous illumination. For chlorophyll formation the active photoreceptor is protochlorophyllide (Rhodes and Yemm, 1966). As in many species the chloroplasts can develop considerably in the dark, when etiolated tissues are illuminated, photosynthetic capacity is related to greening only during the very early phase of chlorophyll accumulation (Tolbert and Gailey, 1955). Similarly, during the normal development of leaves, photosynthetic capacity is not closely related to the chlorophyll content.

III. CHANGES ASSOCIATED WITH AGE

A. Preceding Senescence

The growth processes involve increases in many cell components. Some of these, such as the glycolytic enzymes, the mitochondria and the ribosomes contribute in a catalytic sense to growth. The demand for energy from respiration is greatest during the early stages of growth, and as the growth rate declines, the respiratory rate and the activity of glycolytic enzymes decrease (Smillie, 1962). The ribosomal system functions catalytically in protein synthesis and the rate of growth of the ribosomal system declines appreciably before the rate of accumulation of protein falls (Fig. 3). Once net protein synthesis ceases the demands on the ribosomal system decrease, and so the content of ribonucleic acid (RNA) per cell may decline when expansion growth ceases (Williams and Rijven, 1965; Cherry, 1967; Patterson and Smillie, 1971). A decline in the content of RNA per cell precedes any loss of protein, and so the ratio of protein to RNA per cell becomes larger. A similar widening of the protein to RNA ratio with age is observed in root cells (Heyes, 1959).

The content of protein per cell increases during leaf growth (Fig. 1), and the increase is approximately proportional to the increment in cell size. When growth ceases the protein remains constant for a period which varies greatly with the species and the environmental conditions. There are, however, changes within the enzyme complement of the cell. The activities of some enzymes, for example some of those of the glycolytic sequence, are greatest at the time of maximum growth. Others, as with nitrate reductase of pea leaves

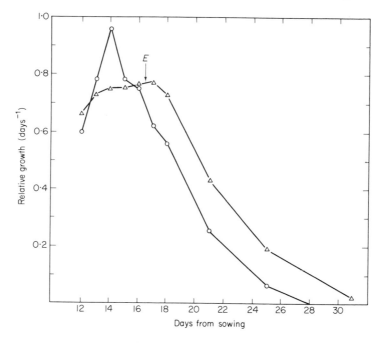

Fig. 3. Rates of accumulation of protein (△) and RNA (○) in the fourth leaf of wheat plants. E denotes the time of leaf emergence (from Williams and Rijven, 1965).

(Wallace and Pate, 1965), show maximum activity as the leaf reaches full expansion. Other enzymes, and particularly a number of hydrolytic enzymes, show maximum activity in the period following the completion of growth. These changes in enzyme activity have generally not been measured as changes in the amount of enzyme protein in the leaf. Evidence of this, which excludes an interaction with an age effect on enzyme recovery, is particularly difficult to obtain.

While non-growing tissues may retain a more or less constant protein content, protein synthesis is not arrested within them.

Protein replacement continues and proteins differ in their rates of turnover. Protein turnover is readily demonstrated as the passage of amino acids labelled with ^{14}C or with ^{15}N into the proteins. Estimates of the extent of protein turnover are, however, difficult to interpret. Amino acids exist in various more-or-less distinct pools within the cells, and it is not certain which of these pools contributes amino acids to protein in particular experimental conditions (Oaks and Bidwell, 1970). Again, when radioactive amino acids are introduced into protein, and their rate of loss measured, the randomness of protein breakdown and absence of amino acid recycling must be assumed. Estimates made by a number of different methods indicate a turnover rate of carbon in the proteins of non-growing leaves in the light to be within the range of 0·2 to 2·0% per hour. Most evidence is that turnover rates decline in the dark and with leaf age, although there are claims that the rate of turnover of protein in senescent barley leaves in the dark is particularly high (Atkin and Srivastava, 1970). Intracellular turnover of proteins is not confined to plants, but occurs also in animal tissues and in bacterial cultures in stationary phase.

RNA molecules also show turnover. In the case of messenger RNA this may be very rapid and provide a primary method for regulating cell function. But the results of Patterson and Smillie in Fig. 2 provide evidence that ribosomal-RNA also shows turnover. In these experiments the rate of incorporation of labelled precursors into ribosomal-RNA showed little correlation with the net increment in RNA, and at least from day 5 to day 7 appreciable turnover of both cytoplasmic and chloroplast ribosomal-RNA can be inferred. When growth of the leaf ceased, turnover of chloroplast ribosomal RNA was no longer apparent, although replacement of cytoplasmic ribosomal RNA was obviously continuing.

The major changes in composition in non-senescent expanded leaves result from the temporary storage of carbohydrates. This storage, as sucrose, fructosan or starch, may cause a progressive increment in the dry matter content of leaves after expansion growth is complete, and a gradual widening of the carbon to nitrogen ratio. Carbohydrate accumulation in leaves will vary greatly with the rate of plant development and with those environmental conditions which influence photosynthesis. However, it is not uncommon to find that starch accumulation, which is usually limited in newly-expanded leaves, increases progressively in older leaves decreasing only in those leaves which have become senescent.

The respiration rate is generally highest in young tissues at the stage of maximum growth. The rate declines later, but the rate of decline in non-growing tissues is gradual. The net photosynthetic rate often changes more dramatically with leaf age. In most species this rate is highest at, or somewhat before, the completion of growth. In general net photosynthesis is not closely correlated with the level of chlorophyll. In some species the photosynthetic rate declines rather sharply once expansion growth has ceased. In other cases, the decline after the completion of growth is gradual and net fixation per unit of leaf area declines rapidly only as senescence commences.

Although there is some evidence of a decline during the period following growth cessation in the activity (Smillie, 1962) or amount (Dorner *et al.*, 1957) of the enzymes involved in carbon fixation, the factors limiting the photosynthetic rate are not yet clearly defined. Physiological evidence (Woolhouse, 1968; Slatyer, 1970; Osman and Milthorpe, 1971) indicates that mesophyll resistance increases as leaves age, and the activity of the carboxylating enzymes is one component of this resistance.

When considering the significance of the declining rate of net photosynthesis, a distinction should be drawn between the observed rates for a leaf at some stage of the plant's development, and the real capacity for photosynthesis within the leaf. Even when measurements exclude interactions with light incidence and with the concentration of carbon dioxide within the leaf, the apparent net photosynthesis rate may be regulated by factors external to the leaf. There is much circumstantial evidence which links the rate of net photosynthesis with the level of assimilate utilization in the "sink" to which the leaf exports carbon. This evidence has often been interpreted as a negative feedback control on photosynthesis as a consequence of assimilate accumulation within the leaf. The evidence for this hypothesis has been critically reviewed by Neales and Incoll (1968).

An illustration of the type of evidence indicating "sink" assimilation regulation of photosynthesis in a source leaf is that of King *et al.* (1967) who studied the flag leaf of the wheat plant (Fig. 4). Two weeks after anthesis this leaf had a high rate of photosynthesis, and about 45% of newly fixed carbon was rapidly translocated to the ear. If the ear was removed, net assimilation in the flag leaf was about halved, but it could be raised again if the lower leaves on the plants were shaded, and these darkened leaves were then a "sink" for assimilates flowing from the flag leaf. As the grain filled in normal plants, the rate of photosynthesis declined; however, this diminished rate increased when photosynthesis in the ear was inhibited by treatment with 3-(3,4-dichlorophenyl)-1,1-dimethylurea (DCMU).

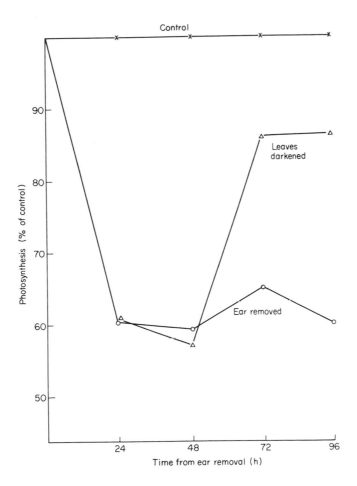

Fig. 4. Photosynthesis by the flag leaf of wheat plants as influenced by removal of the ear at time 0 hours (○). In one set of plants from which the ears were removed, the lower leaves were darkened at time 24 hours (from King *et al.*, 1967).

While such experiments do not establish that photosynthesis is limited by a build up in the leaf of assimilates otherwise transported, they do point to a readily reversible regulation of photosynthesis dependent on the interaction between "source" and "sink" of assimilates (see also Chapter 17).

B. Leaf Senescence

A satisfactory definition of senescence is not easily given, and for practical purposes it is perhaps most convenient to relate senescence and chlorophyll loss in leaves. The leaf changes thoughout its life, and from about the time when the growth rate declines, there are losses in efficiency in some physiological processes. These losses may be accepted by some as aspects of senescence. In many cases, however, the leaves maintain constant nitrogen, phosphorus and chlorophyll contents for some time after growth ceases. This period varies with the species, and with the development of the plant, which is influenced by environment, particularly by light quality, light intensity and nutrition. Eventually, however, the leaves become net exporters of mineral elements and their chlorophyll content declines. At this stage, the leaves are clearly senescent.

In most herbaceous species leaf senescence occurs in a sequential fashion and reflects competition between mature leaves and the regions of the plants where growth persists. Temperate deciduous trees show simultaneous leaf senescence in which the leaves respond to changes within the plant, these changes are imposed by day length and environmental temperature. Shoot senescence, associated with the ripening of fruits in monocarpic species is a further case in which leaf senescence is obviously related to internal factors in the plant.

While chlorophyll loss is a visible sign of senescence, degradation of other macromolecules is also extensive. Ribonucleic acid, protein, lipid and various carbohydrates are degraded. Net losses of RNA and protein may commence before the chlorophyll content declines. In the early literature, leaves in which there was a net loss of protein with associated catabolism of released amino acids were often described as "starving leaves". The implication was that amino acid residues were used as a source of carbon and energy when carbohydrate supply was limiting. This view of senescence is incorrect for there can be substantial breakdown of protein in leaves rich in starch and sucrose and capable of efficient photosynthesis.

Studies of the ultrastructure of senescing leaves have not shown significant changes which precede macromolecule loss measured chemically. At the stage when net losses of chlorophyll and protein can be recorded, local swellings of the lamellar membranes of the chloroplasts are apparent. Droplets, probably of lipid, accumulate within the chloroplasts and these may occur in chloroplasts still containing starch granules. Within the cytoplasm, parts of the endoplasmic reticulum swell and disperse at an early stage. These changes in the membranes of the chloroplast and cytoplasm occur while ribosomes and aggregates or ribosomes are still plentiful. Only in the later stages of senescence, when the plastid membranes are few and disperse, are any disorientations in the tonoplast, plasmalemma or mitochondria detectable (Barton, 1966). Ultrastructure studies suggest that during senescence, structural and storage entities and components associated with photosynthesis are catabolized, while those metabolic entities concerned in mobilization are preserved. If minerals and carbon are to be maximally recovered from older leaves, then cells must retain the capacity for synthesis of molecules suitable for transport and for secretion into conducting vessels. The persistence of active mitochondria in senescent cells is in line with such requirements.

C. Translocation from Senescent Leaves

The proportion of leaf nitrogen which is present in non-protein form normally increases as leaves senesce. While the spectrum of free amino acids present in leaves varies between species, the amides glutamine and asparagine, with γ-aminobutyric acid, glutamic acid, alanine and serine are among the more prominent in most species. When, in the senescent leaves, there is a release of a range of amino acids from protein, the released amino acids tend to be metabolized in such a way that the balance between the free amino acids is maintained. When translocation is inhibited, as in detached leaves, this metabolism results in an accumulation of a large amount of amide in the leaf, though the level of free amino acids also increases. In attached leaves, many of the free amino acid molecules enter the translocation stream without conversion to amide. As a consequence, a wider range of amino acids is apparent in the translocation stream of plants with a number of senescent leaves than is the case in less mature plants.

Relatively little is known of the factors regulating the movement of nitrogen from senescing grass leaves. While an enriched nutrient supply can delay the onset of senescence in wheat leaves, it does not reduce the proportion of nitrogen which is eventually exported (Williams, 1955). Thus factors within the plant eventually induce a flow of nitrogen from the older leaves even when a supply of minerals from the roots is maintained. One effect of this is to supply

the metabolic sinks—the growing points and the developing fruits—
with a more diverse range of protein precursors, since the amides and
the basic amino acids usually dominate root exports (Palfi et al.,
1966).

Nitrogen circulation in the field pea plant has been studied in
some detail (Pate, 1968). These studies reveal that amino acids
exported from the leaves supplement those normally translocated
from the roots so that from these sources combined, the growing
points receive a more balanced amino acid supply. The circulatory
system in the field pea involves the use in the root of carbon derived
from translocated sucrose for the synthesis of asparagine, glutamine,
homoserine and aspartic acid. These amino acids are then translocated to (1) growing points, (2) storage zones, especially in the
stem, and (3) to the leaves where they donate nitrogen to form
glycine, serine, alanine and the aromatic amino acids whose carbon
skeletons are derived from newly fixed carbon. Portions of the amino
acids synthesized in the leaves are exported to younger tissues where
they supplement the supply received more directly from the roots.
When the plants are utilizing nitrate nitrogen, much of the nitrate is
reduced in the root, but some is reduced in the leaves and supplies
nitrogen to the photosynthetically derived amino acids.

Leaves of a number of different species have been shown to
become net exporters of carbon when they have reached one-third to
one-half of full expansion (Doodson et al., 1964; Thrower, 1967).
Carbon movement has most often been measured in terms of
newly-fixed [14]C. When carbon movement has been computed from
dry matter changes (Hopkinson, 1964) substantially similar conclusions have been derived. Carbon export is greatest as the leaves reach
full expansion, and the rate of carbon translocation then decreases
with leaf age. Translocation often decreases more rapidly than does
net photosynthesis and so the ageing leaf increases in dry matter
content and widens in carbon to nitrogen ratio. While the declining
rate of translocation has sometimes been linked with increasing
callose deposition in the phloem, both the pattern and rate of
translocation are closely correlated with the stage of plant development, and there are experiments which demonstrate that both the
supply of assimilate in the leaf, and the demand elsewhere in the
plant are factors influencing carbon translocation.

While carbon translocation may decline as leaves age, remobilization of mineral elements is a feature of leaf senescence. Of the
maximum content, 90% of the nitrogen and phosphorus and 70% of

the potassium may eventually move from the leaves (Williams, 1955). Translocation is more limited when fruit development is prevented but, in these cases, translocation to stems, roots and younger leaves will occur.

The pattern of translocation varies between species and in response to environmental factors including nutrition. A new export of nitrogen and phosphorus may commence before leaves reach full expansion (Fig. 5), or the nitrogen and phosphorus contents of leaves may remain constant for a period after expansion growth ceases. The

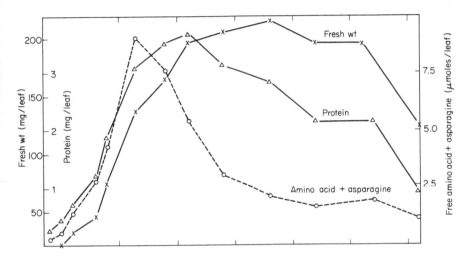

Fig. 5. Changes with time in the contents of protein and the major non-protein components of the fifth leaf of pea plants (from Pate, 1968).

nitrogen and phosphorus contents of leaves commonly decline before the potassium content falls.

While some mineral elements move readily from older to younger tissues, others are less mobile. Of the mineral cations, potassium is the most mobile, but sodium may also be redistributed from ageing leaves to shoots (Bernstein and Pearson, 1956). The alkaline earths and iron are of low mobility; for a long time calcium was considered to be fixed at the site of its initial disposition, but there is evidence that much of the leaf calcium is exchangeable (Millikan and Hanger, 1966); phosphate and chloride are highly mobile, but sulphate is less readily translocated (Biddulph et al., 1958) (see also Chapters 12, Vol. 1, 18, 19).

The ability of cells to accumulate ions is closely related to the growth processes, and declines when growth ceases. Changes in membrane properties in ageing cells (Eilam, 1965; Moore, 1966), a shift in intracellular distribution (Jacoby and Dagan, 1969) and the development of modified polarity interactions within the plant may all contribute to mineral mobilization, but the role of these processes has not been accurately defined.

D. The Senescence Process

While there is a great deal of information on the changes which occur in leaves as they undergo senescence, and there has been a great deal of experimental work on the intracellular processes which are responsible for the observed chemical changes, there is as yet little understanding of the type of regulatory functions which are involved in the initiation and regulation of the degradative changes in leaves. However, there is a general consensus of opinion that leaf senescence is a tightly regulated process, and that it is intimately coupled with the development of the plant as a whole. The older idea that senescence involves only degradative changes, and that the senescing leaves eventually starve as a consequence of a decline in respirable substrates is no longer accepted. Senescence is a stage of differentiation, involving anabolic as well as catabolic metabolism

So many of the changes in leaf senescence are in the function or composition of the plastid that it is a reasonable hypothesis that programmed senescence is confined to regulation of the chloroplast. As nuclear and cytoplasmic functions are involved in chloroplast development, so they may be in the maintenance and ultimate dismantling of the chloroplast. Nor need this dismantling necessarily be complete. In various situations including autumn leaves and developing or ripening fruits, chloroplasts are dedifferentiated to form chromoplasts (Harris and Spurr, 1969) and in some cases this process may be reversible. The chloroplast to chromoplast change involves an extensive dismantling of the lamellar structure and a hydrolysis of large amounts of stroma protein. But these changes stop short of complete disintegration of the plastid, and a new, regulated function is imposed. The selective nature of the hydrolytic processes involved in such cases, suggest that the machinery of the cell imposes limits on degradative changes and the evidence of the persistence of particular enzymes and of functional mitochondria in ageing leaves supports such a view.

Many of the studies of leaf senescence have utilized leaves detached from the plant. Detached leaves are freed from the correlative influences of the plant, and have advantages when treatments involving the introduction of solutions are involved. The detached leaves have commonly been incubated in darkness, and so the experiments have involved three treatments—ageing, detachment and darkness. In a few cases, the influence of separation from the plant on macromolecule changes in leaves have been evaluated with the general conclusion that while there is a similarity in the processes, there may be a considerable shift in the relative time scales. Thus attached wheat leaves retain starch in the plastids when protein breakdown is progressing, while detached leaves, even though held in continuous light, lose all their starch very rapidly and before there is much breakdown of protein. Likewise there is no doubt that the absence of light modifies leaf senescence. Some light effects appear to be dependent upon photosynthesis (Goldthwaite and Laetsch, 1967; Udvardy et al., 1967), but there are, too, effects of light in chloroplast senescence which are not mediated through carbon fixation (Haber et al., 1969).

There remain a host of theories relating to the onset of senescence. In no case is there conclusive evidence to support the theories, but a consideration of them provides information of some of the metabolic factors which change as leaves senesce.

Certain theories relate to carbon metabolism and rely upon a decreased coupling of respiration or a smaller ratio of reduced to oxidized coenzyme (NADH/NAD, NADPH/NAD) in leaves approaching senescence. These theses suggest that synthetic reactions are depressed in this situation and that senescence is a consequence of the resulting metabolic imbalance. Others relate a decline in the rate of synthesis of protein to a deficiency of particular amino acids for protein synthesis. This theory considers that as a net export of nitrogen from the developing leaves is established (see Fig. 5), the level of some free amino acids falls to a level at which protein synthesis is limited and that senescence follows. Such a mechanism cannot apply in detached leaves in which the level of each of the free amino acids increases as senescence is provoked. This thesis is a development of a number of proposals that senescence ensues as a consequence of a decline in the rates of synthesis of RNA and protein. Such proposals assume that the amount of any species of RNA or protein at any time rests upon a balance between its rates of synthesis and breakdown, and that synthesis and breakdown are independent processes. Hence, when the rate of synthesis declines, breakdown predominates, the amount of RNA and protein declines and senescence ensues.

If RNA or protein loss depends upon a decline in the rate of synthesis, then those macromolecules most susceptible to breakdown should be those synthesized most rapidly, or, conversely, those declining most rapidly should be those whose synthesis has declined most dramatically. But experiments reveal no correlation between protein loss and either initial turnover rate or decline in turnover rate (Tung and Brady, 1971), nor are protein or RNA species which have a zero rate of replacement in mature leaves, conserved during senescence. These and other experiments have led to the conclusion that the declining levels of RNA and protein in senescent leaves reflect changes in the activities of ribonucleases and proteases.

There is a great deal of evidence that ribonuclease activity, and the activity of soluble ribonucleases in particular, increases as leaves age, and when senescence is provoked by detaching leaves (Kessler and Engelberg, 1962; Bagi and Farkas, 1967; McHale and Dove, 1968).

However, evidence that links ribonuclease activity with the amount or function of physiologically-limiting RNA is lacking, and there are cases in which a net synthesis of RNA occurs as RNase activity increases. Likewise there is no evidence that protease activity increases before or as net protein hydrolysis is initiated (Anderson and Rowan, 1965). Rather it seems that the potential for RNA and protein breakdown is at all times high, and that it is the regulation of these activities which are of concern. In this regard, Shibaoka and Thimann (1970) propose that an increase in the level of free serine in leaves is responsible for activating proteases, while others invoke lysozyme function (Balz, 1966).

There are common features to many of these theories, but there is as yet no firm foundation for a unifying hypothesis. Such a hypothesis must take regard for the facts that senescence proceeds in an orderly sequential fashion with the conservation of certain intracellular functions and structures, that a wide variety of cell proteins are degraded, but degradation is not related to declining synthesis, and that cells remain capable of RNA and protein synthesis at advanced stages of senescence. That inhibitors of RNA and protein synthesis inhibit the onset or progress of senescence suggests that macromolecule synthesis is involved in the senescence process but not necessarily that macromolecules concerned uniquely with senescence are produced at this time. Nuclear control is certainly suggested and chromatin bound nucleases which increase during senescence may be of significance (Srivastava, 1968).

IV. EFFECTS OF PHYSIOLOGICAL STRESS

The growth and the composition of plants and plant parts is under the influence of a number of interacting environmental factors. Tissue response to environmental change varies with the stages of development of both tissue and plant. In the case of exposure to unfavourable environments, both the degree of change and the time of exposure will affect the response. This complex of interactions between environmental factors, time and plant ontogeny limits the general conclusion which can be drawn from the large literature on environmental influences on plant composition and function. Interpretations in metabolic terms are limited by our scant knowledge of the regulation of intracellular metabolism in cells in a favourable environment.

A. Water Stress

It was noted earlier (Section IIA) that extension growth is dependent upon a positive turgor pressure, and that growth ceases when a critical level of turgor pressure is lacking. There are many other plant functions which are impeded when the water content of the plant is reduced, but in very few cases is the parameter controlling the response known (Greenway, 1970). When the water to dry matter ratio in the tissue narrows, not only does the chemical potential of water fall, and the turgor, osmotic and matrix potential components of water potential vary (Slatyer, 1967), but the concentrations of solutes in the cells and the properties of protoplasm also change. Moreover, the response of individual cells or tissues is influenced by changes in other tissues within the plant, and strong correlative effects can be expected in a cell's response to water stress.

A shift towards the hydrolysis of macromolecules is commonly observed in water-stressed cells. While it is often not apparent whether this is due to a decrease in synthetic reactions or an increase in breakdown, there is evidence that the synthesis of starch, nucleic acid and protein is inhibited in water-stressed tissues. Nonetheless, macromolecules, whose content is not dependent on continued synthesis, are degraded in response to water stress in leaves. This, in turn, may result indirectly from a decline in protein synthesis if hydrolases are regulated by proteins whose concentration is maintained by continued synthesis. Proteins which are protease inhibitors are known, but a role for them in regulating protease function in green tissues has not been demonstrated. There is evidence that ribonuclease activity does increase in stressed leaves (Dove, 1967; Kessler, 1961), though the relation of this increase to RNA content or function is not understood. In other cases, no increase in protease activity has been measured although the protein content declines, and no increase in amylase or phosphorylase although the starch content declines. The potential activity of these enzymes in turgid tissues is normally sufficient to achieve a rapid breakdown of their substrates.

A decline in the contents of protein and RNA is not invariably seen in water-stressed tissues, though this may be the general response in the older leaves of a plant. Gates (1968) has found in *Lupinus albus* that there is no loss of protein from stressed apical leaves, although growth and protein synthesis are severely restricted. The nitrogen content of stressed tissues may increase as a percentage of fresh and dry weight when growth or assimilation is suppressed more than nitrogen uptake. Nitrogen often accumulates as nitrate; nitrate reductase is an inducible enzyme and its activity declines when protein synthesis is inhibited.

An inhibition of protein synthesis in water-stressed tissues (Ben-Zioni *et al.*, 1967), is reflected by a decrease in the proportion of RNA, which is present as polysomes (Clark *et al.*, 1964; Hsiao, 1970). In isolated mitochondria protein synthesis is more readily inhibited by an increase in the sucrose content of the medium than is substrate oxidation, or phosphorylation (Halder and Freeman, 1969). The relative susceptibilities of the plastid and cytoplasmic protein synthesizing systems in water-stressed cells are not known, but both development of apical tissues (Gates, 1968) and the chloroplast development in etiolated tissues are retarded when the water potential is lowered (Bourque and Naylor, 1971).

While protein and RNA synthesis are particularly susceptible to the water status of the tissue, amino acid synthesis is less drastically influenced (Barnett and Naylor, 1966). However, nitrate reduction is reduced, and amide formation may be curtailed. An accumulation of free proline is characteristic of tissues of low water potential (Morris *et al.*, 1969): accumulated proline is rapidly catabolised when the water potential rises. There are claims (Tyankova, 1966) that free proline in wheat plants contributes to drought tolerance.

Associated with the decrease in starch synthesis in water-stressed tissues is a diversion of carbon to sucrose synthesis. Sucrose may accumulate in leaves, stems or roots for its utilization in growth processes is normally impaired in stressed plants (Hartt, 1967; Wardlow, 1969; Boyer, 1970). However, grain filling continues in plants in which photosynthesis and translocation are depressed because of water stress (Wardlaw, 1967). The extent and sites of sucrose and, in grasses, fructosan accumulation is very dependent on the stage of development when stress occurs.

While there is evidence that translocation of carbon and mineral elements occurs more slowly in water-stressed tissues, assimilate and mineral translocation do continue. Rate of flow in the conducting vessels is not slowed in stressed plants, but secretion into them occurs more slowly (Wardlaw, 1967; Greenway and Klepper, 1968). In algae (Greenway and Hiller, 1967) and in non-vacuolated root cells uptake of organic molecules is reduced when the water potential is lowered, but vacuolated cells do not respond similarly (Greenway, 1970). In tomato plants at least, the supply of phosphorus to the shoots is drastically reduced by moderate water stress (Gates, 1957; Greenway et al., 1969).

Photosynthesis and transpiration often decline to comparable extents in water-stressed plants and this suggests that stomatal behaviour is controlling both functions (Brix, 1962; Boyer, 1970). Carbon fixation, however, is often not limited by stomatal resistance, and reduced photosynthesis in stressed leaves persists when the carbon dioxide concentration is raised, and light is saturating (Wardlaw, 1967). Photosynthesis also declines with the osmotic pressure of the cell sap of liverwort thalli in which no stomatal control exists (Slavik, 1965). There is a little evidence to suggest that chloroplasts from water-stressed plants are deficient in Hill activity and photophosphorylation. While this evidence is at present insufficient to relate to plant response, there are strong indications that photosynthesis is inhibited in water-deficient cells by factors additional to carbon dioxide supply.

Respiration has generally been found to fall gradually with the water content of plants, though short term observations may reveal a period of increased respiration when stress is applied suddenly. This short burst of increased respiration may result from cell damage resulting from the experimental procedures rather than from the influence of cell water status on respiration rate. The respiration of isolated mitochondria is lowered when they are placed in media of high salt or sucrose content (Flowers and Hanson, 1969; Campbell *et al.*, 1972) but, except in extreme cases, oxidation remains coupled to respiration. Mitochondria recovered from water-stressed tissues also have oxidation coupled to phosphorylation. Nonetheless, there is evidence that phosphorylation is depressed in stressed tissues (Henckel, 1964; Zholkevich and Rogacheva, 1968), but there is no information on the mechanisms involved. Efficient synthesis and translocation of sucrose in the stressed leaf cells is evidence of available energy in these cells.

Many of the responses to water-stress are parallel to those of senescence, and, in many tissues, water stress induces premature senescence rather than arrested development. This is particularly so in more mature tissues, and these are less likely to recover from a period of stress. This may be a consequence of a re-established polarity in the plant. Apical regions are tolerant of drought in the sense that, although their development and metabolism are inhibited, they recover more readily than do mature tissues, when water availability improves. The drought tolerance of apical tissues appears to derive from their non-vacuolated condition (Milthorpe, 1950; Greenway, 1970).

B. Nutrition

The mineral composition of plants, and the biochemical roles of mineral elements within plants are discussed in Chapters 12, Vol. 1, 18 and 19. Because of the strong interactions between groups of mineral nutrients e.g. between potassium, calcium and boron, because minerals affect catalytic components of the plants, and because the pattern of plant development is altered when mineral components are limiting, changes in plant composition are not confined to those substances which themselves contain the limiting element. This is well illustrated in the case of potassium. Potassium influences the activities of a large range of enzymes (Evans and Sorger, 1966), and is also involved with magnesium in ribosome stability. In potassium deficiency, the contents of sucrose and reducing sugars, and of free amino acids and amides within leaves are high. These changes are the consequence of reduced rates of sugar and amino acid utilization coupled with reduced translocation (Amir and Reinhold, 1971). An accumulation of agmatine and putrescine is also a feature of potassium deficiency, and this may be a reaction to lowered pH (Smith, 1968). Although zinc is present in a number of enzymes, the major effect of zinc deficiency in plants may be exercised through an inhibition of auxin synthesis. In zinc deficient plants, differentiation is retarded, and plastid structure is modified.

The pattern of response to a limiting nutrient is very much influenced by the mobility of the element in the plant. With the minerals which are readily retranslocated—phosphorus, nitrogen and potassium—early formed leaves have no symptoms of deficiency. As the plant develops, minerals move to the developing leaves and the older leaves become deficient. When the deficient mineral is not readily translocated, as with calcium, iron and sulphate, symptoms appear in the younger leaves of plants whose older leaves may appear normal.

Chlorosis is a symptom common to most mineral imbalances. The pattern of chlorosis on the leaf and through the plant varies, and there is a little evidence that chloroplast structure is influenced in different ways by particular deficiencies (Machold, 1971). The fact that chlorosis is so generally a deficiency symptom may indicate the range of cell functions which are involved in chloroplast formation, or it may be related to the fact that within the cell, the highest concentrations of many minerals—potassium, iron, magnesium, nitrogen—are within the chloroplast. When an element is limiting within a cell, some priority between the various components influenced can be expected; for example, in iron deficiency, the content of the non-haem iron, chloroplast protein, ferredoxin, is drastically reduced while the content of cytochrome oxidase is scarcely influenced. The factors which regulate priorities in limiting situations are not defined (Price, 1968).

As chloroplast senescence is hastened when the supply of key minerals is sub-optimal, so it may be delayed when these are supplied in large amount. A plentiful supply of phosphorus and nitrogen delays senescence in the leaves of annual plants, but the older leaves do eventually senesce. The percentage export of nitrogen and phosphorus from senescing leaves is as great when these elements are available in luxuriant amount, indicating the importance of internal factors in the plant in regulating senescence and mineral translocation (Petrie, 1937). How the internal regulation is modified by nutrient supply so as to delay leaf senescence is not understood.

When the nitrogen supply is plentiful, the protein content of leaves is increased, and only when very large amounts of nitrogen are present does the proportion of non-protein, and particularly of amide, nitrogen increase. From limited evidence, it seems that the content of most proteins in leaves is increased, and there is no indication of specific storage proteins.

The growth of plants is depressed when the root medium has a high content of sodium chloride (Hayward and Bernstein, 1958). Growth depression may be a consequence of a lowered water potential, a modified internal ionic environment, or a combination of both. While plants vary in their capacities to achieve internal osmotic adjustment in these conditions (Gale *et al.*, 1967), in most plants the high internal ion concentration has the greater influence on growth and metabolism (Slatyer, 1967). Nonetheless, the responses to salinity are in many respects similar to those of water stress. Photosynthesis, and RNA and protein synthesis are depressed, and free amino acids and especially proline accumulate. There are, however, specific responses to a high salt content, and the response to sodium chloride differs from that to sodium sulphate (Kahane and Poljakoff-Mayber, 1968). Sodium chloride has been shown to inhibit the pentose phosphate pathway of carbohydrate oxidation, to modify the spectrum of isozymes of malate dehydrogenase, and to alter the pattern of protein synthesis (Hason-Porath and Poljakoff-Mayber, 1969). Some plants become tolerant of saline conditions by adjustments to metabolism. The nature and mechanisms of these adjustments are of great interest but are not presently understood.

C. Other Environmental Factors

Of the array of environmental factors which influence the growth and composition of plants, few can be considered here. The influence of modifications made by man as with spray residues or atmospheric fluoride, ozone or peroxyacetyl nitrate has been reviewed recently by Treshow (1970). The effects of light quality and intensity is considered in Chapter 17 of this book, while the role of phytochrome has been mentioned (Section II B), and some effects of daylength were reviewed recently (Evans, 1971).

Effects of high temperature appear to be on the integrated plant response; thus polysaccharide and protein hydrolysis in cotyledons is inhibited at temperatures much below those needed to inhibit the hydrolysing enzymes, and the primary site of inhibition is the shoot axis and not the cotyledons themselves. In the field, inhibition by high temperature will interact with that due to moisture stress, while reduced growth with low soil temperatures may be primarily due to a reduced rate of mineral uptake. A range of plants is damaged when they experience low, but non-freezing temperatures for some days. Raison and Lyons (1970), have concluded that damage in the susceptible species results from changes at low temperature in the mitochondrial membranes. These changes are dependent on the composition of the lipids of the membranes, and other membranes of the cell as well as the mitochondrial membranes are probably involved. Freezing injury in cells is of a quite different nature (Mazur, 1969).

D. Disease

The response of host cells to infection varies immensely according to the host and its environment, and to the nature of the invading organism. The range of response varies from that of cells hosting viruses, when carbon metabolism and the protein synthesizing system of the host are utilized in the synthesis of virus molecules, to that of resistant tissues, which isolate infected regions by a diversion of carbon to form lignin around about infection sites. When galls are formed, increases in the amounts of nucleic acid and protein occur, and there are cases in which starch accumulation occurs in response to specific infections. In contrast, an increase in the permeability of cell membranes associated with leakage of cell contents, hydrolase action and a higher respiration rate are symptoms of other infections. These changes may occur in response to diffusable toxins in cells distant from fungal hyphae. Premature senescence in cells distant from the zone of infection commonly occurs, yet in tissues immediately adjacent to infected cells, photosynthesis may increase in response to an increased "sink" in the cells supporting microbial growth.

Responses to infection are often similar to responses to mechanical injury. Ethylene production (see Section V, E) and the respiration rate increase, and the proportion of carbon moving through the pentose phosphate pathway also increases. The activities of the enzymes, phenylalanine ammonia lyase, phenolase, peroxidase and in grasses, tyrosine ammonia lyase, increase. In infection, higher contents of phenylalanine and tryptophane may occur before infection is obvious, and before there is a net loss of protein from the tissue. Such changes have a number of effects on metabolism. Phenylpropane carbon skeletons are provided for the synthesis of hydroxylated cinnamic acid derivatives, lignin and the flavanoid phytoalexins, such as pisatin and phaseolin. Quinone synthesis increases with phenolase activity and an increased supply of hy-droxylated phenols, and shifts in the concentrations of such com-pounds as p-coumaric, caffeic and ferulic acids alter the regulation of peroxidases thus modifying the synthesis and catabolism of indoleacetic acid (see Section V, A).

V. THE ROLE OF PLANT HORMONES

At several points in this chapter, emphasis has been placed on the internal regulation of organ response in plants. While we are far from understanding the interactions of organs within the plant, we recognize the regulatory role of groups of plant hormones in this regard. Responses within cells, and interactions between organs depend, at least in part, on their contents of auxins, gibberellins, cytokinins and abscisic acid, and on their production and content of ethylene. Responses depend on the relative content of each of these growth regulators both within and between tissues, and also on the stage of development of the organ considered, and thus on the history of contact with the hormones.

The term "plant hormone" has been assigned to the plant growth regulators by derivation from the hormones of animals. In the latter case, the hormones are produced in one tissue and exert their effect after translocation to other tissues. While translocation—or, with ethylene, diffusion—is involved in plants, the plant hormones are also active in the cells in which they are produced, and often the endogenous compounds have the major influence. Besides the five groups of plant hormones which are considered below, there are a number of other chemicals with growth regulating properties, especially as growth inhibitors. Gallic acid, p-hydroxybenzoic acid, coumarin and a number of related cinnamic acid derivatives, and the flavonoids, naringenin and phloridzin are included among these (Kefeli and Kadyrov, 1971).

Applications of plant hormones produce a diversity of responses according to the tissue treated, and its developmental stage. The response to one hormone in one species may be measured in terms of water uptake, cell division or cell elongation, stomatal movement, plastid development or the production of a particular enzyme. Again, any one response may be influenced by a number of hormones. Thus stomates may open in response to cytokinins or gibberellic acid, and close in response to auxins or abscisic acid, and the magnitude of the response will vary with leaf age (Tal and Imber, 1970). Leaf senescence is delayed in particular species by cytokinins, by gibberellins or by auxins, and in other species by combinations of these; leaf senescence is accelerated by abscisic acid and ethylene. Observations of this type have led to the suggestion that auxins, gibberellins and cytokinins are positive regulators of development, and that abscisic acid and ethylene modulate their influences, but this reasoning is difficult to apply to some responses as to effects on water uptake and transpiration.

Responses to the plant hormones are usually inhibited when nucleic acid and protein synthesis are inhibited. This fact and the fundamental nature of many of the changes induced by hormone treatment have led many to the conclusion that plant hormones regulate gene expression in the cells. How directly the hormones interact with DNA, how the interaction between the different hormones is effected, and how, at a particular stage of cell ontogeny the portion of the genome to be activated or repressed is selected are unanswered questions.

A multitude of experiments describe tissue response to additions of plant hormones. From these the inference is often made that in the untreated tissue metabolism is regulated by the hormone which provokes a response. Thus, if senescence of a leaf is retarded when a cytokinin is added, the inference is that senescence is the result of a deficiency of cytokinin in the leaf. This may be so, but in a number of cases where the endogenous concentrations have been measured, the assumption has not been substantiated, and it is either the balance with other hormones in the tissues, or the ability, for some other reason, of the tissue to respond to added hormone which has changed.

A. Auxins

As the name implies, auxins were originally considered to be concerned primarily in growth, and particularly in cell elongation. Assay of auxins by their ability to promote the elongation of decapitated *Avena* coleoptiles is still widely practised, but auxins are now known to be involved in many other plant functions. Many synthetic auxins, for example, 2,4-dichlorophenoxyacetic acid (2,4-D) are known, but the two compounds recognized as natural auxins are indoleacetic acid (1) and indole-acetonitrile (2).

Both natural auxins are formed from L-tryptophan. The initial step in the formation of indoleacetic acid may be decarboxylation to

(1) Indole-3-acetic acid   (2) Indole-3-acetonitrile

tryptamine or transamination to yield indolepyruvic acid. In either case, further oxidation proceeds through indoleacetaldehyde. Kinetic experiments and the low activity or limited distribution of trypro-phane decarboxylase indicate that the more important route is through indolepyruvic acid. Indoleacetaldoxime is an intermediate in the conversion of tryptophan to indoleacetnitrile.

Indoleacetic acid is broken down by photo-oxidation and enzymic oxidation. Some isozymes of peroxidase are particularly active in the oxidation of indoleacetic acid. The level of auxin in cells is also lowered by conjugation to yield esters, glycosides or indoleacetyl aspartate, which are all inactive as auxins.

Synthesis of auxins exceeds degradation and inactivation in meri-stematic tissues, and the auxin content of stem and root apices, and developing leaves, flowers and fruits is high. Destruction of auxin is rapid in roots, older leaves and damaged tissues which have high contents of auxin-specific peroxidases. Auxin in tissues may be measured as that extractable from the tissue by solvents or that diffusing from the tissue. The diffusible auxin content often cor-relates well with the growth potential of the tissue.

Auxin has a regulatory function in cell elongation, in apical dominance, in metabolite movement and in responses to light and gravity. Other hormones interact in these functions, sometimes by stimulating the synthesis of auxin. High concentrations of auxin stimulate ethylene production by most tissues, and some responses to auxin may indeed be to the induced ethylene. However, at lower concentrations auxin responses are independent of ethylene.

B. Gibberellins

The gibberellins are a group of terpenoid derivatives, originally isolated from fungi, but now known to be widely distributed through higher plants. Two systems of nomenclature have been used for the gibberellins. In the older system, gibberellins were treated as deriva-tives of the gibbane (3) skeleton. More recently gibberellins have been described as derivatives of gibberellane (4). More than 30 naturally occurring gibberellins have been described, but gibberellin A_3, also known as gibberrellic acid (5), has been used most widely in physiological experiments.

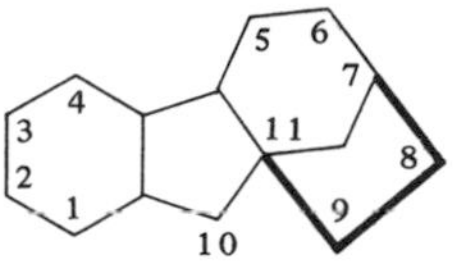

(3) Gibbane

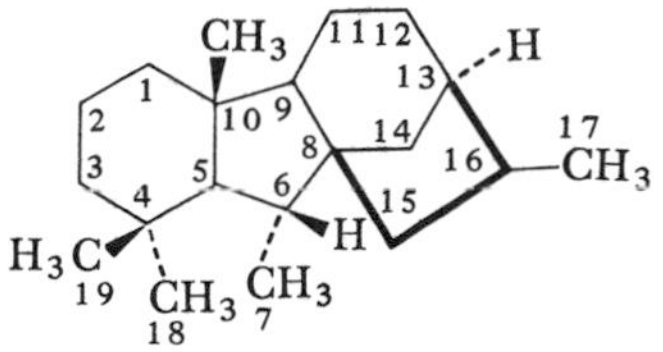

(4) Gibberellane

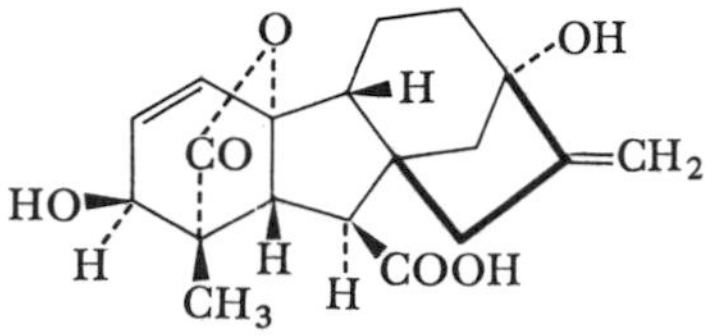

(5) Gibberellic acid (Gibberellin A_3)

In gibberellin biosynthesis, carbon from acetate and mevalonate flows to geranylgeraniol in a pathway common to the terpenes. The cyclic diterpenoid, kaurene is a key intermediate between the gibberellins and geranylgeraniol. The reactions from acetate to kaurene are catalyzed by soluble enzymes and require ATP and magnesium ions. Oxidation of kaurene is accomplished by components of the microsomal fraction which utilize pyridine nucleotides and oxygen. A little is known of the metabolic reactions between the various gibberellins, but scarcely anything is known of their physiological interactions. Bound gibberellins, some of which are glycosides, have been described.

The young leaves of the apical bud, the root tip and developing seeds are sites of gibberellin synthesis. Translocation occurs in the xylem and the phloem, and there is evidence of directed transport. The content of active gibberellins declines with leaf age, but the roles of declining synthesis, transport, catabolism and inactivation in this are not defined.

The growth retardants CCC (B-chlorethyltrimethyl ammonium chloride), AMO-1618 (2'-isopropyl-4'-(trimethylammonium chloride)-5'-methylphenyl piperidene carboxylate), and phosphon D (tributyl-2,4-dichlorobenzylphosphonium chloride) inhibit the synthesis of gibberellins by inhibiting the reactions between geranylgeranol and kaurene (Cathey 1964). These inhibitors may influence plant growth in other ways, but in some plants at least their inhibitory effect stems from an inhibition of the synthesis of gibberellins.

Gibberellin treatment increases stem elongation in many species and this effect in dwarf strains of peas and maize may be used as an essay. But, like other plant hormones, gibberellins influence many aspects of growth, development and metabolism. Their effect on the synthesis of enzymes in the aleurone layer of embryoless half seeds of barley has been studied in great detail. That gibberellin treatment induces *de novo* synthesis of a number of enzymes is proven, but the mechanism of regulation is not known.

C. Cytokinins

Plant cytokinins are a group of substances which promote cell division and regulate growth and metabolism in a manner analogous to kinetin (6) (6-furfurylaminopurine). A very large number of compounds with cytokinin activity have now been described. Many are purine derivatives, but others, like the substituted phenylureas, are not. Included among the cytokinins which have been well characterized are zeatin (7), a purine derivative isolated from corn

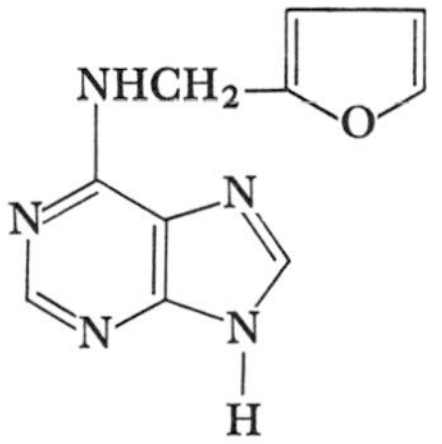

(6) Kinetin

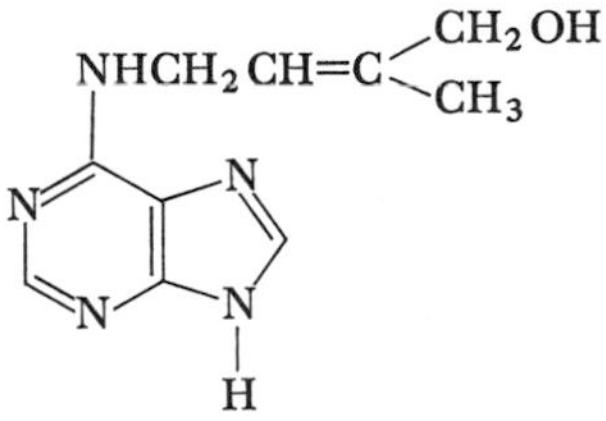

(7) Zeatin

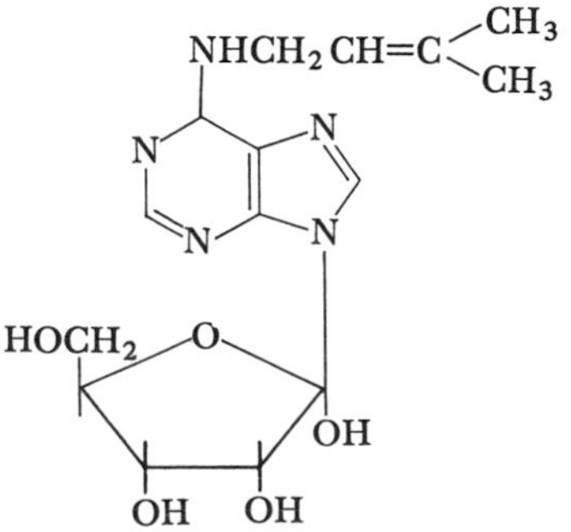

(8) Isopentenyladenosine

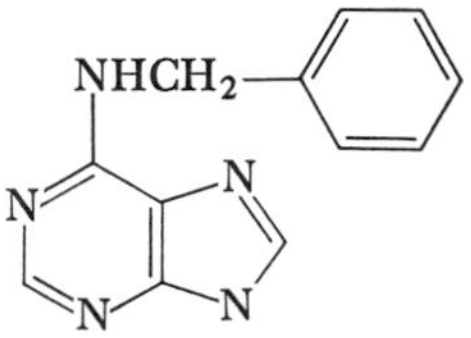

(9) 6-Benzylaminopurine

kernels and occurring as the free base, the riboside and the ribotide, and isopentenyladenosine (8) which has been isolated from a number of plants and which occurs also in transfer RNA in plants, animals and bacteria. The structure-activity relationships of plant cytokinins are discussed by Bruce and Zwar (1966) and Skoog and Armstrong (1970). In studies of plant development and metabolism in response to added cytokinins, kinetin and 6-benzylaminopurine (9) have been used most commonly.

Cytokinin activity was initially defined in terms of a stimulation of cell division, but they affect plant function and development in many ways. Effects on nutrient uptake, transpiration, organelle development and enzyme production have been described. An application of a cytokinin delays the senescence of the leaves of many species, and in a few cases the cytokinin effect is enhanced when a gibberellin is also applied. Many responses to cytokinins are enhancements of functions induced by other regulatory systems, and such an enhancement may limit, or appear to limit, further development (Mann *et al.*, 1967). This may mean that cytokinins influence translation rather than transcription processes. While there is some evidence to support this view, and the occurrence of cytokinins in transfer-RNA also suggests a role in the regulation of translation, the evidence is at best circumstantial and the mechanism of cytokinin action is not understood.

Cytokinins are produced in the roots and translocated to the shoots (Kende, 1965). Cytokinin activity in root exudate is lowered rapidly when the roots are water-stressed, and this has led to the thesis that the premature senescence of leaves in such plants is a consequence of a decreased supply of cytokinins from the roots (Itai and Vaadia, 1965). In the same way, the senescence of the older leaves of turgid plants could be a consequence of their inability to attract sufficient cytokinins from the roots in the face of increasing competition from younger leaves. Evidence for these views is incomplete and other factors are involved in the response to water stress. Direct measurements of the cytokinin content of ageing leaves have seldom been made. Because of the small amounts present, the measurements are difficult to make, and because of uncertainties as to which compounds are relevant, they are difficult to interpret. Whether roots are the only sites of synthesis of cytokinins is also uncertain. Young shoots and developing fruits are good sources of cytokinins, but whether this is from synthesis *in situ* or from accumulation is unknown.

D. Abscisic Acid

Of a number of abscisins or dormins which occur in plant tissues, the most widespread is (S) — (+) abscisic acid (10) (3-methyl-5   (1'-hydroxy-4'-oxo-2',6',6,'-trimethyl-2'-cyclohexen-1'-yl)-cis,trans-2,4-pentadienoic acid). Besides being active in promoting abscission, depressing growth, and imposing dormancy, abscisic acid promotes senescence and fruit ripening, influences stomatal closure and depresses enzyme production as in the giberellin stimulated production of α-amylase by barley seed aleurone cells.

(10) (S)-Abscisic acid

Responses to abscisic acid are subject to interaction with the other plant hormones, and in several bioassays there is a negative interaction between abscisic acid and either gibberellins or cytokinins. In senescent leaves, and particularly in water-stressed leaves there is a rapid accumulation of abscisic acid (Wright and Hiron, 1969; Mizrahi et al., 1970). Since added abscisic acid promotes leaf senescence, senescence may be thought to be a direct response to the accumulation of abscisic acid. However, the senescence response to added abscisic acid is dependent upon the stage of leaf and plant development, and leaves recovering from water stress retain high concentrations of abscisic acid. The latter compound is undoubtedly part of the syndrome of leaf senescence, but it is not a sole causative agent.

E. Ethylene

Of the known growth regulators, ethylene (C_2H_4) is the most simple chemically. Other unsaturated hydrocarbons, for example propylene, have some activity, but ethylene is by far the most active of the series. Concentrations of ethylene above 0·1 parts per million of air produce a range of responses, including the induction of fruit ripening, stem thickening, shoot epinasty, seed germination, regulation of flowering in monoecious plants, leaf abscission and the inhibition of the polar transport of auxin (Pratt and Goeschl, 1969). There is doubt as to which of the cell functions involved in these responses is normally regulated by ethylene. Most cells produce some ethylene, and ethylene production can be increased in a number of ways, for example, in response to injury, to water stress or to various chemicals including auxins, gibberellins, cytokinins and abscisic acid. Ethylene produced in such circumstances may provoke a physiological response, such as leaf or fruit abscission. However, some responses to ethylene occur without any increase in ethylene production, but reflect an increased susceptibility of the tissue to the regular endogenous concentration. When ethylene has a regulatory role in unstressed tissues, it may function in this manner. Since increasing the content of auxins or cytokinins in tissues decreases the responses to added ethylene, an increased response to a constant concentration of ethylene may result from a declining content of these other plant hormones.

REFERENCES

Amir, S. and Reinhold, L. (1971). *Physiologia Pl.* 24, 226-231.
Anderson, J. W. and Rowan, K. S. (1965). *Biochem. J.* 97, 741-746.
Atkin, R. K. and Srivastava, B. I. S. (1970). *Physiologia Pl.* 23, 304-315.
Bagi, G. and Farkas, G. L. (1967). *Phytochemistry* 6, 161-169.
Balz, H. P. (1966). *Planta* 70, 207-236.
Barnett, N. M. and Naylor, A. W. (1966). *Pl. Physiol., Lancaster* 41, 1222-1230.
Barton, R. (1966). *Planta* 71, 314-325.
Ben-Zioni, A., Itai, C. and Vaadia, Y. (1967). *Pl. Physiol., Lancaster* 42, 361-365.
Bernstein, L. and Pearson, G. A. (1956). *Soil Sci.* 82, 247-258.
Biddulph, O., Biddulph, S., Cory, R. and Koontz, H. (1958). *Pl. Physiol., Lancaster* 33, 293-300.
Bourque, D. P. and Naylor, A. W. (1971). *Pl. Physiol, Lancaster* 47, 591-594.
Boyer, J. S. (1970). *Pl. Physiol., Lancaster* 46, 233-235.
Brix, H. (1962). *Physiologia Pl.* 15, 10-20.
Bruce, M. I. and Zwar, J. A. (1966). *Proc. R. Soc. B.* 165, 245-265.
Campbell, L. C., Raison, J. K. and Brady, C. J. (1972). *Biochim. biophys. Acta* (in press).
Cathey, H. M. (1964). *A Rev. Pl. Physiol.* 15, 271-302.
Cherry, J. H. (1967). *Symp. Soc. Exp. Biol.* XXI, 179-214.
Clark, M. F., Mathews, R. E. F. and Ralph, R. K. (1964). *Biochim. biophys. Acta* 91, 289-304.
Cleland, R. (1971). *A. Rev. Pl. Physiol.* 22, 197-222.
Doodson, J. K., Manners, J. G. and Myers, A. (1964). *J. exp. Bot.* 15, 96-103.
Dorner, R. W., Kahn, A. and Wildman, S. G. (1957). *J. biol. Chem.* 229, 945-952.
Dove, L. D. (1967). *Pl. Physiol., Lancaster* 42, 1176-1178.
Eilam, Y. (1965). *J. exp. Bot.* 16, 614-627.
Evans, H. J. and Sorger, G. J. (1966). *A Rev. Pl. Physiol.* 17, 47-76.
Evans, L. T. (1971). *A. Rev. Pl. Physiol.* 22, 365-394.
Feiererabend, J. and Pirson, A. (1966). *Z. Pflanzenphysiol.* 55, 235-245.

Flowers, T. J. and Hanson, J. B. (1969). *Pl. Physiol., Lancaster* **44**, 939-945.
Gale, J., Kohl, H. C. and Hagan, R. M. (1967). *Physiologia Pl.* **20**, 408-420.
Gates, C. T. (1957). *Aust. J. biol. Sci.* **10**, 125-146.
Gates, C. T. (1968). *In* "Water Deficits and Plant Growth" (T. T. Kozlowski, ed.), Vol. II, pp. 135-190. Academic Press, New York and London.
Goldthwaite, J. J. and Laetsch, W. M. (1967). *Pl. Physiol., Lancaster* **42**, 1757-1762.
Graham, D., Grieve, A. M. and Smillie, R. M. (1968). *Nature, Lond.* **218**, 89-90.
Graham, D., Hatch, M. D., Slack, C. R. and Smillie, R. M. (1970). *Phytochemistry* **9**, 521-532.
Greenway, H. (1970). *Pl. Physiol., Lancaster* **46**, 254-258.
Greenway, H. and Hiller, R. G. (1967). *Planta* **75**, 253-274.
Greenway, H. and Klepper, B. (1968). *Planta* **83**, 119-136.
Greenway, H., Hughes, P. G. and Klepper, B. (1969). *Physiologia Pl.* **22**, 199-207.
Haber, A. H., Thompson, P. J., Walne, P. L. and Triplett, L. L. (1969). *Pl. Physiol., Lancaster* **44**, 1619-1628.
Halder, D. and Freeman, K. B. (1969). *Biochem. J.* **111**, 653-663.
Harris, W. M. and Spurr, A. R. (1969). *Am. J. Bot.* **56**, 369-389.
Hartt, C. E. (1967). *Pl. Physiol., Lancaster* **42**, 338-346.
Hason-Porath, E. and Poljakoff-Mayber, A. (1969). *Pl. Physiol., Lancaster* **44**, 1031-1034.
Hayward, H. E. and Bernstein, L. (1958). *Bot. Rev.* **24**, 584-635.
Henckel, P. A. (1964). *A. Rev. Pl. Physiol.* **15**, 363-386.
Heyes, J. K. (1959). *Symp. Soc. exp. Biol.* **XIII**, 365-385.
Hopkinson, J. M. (1964). *J. exp. Bot.* **15**, 125-137.
Hsiao, T. C. (1970). *Pl. Physiol., Lancaster* **46**, 281-285.
Itai, C. and Vaadia, Y. (1965). *Physiologia Pl.* **18**, 941-944.
Jacoby, B. and Dagan, J. (1969). *Physiologia Pl.* **22**, 29-36.
Juniper, B. E. and Clowes, F. A. L. (1965). *Nature, Lond.* **208**, 864-865.
Kahane, I. and Poljakoff-Mayber, A. (1968). *Pl. Physiol., Lancaster* **43**, 1115-1119.
Kefeli, V. I. and Kadyrov, C.Sh. (1971). *A Rev. Pl. Physiol.* **22**, 185-196.
Kende, H. (1965). *Proc. natn. Acad. Sci. U.S.A.* **53**, 1302-1307.
Kessler, B. (1961). *Recent Advan. Bot.* **2**, 1153-1159.
Kessler, B. and Engelberg, N. (1962). *Biochim. biophys. Acta* **55**, 70-82.
King, R. W., Wardlaw, I. F. and Evans, L. T. (1967). *Planta* **77**, 261-276.
Kirk, J. T. O. (1970). *A. Rev. Pl. Physiol.* **21**, 11-42.
Lamport, D. T. A. (1970). *A. Rev. Pl. Physiol.* **21**, 235-270.
Lockhart, J. A. (1965). *In* "Plant Biochemistry". (J. Bonner and J. E. Varner, eds), pp. 826-849. Academic Press, New York and London.
Machold, O. (1971). *Biochim. biophys. Acta* **238**, 324-331.
Mann, J. D., Kung-Hing, Y., Storey, W. B., Pu, M. and Conley, J. (1967). *Pl. Cell Physiol. Tokyo* **8**, 613-622.
Mazur, P. (1969). *A. Rev. Pl. Physiol.* **20**, 419-448.
McHale, J. S. and Dove, L. D. (1968). *New Phytol.* **67**, 505-515.
Mego, J. L. and Jagendorf, A. T. (1961). *Biochim. biophys. Acta* **53**, 237-254.
Millikan, C. R. and Hanger, B. C. (1966). *Aust. J. biol. Sci.* **19**, 1-14.
Milthorpe, F. L. (1950). *Ann. Bot. N.S.* **14**, 79-89.
Mizrahi, Y., Blumenfeld, A. and Richmond, A. E. (1970). *Pl. Physiol., Lancaster* **46**, 169-171.
Moore, K. G. (1966). *A. Bot. N.S.* **30**, 683-699.

Morris, C. J., Thompson, J. F. and Johnson, C. M. (1969). *Pl. Physiol., Lancaster* **44**, 1023-1026.

Mühlethaler, K. (1967). *A. Rev. Pl. Physiol.* **18**, 1-24.

Neales, T. F. and Incoll, L. D. (1968). *Bot. Rev.* **34**, 107-125.

Oaks, A. and Bidwell, R. G. S. (1970). *A. Rev. Pl Physiol.* **21**, 43-66.

Osman, A. M. and Milthorpe, F. L. (1971). *Photosynthetica* **5**, 61-70.

Palfi, G., Vamos, R. and Barkoczy, M. (1966). *Adv. Frontiers Pl. Sci.* **15**, 149-155.

Pate, J. S. (1968). *In* "Recent Aspects of Nitrogen Metabolism in Plants" (E. J. Hewitt and C. V. Cutting, eds), pp. 219-240. Academic Press, London and New York.

Patterson, B. D. and Smillie, R. M. (1971). *Pl. Physiol., Lancaster* **47**, 196-198.

Petrie, A. H. K. (1937). *Aust. J. exp. Biol. med. Sci.* **15**, 385-404.

Pratt, H. K. and Goeschl, J. D. (1969). *A. Rev. Pl. Physiol.* **20**, 541-584.

Price, C. A. (1968). *A. Rev. Pl. Physiol.* **19**, 239-248.

Raison, J. K. and Lyons, J. M. (1970). *Pl. Physiol., Lancaster* **45**, 382-385.

Rhodes, M. J. C. and Yemm, E. W. (1966). *New Phytol.* **65**, 331-341.

Siegelman, H. W. and Butler, W. L. (1965). *A Rev. Pl. Physiol.* **16**, 383-392.

Shibaoka, H. and Thimann, K. V. (1970). *Pl. Physiol., Lancaster* **46**, 212-220.

Skoog, F. and Armstrong, D. J. (1970). *A Rev. Pl. Physiol.* **21**, 359-384.

Slatyer, R. O. (1967). "Plant Water Relationships," Academic Press, London and New York.

Slatyer, R. O. (1970). *Planta* **93**, 175-189.

Slavik, B. (1965). *In* "Water Stress in Plants" (B. Slavik ed.), Proc. Symp. Prague, 1963. pp. 195. Czech. Acad. Sci. Prague.

Smillie, R. M. (1962). *Pl. Physiol., Lancaster* **37**, 716-721.

Smith, T. A. (1968). *In* "Recent Aspects of Nitrogen Metabolism in Plants" (E. J. Hewitt and C. V. Cutting, eds), pp. 139-146. Academic Press, London and New York.

Srivastava, B. I. S. (1968). *Biochem. J.* **110**, 683-686.

Tal, M. and Imber, D. (1970). *Pl. Physiol., Lancaster* **46**, 373-376.

Treshow, M. (1970). "Environment and Plant Response" McGraw-Hill, New York.

Thrower, S. L. (1967). *Symp. Soc. exp. Biol.* **XXI**, 483-506.

Tolbert, N. E. and Gailey, F. B. (1955). *Pl. Physiol., Lancaster* **30**, 491-499.

Tung, H. F. and Brady, C. J. (1971). Proceedings 7th Intern. Conf. on Plant Growth Regulators (in press).

Tyankova, L. A. (1966). *C.r. Acad. bulg. Sci.* **19**, 847-850.

Udvardy, J., Farkas, G. L., Marre, E. and Forti, G. (1967). *Physiologia Pl.* **20**, 781-788.

Villemez, C. L., NcNab, J. M. and Albersheim, P. (1968). *Nature, Lond.* **218**, 878-880.

Wallace, W. and Pate, J. S. (1965). *Ann. Bot.* **29**, 655-671.

Wardlaw, I. F. (1967). *Aust. J. biol. Sci.* **20**, 25-39.

Wardlaw, I. F. (1969). *Aust. J. biol. Sci.* **22**, 1-16.

Williams, R. F. (1955). *A. Rev. Pl. Physiol.* **6**, 25-42.

Williams, R. F. and Rijven, A. H. G. C. (1965). *Aust. J. biol. Sci.* **18**, 721-743.

Woolhouse, H. W. (1968). *Hilger J.* **11**, 7-12.

Wright, S. T. C. and Hiron, R. W. P. (1969). *Nature, Lond.* **224**, 719-720.

Zholkevich, V. N. and Rogacheva, A. Y. (1968). *Fiziologiya Rast.* **15**, 537-545.

CHAPTER 25

Organic Reserves and Plant Regrowth

R. W. SHEARD

Department of Land Resource Science, Ontario Agricultural College,
University of Guelph, Guelph, Ontario, Canada

The survival of a plant species in its environment depends to some degree on its regrowth ability following a period of stress. The stress may be caused by a dry season in a Mediterranean climate, a cold season as in a north temperate climate, or an induced stress resulting from defoliation by a grazing animal or mechanical harvesting by man. Following a period of stress, regrowth of the plant may be from seeds or vegetatively. In both instances the initial growth is largely dependent on substances which have accumulated in the perennating organs of the previous generation or in the seed. With perennial species the regeneration of the plant under normal systems of pastoral management occurs by vegetative reproduction with the utilization of substances stored in the more permanent plant tissues. These substances are commonly referred to as "food reserves."

This chapter deals with the storage and mobilization of substances involved in regrowth following a period of stress, and should be read in conjunction with Chapters 17 and 24.

I. DEFINING ORGANIC RESERVES

The most comprehensive outline of the role or organic reserves in regrowth is that originally stated by Graber *et al.* (1927): "that new top growths, especially in the early stages, are initiated and developed largely at the expense of previously accumulated organic reserves, that the roots of alfalfa or rhizomes of grasses are not only organs of absorption and translocation but are organs of storage for such reserves, that such storage occurs principally during the maturity of top growth, that these organic reserves are essential to normal top and root development, that their quantity and availability sharply limit the amount of both top and root growth that will occur, and that the progressive exhaustion of such reserves by early, frequent and complete removal of top growth results ultimately in the death of the plant regardless of the most favourable climate and soil environment." The many points included in Graber's summary of the role of organic reserves in herbage growth have received considerable attention in the past half century. The subject has been reviewed by Archbold (1940), Weinmann (1948), Troughton (1957) and May (1960).

Graber *et al.* (1927) originally defined organic reserves as "those carbohydrate and nitrogen compounds elaborated, stored and utilized by the plant itself as food for maintenance and the development of future top and root growth." Wienmann's (1952) addition of the words "stored in the more permanent organs of the plant body" further refined the definition. Thomas *et al.* (1956) considered that prior to classifying a substance as an organic reserve it must be shown that there is a period of accumulation *in situ* of the substance to a relatively high concentration, and later, as a result of some physiological stress, a period of diminution in concentration. May (1960) presented arguments for the use of the word "accumulate" in preference to "reserves" since the former is non-committal in terms of purpose. Furthermore, the word "food" has primarily a zoological implication, hence it could be considered inappropriate in the botanical field. Tradition, however, has accepted the term "reserve" and its use will be in this context during the discussions which follow.

In the light of the preceeding discussion a satisfactory definition would state that organic accumulates are those carbohydrate and nitrogen compounds elaborated and stored by the plant in the more permanent organs of the vegetative reproductive system to be used as substrate for maintenance during periods of stress and development of future top and root growth.

II. CHEMICAL COMPOSITION OF ORGANIC RESERVES

Since a more comprehensive study of the carbohydrate composition of the organic accumulates is presented in Chapter 3, Vol. 1, only brief reference will be made to them here to illustrate the differences between species.

De Cugnac (1931), studied the carbohydrates of 38 species of Gramineae for carbohydrates and classified the grasses into two groups. The first group contained species indigenous to cool, temperate climates and accumulated a preponderance of sucrose and fructosans. The second group were indigenous to warm, tropical climates and accumulated sucrose, sometimes starch, but never fructosans. Substances such as pentosans, lignin and cellulose change only slightly, if at all, when plants are defoliated (Wienmann, 1948; Sullivan and Sprague, 1943; Mackenzie and Wylan, 1957) whereas hemicellulose changes at a slow rate (McCarty, 1938) and may have an organic reserve function.

Legume species are characterized by an accumulation of sucrose and starch (Graber *et al.*, 1927; Ruelke and Smith, 1956; Smith, 1962) in conjunction with the monosaccharides, glucose and fructose. The species examined included alfalfa (*Medicago sativa*), red clover (*Trifolium pratense*), Ladino clover (*T. repens*), and birdsfoot trefoil (*Lotus corniculatus*). Similar characterization of carbohydrates in five tropical grass species and five tropical legume species has been reported by Hunter *et al.* (1970). The tropical grasses stored carbohydrate principally in the form of sucrose with very little fructosan and also as starch in basal stems (see Chapter 3, Vol. 1 and Smith, 1968a) The legume species were characterized by sucrose, starch and pectin.

The definition of organic reserves includes nitrogenous materials yet there are few reports of such compounds being measured.

Weinmann (1948) recognized that nitrogen in organic combination might serve a reserve function. Davidson and Milthorpe (1966), in a detailed study with orchard grass (*Dactylis glomerata*) suggested the decrease in soluble carbohydrates could account for only a portion of the total respiratory loss and that protein may be a factor in supplying substrate.

Storage organs of grass species have been shown to lose total nitrogen at times of rapid herbage growth in the spring and early summer and to gain nitrogen in the autumn (Arny, 1932; Richardson *et al.*, 1932; Wienmann, 1948; McIlvanie, 1942). Peters (1956) reported the total nitrogen content in the haplocorm of timothy (*Phleum pratense*) decreased with increased frequency of cutting indicating a utilization of nitrogen compounds for regrowth following frequent cutting. Smith and Silva (1969) found proportionally less total nitrogen than nonstructural carbohydrate was translocated from the roots of alfalfa for the production of new growth.

Although total nitrogen determinations (Kjeldahl nitrogen) may show no significant gain or loss of nitrogen from the storage organs, major alterations of the nitrogen compounds within the tissue may have occurred during regrowth. Since plants do not excrete nitrogen, as they excrete CO_2, total nitrogen would not reflect such alterations. The carbon skeletons of amino acids from proteins may function as respiratory substrates (Steward, 1959). the nitrogen being stored as amides and the carbon skeletons lost as CO_2. A true evaluation of the role of nitrogen compounds as organic reserves would necessitate the separation of amides, amino acids, structural and non-structural proteins, as has been performed for carbohydrates.

Sullivan and Sprague (1943) reported a negative correlation between aftermath yields and ethanol-insoluble nitrogen in the stubble of orchard grass, with no relationship between the ethanol-soluble nitrogen and regrowth. Dilz (1966) reported that the amount of N in perennial ryegrass gave a better estimate of reserves involved in herbage regrowth than the total soluble carbohydrates. Sheard (1968b) found a positive correlation between spring growth in the field and the nitrogen fractions in timothy. Later studies (Sheard, 1970) further indicated there were positive correlations between amide, amino and nitrate nitrogen and the regrowth of timothy.

Few studies with nitrogenous organic reserves have been conducted to demonstrate a primary requirement of the definition of organic reserves, namely that the compounds go through a cycle of accumulation and depletion, prior to and during a period of stress. In an attempt to fulfil the requirement, however Sprague and Sullivan (1950) found that both alcohol-soluble and -insoluble nitrogen in the stubble of orchardgrass increased following defoliation and then decreased as the foliage developed. Nielsen and Lysgaard (1956) reported a similar trend for total nitrogen in the tap root of alfalfa.

III. PLANT MORPHOLOGY AND ORGANIC RESERVES

The particular region of organic reserve accumulation in the plant varies from species to species and its delineation should be a prerequisite of all studies. By definition the region includes the more permanent parts of the plant which are associated with vegetative reproduction. Hence consideration should be given to roots, haplocorms, rhizomes, stubble or stem base and other parts remaining after defoliation.

TABLE I

Concentration of non-polymer and fructosan fructose in several morphological sections of mature timothy (*Phleum pratense*).

Plant section	Non-polymer fructose	Fructosan fructose
	(% of dry weight)	
Stubble	3·0	36·5
Haplocorm	3·0	48·7
Root	1·2	1·2

Sheard (1967) established that the enlarged stem base was the primary area of fructosan accumulation in timothy (Table I). Whereas the haplocorm contained 25% more fructosan than the stubble, there was still a high concentration in the stubble while the fibrous roots were essentially devoid of fructosan. Sullivan and Sprague (1943) reported 36% of fructosan in orchard grass was located in the lower one-half of the stubble whereas the roots contained only eight per cent. The fructosans in smooth brome (*Bromus inermis*) also appear to be concentrated in the stubble (Smith, 1967) although no data is presented for rhizomes and roots.

The perennial grass plant may be considered as a succession of comparatively short-lived individuals, each individual capable of a separate existence and the reproduction of itself (Langer, 1963).

Following defoliation, the appropriate tissue to sample for organic reserves would be from that individual which will produce the next generation of tillers. This reasoning formed the basis for the recognition of a sequence of development of the storage organ in timothy by the author (Sheard, 1968a).

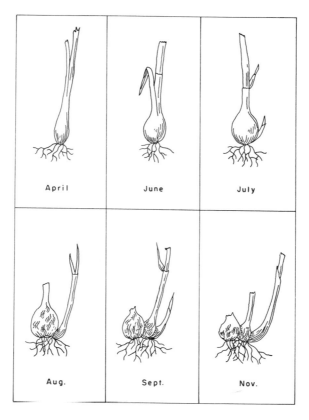

Fig. 1. Development sequence of the haplocorm of timothy. April—Shoot of the seedling year or tertiary shoot of an established stand; June—The development of the primary haplocorm from the tertiary shoot; July—Axillary bud development at nodes below and above the primary haplocorm; Aug.—Development of secondary haplocorms on tillers originating from buds on the primary haplocorm and the beginning of the decay of the primary haplocorm; Sept.—Tertiary shoot development from axillary buds on the secondary haplocorm; Nov.—Further decay of the primary and secondary haplocorm and development of the tertiary shoot to the overwintering stage (reproduced with permission from Sheard, 1968a).

The sequence (Fig. 1) begins with vegetative shoot from the seedling year or the tertiary shoot of the previous harvest year. Upon stem elongation the basal internode enlarges to form a primary haplocorm during May. Following removal of the first hay crop, axillary buds on the primary haplocorm commence rapid development into vegetative tillers that have elongated by August, permitting secondary haplocorms to develop at their base. Following removal of the aftermath growth, a new set of axillary buds at the base of the secondary haplocorms develops into tertiary shoots. During the fall the tertiary shoots do not elongate, thus no true haplocorms can develop although the swollen appearance of the shoot base may resemble haplocorms. Under conditions of adequate nitrogen the primary and secondary haplocorms have largely disintegrated before the winter stress period. With this sequence the primary haplocorm would be the appropriate plant part to sample for organic reserves in May through July. However, sampling of the new shoot bases and the

TABLE II

The percentage reducing sugar, total sugar and starch in the crown, woody cortex and bark segments from the tap root of medium size alfalfa (*Medicago sativa*) plants twenty-one days after cutting (reproduced with permission from Ueno and Smith, 1970)

Carbohydrate	Crown	Segment Woody Cortex (all % of dry weight)	Bark
Reducing sugars	1·5	1·0	1·7
Total sugars	8·3	12·1	14·3
Starch	4·0	18·5	8·3
Total non-structural carbohydrate	12·3	30·6	22·6

secondary haplocorm would be appropriate for studying aftermath regrowth. Studies of organic reserves in timothy in relation to the winter stress period would involve sampling the tertiary shoot only.

A sequence of tissues to sample does not apply to the perennial legume. The tap root of alfalfa has been shown to be the primary area of carbohydrate accumulation (Graber *et al.*, 1927; Smith, 1962). The crown or stubble also serves as an important storage organ (Graber *et al.*, 1927; Jung and Smith, 1961; Sonneveld, 1962).

The tissue within a plant part which serves as the point of accumulation of organic reserves has not received much attention. Ueno and Smith (1970), however, attempted to differentiate between carbohydrate changes in the woody cortex and bark segments of the tap root and of the crown segment of the plant (Table II). While only minor changes occurred in reducing sugars between the three segments, the total sugar percentage was highest in the bark and least in the crown. The woody cortex segment of the tap root was the primary site of starch storage. The highest percentage of total nonstructural carbohydrates (total sugar plus starch) occurred in the same segment. During regrowth 46 to 52% of the total nonstructural carbohydrate from the woody cortex of the tap root was estimated to be utilized, as compared with 21 to 26% from the root bark and 24 to 28% from the crown segment.

Regrowth after cutting is not only dependent on the carbohydrate root reserves, but may also depend partly on the morphological development of the plant at the time of cutting. For example, alfalfa regrowth has been shown to depend to some extent on the number and size of buds on the stubble at the time of harvest (Leach, 1968; Langer and Steinke, 1965). Regrowth of timothy and bromegrass, defoliated above the shoot apex, is rapid in contrast to regrowth following stem elongation where the stem apex is removed and new photosynthetic tissue originates from axillary bud growth (Milthorpe and Davidson, 1966; Sheard and Winch, 1966). The requirement for organic reserves in the latter case would be greater as regeneration by axillary bud development was slow (Sheard and Winch, 1966).

With grasses and many legumes the youngest and most photosynthetically active leaves are produced at the top of the sward. Leaves of many *Trifolium* species, however, arise near the soil surface and grow up through a canopy of older leaves, which at early stages intercept much of the light. These newly emerging leaves may draw on a reserve of different origin than newly emerging leaves of other species. An attempt to demonstrate the movement of carbohydrate during the ontogeny of the leaves of these species was made by Hoshino *et al.* (1964) using ^{14}C. The emerging leaf (Leaf No. 1) of Ladino clover (*T. repens* var. *ladino*) translocated labelled assimilate to the roots, growing points and lateral buds, with some translocation to older leaves. Leaves 4 to 8 were active in supplying assimilate to all plant parts whereas the ability of Leaf 11 to assimilate $^{14}CO_2$ was low and translocation was basipetal. These studies were not carried out under sward conditions, however, where light interception would influence the need for organic accumulates.

IV. SIMPLIFIED PROCEDURES FOR DEMONSTRATING STORAGE
AND MOBILIZATION

The more common procedures for demonstrating storage and mobilization of organic reserves are chemical analysis for a variety of carbohydrate constituents before and after a stress period or defoliation (Sullivan and Sprague, 1943; Okajima and Smith, 1964; Sheard, 1968a; Smith and Silva, 1969; Nelson and Smith, 1968). The reader is referred to the work of Dale Smith (see Chapter 3, Vol. 1), a contributor of many papers on methodology for carbohydrate constituents and their relationship to herbage regrowth and production.

During the past decade simplified procedures have been developed for evaluating organic reserves which do not necessitate that the investigator has knowledge of the appropriate plant part to analyze or the chemical component to analyze for. Furthermore, the procedures require a minimum of special equipment or facilities.

A. Etiolated Regrowth

Burton and Jackson (1962) proposed the regrowth of a defoliated plant in darkness would give an integrated measure of all energy-producing materials in the plant regardless of the plant part associated with the materials. The technique involves the removal of the sod, usually a 6 in. diameter plug, and allowing the defoliated plants to regrow in darkness with adequate moisture until the energy producing materials are exhausted. The weight of the clippings produced during the period form an index of the regrowth potential of the plant.

Macleod (1965) employed etiolated regrowth to evaluate the influence of fertilizer treatments on organic reserves in alfalfa and three grasses (Table III). He obtained some degree of correlation within a species between etiolated regrowth and the concentration of 0.2 N H_2SO_4 extractable carbohydrates. There was considerable difference in the weight of regrowth produced by the different species, partially a reflection of the weight of the original tissue and partially a reflection of the concentration of carbohydrate.

Etiolated regrowth procedures do not define the source or type of organic reserve. Therefore in correlation studies it would be appropriate to consider the total pool of organic reserves defined as the weight of the storage organ multiplied by the concentration of carbohydrate. Raese and Decker (1966) obtained a significant increase in correlation when they combined weight of the storage organ and concentration of carbohydrate to establish a relationship with etiolated regrowth of orchard grass, bromegrass and reed canarygrass (*Phalaris arundinacea*).

TABLE III

Influence of nitrogen and potassium on the etiolated regrowth (mg dry matter/plant) and the concentration of 0·2 N H_2SO_4 extractable carbohydrate (% CHO of dry weight) from the storage organs of four forage species (reproduced with permission from MacLeod (1965)

Treatment (lb/acre)	Alfalfa regrowth	CHO	Bromegrass regrowth	CHO	Orchard grass regrowth	CHO	Timothy regrowth	CHO
Rate of K								
0	283a[a]	37·2a	15a	16·7a	48a	27·6a	34a	24·3a
50	430b	47·0b	28a	17·1a	68a	28·1a	48a	26·0a
100	463bc	48·4b	24a	16·0a	73ab	27·8a	40a	24·2a
200	520c	48·5b	32a	17·8a	103b	27·7a	64a	24·5a
Rate of N								
0	466b	44·4a	6a	17·3a	20a	26·0a	14a	22·1a
25	352a	46·4a	13a	17·6a	40ab	26·8ab	40b	25·8b
50	456b	47·8a	27b	17·6a	68ab	28·3b	52bc	27·0bc
100	423b	45·1a	37bc	16·0a	118c	29·0b	62bc	25·5b
200	415b	42·6a	40c	16·3a	120c	28·8b	66c	23·3ab
Corr. coefficient	0·511		−0·183		0·720		0·500	

[a] Values followed by a different letter are significantly different at P = 0·05.

Blacklow and McGuire (1971), in studies with tall fescue (*Festuca arundinacea*), proposed a more rigorous, quantitative treatment of etiolated regrowth as a measure of organic reserve. More recently Blacklow* (pers. commun.) has further expanded the mathematical treatment of data obtained from etiolated regrowth. If regrowth in the dark is clipped frequently and the cumulative weight, W, is plotted against time, t, a curvilinear relationship is established. The rate of regrowth, dW/dt, decreased as the cumulative weight approached the maximum weight of regrowth, A, that is:

$$dW/dt = k\,(A\text{-}W) \tag{1}$$

where k is the proportionality constant with the dimension 1/time. On integration (1) gives the first order function:

$$\ln A - W/A = -kt \tag{2}$$

Consequently, if the rate of regrowth declines and the substrate is depleted the data plotted as (2) should be a straight line with slope k.

*Personal Communication with W. M. Blacklow, Agron Dept., Univ. of Western Australia, Nedlands, Australia 6009.

The exponential form of (2) is:

$$W = A\ (1 - 1/e^{kt}) \qquad (3)$$

and for $t = 1/k$ it can be seen that $W = 0·63A$. Consequently a useful basis for comparing treatments is to define a Response Time (RT) as the reciprocal of k; it is the time required to attain 63% of the maximum regrowth.

The proportionality constant, k, is also a measure of the rate of utilization of organic reserves, that is:

$$k = \frac{dW/dt}{A-W} \qquad (4)$$

and can be expressed as mg regrowth per day per g of organic reserves. It is, therefore, a measure of the Relative Regrowth Rate (RRGR) and is suitable for comparing plants of different size and amount of accumulated reserves for their ability to utilize reserves in regrowth.

Some precautions should be exercised in the use of etiolated regrowth procedures. Where different fertilizer elements and rates have been used, plant nutrient levels may exist in the soil core which will influence regrowth—particularly nitrogen. The addition of the appropriate elements would be necessary to equalize any differences. Matches (1969) found a cutting height of six to nine cm was necessary to differentiate between low and high levels of organic reserves in tall fescue. The procedure is subject to large sampling errors, particularly where the stand density is low. Hence cores removed from a bluegrass turf (*Poa pratensis*) may show less variability than cores from an orchard grass sward.

The author would suggest a variation of the procedure to be used for measuring etiolated regrowth of alfalfa and other tap-rooted legumes. Following removal from the field the roots are trimmed to a length of 15 cm and all major lateral roots and stubble to a height of 2·5 cm are removed. The roots are transplanted into sand or expanded vermiculite, then treated in a manner similar to that used for conventional etiolated regrowth studies.

B. Root Density

The total root weight of alfalfa has been shown by Baker and Garwood (1959) to be largely due to the weight of the tap root. Nielsen and Lysgaard (1956) suggested the volume-weight of dried and finely ground root material from alfalfa would serve as a reliable index of the quantity of root dry matter, an index which paralleled the changes in sugars and starches. Fulkerson (1970) and his students have further developed the technique to obtain empirical measurements of organic reserves in their studies of harvest management and winter survival of alfalfa.

The procedure is based on the assumption that changes in the density of the whole root reflect the loss or gain of reserve materials from the cells of the root. Root samples are washed free of adhering soil, dried at 100°C for one hour followed by 80°C until constant weight is attained. The dried roots are then stripped of fine lateral roots, weighed in air and then weighed when immersed in water. Root density may then be calculated from the formula:

$$\text{Root density} = \frac{\text{Weight in air}}{\text{Weight in air} + (\text{weight in air} - \text{weight in water})}$$

where the denominator provides a measure of the volume of the roots.

The change in root density of three alfalfa cultivars following an August 25th defoliation is shown in Fig. 2. The pattern established

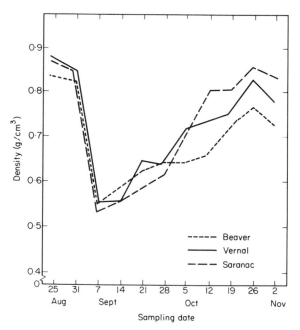

Fig. 2. Post-harvest density patterns of alfalfa cultivars (reproduced with permission from Stauffer, 1969).

for root densities is very similar to that reported by Nelson and Smith (1968) and Kust and Smith (1961) for carbohydrate depletion and accumulation in alfalfa which would substantiate the applicability of the procedure.

Root density measurements have been related to winter survival and regrowth measured as subsequent hay yields of alfalfa (Table IV). Subjecting alfalfa to a series of fall cutting dates altered the root density as measured at the beginning of the winter stress period. The low root densities were associated with a reduced plant population and vigour of the surviving plants which in turn was reflected in hay yields.

TABLE IV

The relationship of fall cutting date to root density at the end of the growing season (Nov. 8), plant population and hay yeild during the subsequent season (reproduced with permission from Stauffer, 1969).

Fall cutting date[a]	Root density on Nov. 8 (g/cm^3)	Plant population (plants per 0·09 m^2)	Hay yield (kg/ha)
September 8	0·78	23·8	4910
September 15	0·76	19·0	4320
September 22	0·62	6·7	2530
September 29	0·64	12·2	2860
October 6	0·69	15·5	3520
Uncut	0·77	26·8	4950
L.S.D. @ P = 0·05	0·05	4·8	430

[a]Following three harvests during the summer at the first flower stage, the last hay harvest occurring on August 30.

Fulkerson (1970) employed the technique to establish the critical fall harvest date at 24 representative locations throughout Southern Ontario, Canada. From the dates he developed a fall harvest management map for alfalfa in the region, a map delineating the areas where a fall harvest should not occur later than three weeks before the critical date.

The root density procedure as described is limited to use with tap-rooted legumes. Errors in measurement can occur from air bubbles trapped on the dried root and from water entering the tissue during immersion. These errors are largely avoided by the use of fresh roots or roots which have been stored under frozen conditions until the time of measurement.

V. STUDIES OF ORGANIC RESERVES USING CARBON-14

Without doubt, studies of movement and distribution of organic compounds in herbage has made considerable advances through use of ^{14}C which serves as a marker for carbon assimilated during photosynthesis by the leaves (Wardlaw, 1968). The technique has been used to assess the relative importance of leaf position and function, to trace vascular connections and to study accumulation and depletion of organic reserves during the development of the plant.

Basically the procedure involves enclosing plants in a plastic chamber, preferably a chamber where there is some air turbulence and humidity and temperature is controlled (Musgrave and Moss, 1961; Wolf, 1967a). ^{14}C is generated by dropping lactic acid or dilute HCl on $Ba^{14}CO_3$ or $Na_2^{14}CO_3$ within the chamber (Hodgkinson, 1969; Pearce et al., 1969; Smith and Marten, 1970). Following exposure of the plants under sunlight or strong artificaal light for periods ranging from 15 min (Wolf, 1967a) to two hours (Pearce et al., 1969) the unfixed $^{14}CO_2$ is trapped by flushing the enclosed air through KOH. Levels of initial activity employcd in the carbonate salt depend on the volume of material in the chamber and range from 5 μc (Marshall and Sager, 1968a) to 2 mc (Pearce et al., 1969).

At sampling time the ^{14}C in the plant parts under examination may be determined by: (1) combustion of the ground samples and absorption of the $^{14}CO_2$ in ethanolamine followed by liquid scintillation counting (Hodgkinson and Veale, 1966); (2) counting directly as ground, 40-mesh plant material on planchets using a windowless gas-flow detector (Wolf, 1967a); (3) extracting organic constituents followed by liquid scintillation spectrometry using Bray's solution (Bray, 1960); (4) using a wet oxidation procedure (Jeffray and Alverez, 1961) followed by liquid scintillation counting and (5) suspending ground plant material in a thixotropic gel suspension of Cab-O-sil* (Ott et al., 1959).

A. Utilization During Regrowth

During regrowth the organic materials in the storage organs may be redistributed by three processes: translocation to new root growth, translocation to new shoot growth, and respiration. Some estimates

---

* 4% w/v Cab-O-Sil, 4 g 2,5-diphenyloxazole (PPO) and 50 mg 1-4-bis-[2-(4-methyl-5-phenyloxazolyl)]-benzene (POPOP).

of the relative importance of organic reserves in the three processes were obtained from ^{14}C studies. The transfer of ^{14}C-labelled organic reserves from those portions of the plant remaining after defoliation was investigated by Pearce *et al.* (1969), Hodgkinson (1969), Hough (1970) and Smith and Marten (1970). Their studies have firmly established the importance of organic reserves in the storage organs

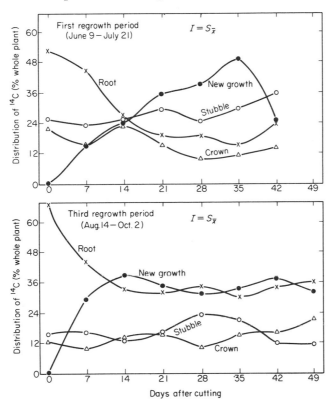

Fig. 3. Distribution of ^{14}C in the plant parts of alfalfa during the first and third regrowth periods (reproduced with permission from Hough, 1970).

as material that is mobilized and utilized in the synthesis of new tissue during regrowth.

Unpublished data (Fig. 3) from the thesis of Hough (1970) clearly showed the decline of ^{14}C-labelled material in the root and the transfer of the material to the new top growth. The change in ^{14}C in the crown and stubble, however, did not indicate a contribution of stored organic material from this region to the regrowth of alfalfa either in the summer or fall. The ^{14}C in the regrowth increased steadily for 21 days in the first and for 14 days in the third regrowth periods, increases that corresponded to a decrease in the proportion of ^{14}C in the root. Hodgkinson (1969) observed translocation to the new shoots to continue for the first 20 days of regrowth of alfalfa. Pearce et al. (1969) found the greatest changes in alfalfa to occur between three and fifteen days. Smith and Marten (1970) observed the redistribution of ^{14}C-labelled carbohydrate to be at a maximum during early vegetative growth and to decrease with advancing maturity. A shorter period of dependency on organic reserves is reported for Ladino clover. Hoshino and Oizumi (1968) concluded new leaves were dependent on reserve assimilates for the first week after defoliation, and although they continued to utilize reserves during the second week, they were themselves exporting assimilates and hence were not solely dependent.

Measurement of redistribution 28 days after defoliation by Pearce et al. (1969) suggest respiration may account for 45% of the ^{14}C-labelled material, new top may utilize 19% with the remaining ^{14}C-labelled material staying in the storage organ. Hough (1970) reported a net loss of 56% of the ^{14}C present in the storage organ at the time regrowth commenced, presumably a loss due to respiration; the majority of the respiration loss occurring during the first 21 days of regrowth. Hodgkinson (1969), however, reported high respiration losses of 81% in 20 days and 19% utilization in new top growth. Smith and Marten (1970) reported a respiratory utilization of ^{14}C-labelled carbohydrate of 26% by the entire alfalfa plant at the vegetative stage, increasing to 71% at the 50% bloom stage. It is of interest to note that on the basis of weight and carbohydrate analysis Smith and Silva (1969) reported respiration losses of only 23% in the tap root of alfalfa and 68% utilization in regrowth.

Incorporation of ^{14}C initially stored in the roots as non-structural carbohydrate accounted for 25% of the structural tissue in the shoot of vegetative alfalfa (Smith and Marten, 1970). A minimum incorporation of 3% occurred at the 50% bloom stage, the decrease resulting from the increasing contribution from concurrent photosynthesis in the formation of the shoot.

An attempt was made by Pearce et al. (1969) through the use of chemical extractants to identify the particular ^{14}C-labelled compounds redistributed during regrowth. Their estimates, obtained by ^{14}C measurements, of starch removal from the large roots during regrowth were remarkably similar in shape and magnitude to that obtained by conventional carbohydrate analysis. The evidence lends support to previous literature on changes in the concentration of various carbohydrates in the storage organs during regrowth but unfortunately has not included a consideration of the nitrogen economy of the plant during regrowth.

B. Restoration of Organic Reserves

At some stage during regrowth, recovery of the organic reserves must occur to enable the plant to survive the next stress period. At this stage, the age of the leaf influences the amount and direction of the movement of ^{14}C-labelled assimilates. Webb and Gorham (1964) found very young leaves of wheat (*Triticum vulgare*) received carbohydrate for their growth from older leaves and their growth requirements are met by photosynthesis with export of assimilate commencing only when the leaf has reached $\frac{1}{3}$ to $\frac{1}{2}$ of its final area.

In a study with timothy, Williams (1964) followed the movement of ^{14}C-labelled assimilates as measured by autoradiographic techniques. Young, growing leaves retain all their assimilates, using them for growth. These leaves import assimilates from older leaves. Import ceases, and export begins, before leaf expansion is complete. Export is first upwards to younger leaves and the growing point; later, some moves downwards, and finally movement is entirely downward to the root. The older the leaf, the greater the downward movement of assimilates in contrast to upward movement.

Ryle (1970) further elaborated on the influence of leaf age on the distribution of ^{14}C-labelled assimilates in perennial ryegrass. During early vegetative growth, the terminal meristem, tillers and roots received most of the ^{14}C exported from the youngest, fully-expanded leaf. At the transition from vegetative to reproductive growth there was an abrupt increase in the export of ^{14}C to the stem from the upper leaf, but there was little change in the proportion moving to developing leaves or the incipient inflorescence. As a result, less ^{14}C-labelled assimilates moved to daughter tillers and much less to the roots.

The proximity of the leaf to actively growing parts of the shoot which may function as sinks may be an important factor in the distribution of carbohydrate in graminaceous plants (Ryle, 1970). In vegetative plants, without elongated internodes, terminal meristems, tillers and roots are all close to the exporting leaves, and may thus have equal access to the assimilates. When the stem elongates, ^{14}C from the upper leaves must move through the extended internodes to reach tillers and roots giving the stem a clear advantage in terms of proximity. Other factors, such as hormonal control, have been suggested as influencing the pattern of distribution between the terminal meristem and the stem. Data to support these suggestions, however, has not been obtained.

Studies with legume species have shown Ladino clover exposed to ^{14}C three days after cutting, translocated assimilates to opening leaves at the growing point (Hoshino *et al.*, 1967). Three days later translocation to stolon and roots began, reaching a maximum on a

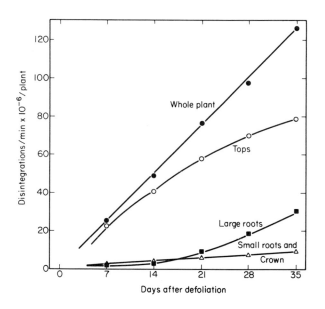

Fig. 4. ^{14}C uptake of the whole plant and plant parts when exposed to $^{14}CO_2$ at intervals after defoliation. (reproduced with permission from Pearce *et al.*, 1969).

whole plant basis between six and nine days. Vegetative or early growth of alfalfa and reed canarygrass translocated less ^{14}C-labelled material to the tap root or rhizomes than did more mature plants (Wolf, 1967b).

In a more detailed study Pearce *et al.* (1969) exposed freshly defoliated alfalfa to ^{14}C at weekly intervals over five weeks. They found most of the early photosynthate—83% at one week—remained in the top, with only a small amount translocated to the crown and small roots (Fig. 4). When labelled thirty-five days after regrowth commenced 64% of the ^{14}C remained in the tops, indicating that one-third of the assimilated ^{14}C was translocated to the storage organs. The crown and small roots accumulated photosynthate at a constant rate; however, the accumulation in the large roots (tap root plus main laterals) increased as the plant developed a greater photosynthetic area. Accumulations in the tap root at early stages of growth tended to be in the ethanol-soluble materials whereas by the third week significant amounts of ^{14}C were found in the starch fraction.

The removal of top growth by a grazing animal is more selective than cutting as younger leaves and tillers are preferred to older leaves and stems (Moore and Bibbiscome, 1964). Marshall and Sager (1968a, b) and Sager and Marshall (1966) employed ^{14}C techniques to determine the interdependence of tillers of Italian ryegrass (*Lolium multiflorum*). They demonstrated a mutual exchange of assimilates between parent and daughter tillers and between sister tillers. Six hours after exposure to ^{14}C, 45% of the isotope exported by a mature leaf was recovered from the tiller to which it was attached and 45% was recovered from the root system. The remainder was distributed between daughter and sister tillers. If the entire main shoot was exposed to ^{14}C the greater proportion of the exported material (*ca.* 63%) was recovered from the root, the remainder being distributed among daughter tillers, the youngest receiving a greater proportion than the oldest. Exposure of a daughter tiller, however, resulted in a lower export to the root and a significant export to the main tiller. Marshall and Sager (1968b) reported that during partial defoliation of Italian ryegrass the flow of ^{14}C-labelled assimilate to the root sink remained constant whereas the flow to the defoliated tiller increased through an increased flow of carbon from the intact tiller.

During restoration of the carbohydrates, attention has not been given to the nodule on the root of the legume species as a potential sink for photosynthate. Hoshino *et al.* (1964), using autoradiographs, demonstrated a strong movement of ^{14}C-labelled assimilate into the nodule of Ladino clover. Small and Leonard (1969) found 24% of the radioactivity to be evenly divided between the roots and nodules of field pea (*Pisum sativum*) 20 h after exposure to $^{14}CO_2$. Subterranean clover (*Trifolium subterraneum*), by contrast, translocated 5·9% of the ^{14}C-labelled assimilate to the nodule and 16% to the root. Interestingly, nitrate nitrogen decreased the translocation of photosynthate to the nodule and resulted in a corresponding increase in the translocation to the root of both species.

VI. LEVEL OF ORGANIC RESERVES AND REGROWTH

While it has been established that organic reserves function in regrowth, "critical levels" for the various types of compounds involved and conditions for regrowth have not been established. May (1960) stated that, provided sufficient organic reserves exist to support early regrowth and sustain the respiratory requirement of the underground organs while there is no foliage, there is no evidence to suggest that fluctuations in concentration above a critical level has any bearing on regrowth.

Milthorpe and Davidson (1966) reported tiller bases of perennial ryegrass (*L. perenne*) defoliated at flower initiation contained 4 mg/pot of water soluble carbohydrate compared to 105 mg/pot at ear emergence and 278 mg/pot at grain development. New growth after three weeks, however, was greatest where the lesser amount of carbohydrate was measured, particularly at low levels of visible radiation. As the amount of carbohydrate was more than adequate, growth was dependent on the growth rate of intact flowering tillers in contrast to vegetative tillers.

Timothy, containing 24·5 to 42·7% water soluble carbohydrate in the stem bases underwent only minor depletion of reserves during regrowth, indicating that the reserves were present in levels far in excess of the requirements of the plant (Jewiss and Powell, 1965). Sheard (1968a) also questioned the value of high levels of carbohydrate often observed in timothy, considering them to be in excess of the critical level for survival of the winter dormancy period and spring regrowth. Following establishment of the tertiary shoot (see Fig. 1) as the appropriate plant part for sampling, he found a fructosan concentration of 16 to 20% in November was sufficient to provide respiratory substrate during winter dormancy, leaving a concentration of 5 to 7% to initiate spring growth. Concentrations in excess of 40%, measured in the large primary haplocorm at the end of the first growth cycle, were considerably greater than was required for survival of a normal midsummer dormancy. Where organic reserves reach such high levels they must be considered as representing growth not being made, rather than contributing to the organic reserve accumulation required for the survival of the plant.

The cause of the high values—lack of nitrogen nutrition, low temperatures, or other factors decreasing the potential for top growth—should be investigated.

VII. ORGANIC RESERVES IN WEED CONTROL

A contrast to the objective of reserve accumulation in herbage production is the objective employed in the control and elimination of perennial weeds where management procedures are designed to minimize the level of organic reserves at the time of herbicide application (Åberg, 1964). Hence Arny (1932) and Granström (1954) found the most suitable time for eradication of thistle (*Cirsium arvense*) by cultivation or chemical means was immediately before flowering when the reserve carbohydrates were at a minimum.

LeBaron (1962) reported a short-term decline in fructosans in the rhizomes of quackgrass (*Agropyron repens*) from nitrogen applications but found the nitrogen to eventually stimulate the recovery of fructosans. Repeated cultivation was also an effective treatment for the reduction of fructosans in the rhizomes. Atrazine was superior to dalapon when used in conjunction with nitrogen as it caused a rapid and complete decline in the carbohydrates.

In contrast Linscott and McCarty (1962) did not find the herbicidal action of 2,3-D on iron weed (*Vernonia baldwini*) to depend on the changes in the carbohydrate content of the roots. Hull (1969) also failed to show a significant effect of repeated clippings on the translocation of ^{14}C-labelled assimilate into the rhizome of Johnson grass (*Sorghum halepense*). Translocation of the assimilates to the rhizomes was greater at the full flower than at earlier stages whereas ^{14}C-dalapon was translocated to the rhizomes prior to stem elongation.

VIII. SURVIVAL AND REGROWTH

Regrowth of the perennial herbage plant must occur not only following defoliation but also following climatic conditions which enforce a period of dormancy. Under Mediterranean and semi-arid to arid conditions drought enforces the period of dormancy, the drought frequently being associated with high temperatures and high respiration rates. In contrast, a northern continental climate enforces a period of dormancy due to low temperatures with a relatively low respiration rate (Jung and Smith, 1960). A feature of the low temperature survival is the changes in the permanent parts of the plant as it proceeds through various stages of cold acclimation (Weiser, 1970).

Smith (1968b) reviewed the chemical changes associated with freezing resistance in forage plants. Although no positive correlations occur between various carbohydrate constituents and low tempera-ture survival in legumes and grasses, a starch to sugar conversion during late autumn and early winter occurred in all starch-stored legumes and grasses. The exact role of the conversion has not yet been established. Data for alfalfa have shown positive correlations, however, between frost resistance and water soluble protein nitrogen. The role of protein in frost resistance is generally associated with the influence of water on the macromolecular configuration of proteins (Weiser, 1970).

Julander (1945) found plants of couch grass (*Cynodon dactylon*) to make little or no growth during a drought and very little carbohydrate change to occur. A decrease in carbohydrates occurred, however, when growth was stimulated by partial defoliation or water additions. He also associated low organic reserves with a decrease in heat resistance.

IX. A CARBON CYCLE IN HERBAGE

The carbon compounds in a herbage plant may be considered to be in a dynamic system strongly influenced by the development of the plant and the environment in which the plant is growing. The concentration of a compound in a plant part at any particular moment is the resultant of past and current photosynthesis, trans-location, respiration and synthesis of new material. Whereas diurnal variations may be appreciable, the long-term effect is for high concentrations to arise if conditions allow moderate rates of photo-synthesis but growth is restricted by low temperature or nitrogen deficiency (Davidson and Milthorpe, 1965). Furthermore, import or export of carbon from a specific plant part depends on the ontogeny of the plant part (Williams, 1964).

A schematic diagram may be used to provide a qualitative summary of the carbon cycle in herbage prior to defoliation and during regrowth (Fig. 5). In an intact, mature plant the uppermost leaves actively produce photosynthate, a portion of which is used in growth of new tissue, a portion functions as substrate for respiration and the remainder is exported to the lower regions of the plant. At an intermediate level in the canopy of leaves photosynthesis will be somewhat less, growth of leaves is nearly complete and respiration may be slightly less than in the more active leaves at the surface of the leaf canopy. As a result the export of photosynthate should be at a maximum. At the lower levels in the canopy there will be little, if any, photosynthetic activity, growth has essentially ceased and respiration may be slightly reduced. If import into the region exceeds respiration, accumulation of organic reserves begins. Within the root system growth and respiration are dependent on transport from the leaves, accumulation of organic reserves only occurring when the transport into the region exceeds growth and respiration.

Following defoliation root growth ceases for a brief period of time; however, respiration continues to draw on organic reserves in both the root and stubble. As new root and top growth is initiated

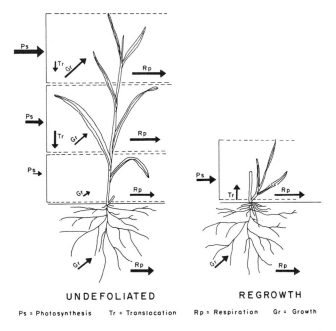

UNDEFOLIATED   REGROWTH

Ps = Photosynthesis   Tr = Translocation   Rp = Respiration   Gr = Growth

Fig. 5. A schematic representation of the movement of carbon during undefoliated growth and the regrowth of a herbage plant.

from axillary buds the export of food reserves from the root and stubble to the new growth for respiration and building material for new protoplasm is accelerated. This export probably continues until at least one fully expanded leaf has been formed in grasses (Williams, 1964) and for 21 days in alfalfa (Fig. 4, pp. 370).

Sufficient information is not available at this time to provide quantitative estimates of the various components in Fig. 5 or to illustrate how the components differ between species. The techniques of CO_2 fixation studies in leaf chambers, $^{14}CO_2$ labelling and chemical procedures for extraction and identification of organic compounds can provide the necessary data. An attempt to provide estimates of the compounds was made by Marshall and Sager (1968b) who proposed a "reciprocity diagram" showing the movement of ^{14}C-labelled material between tillers of ryegrass which act as "source" and "sink". Further provision of quantitative estimates of both the amount and direction of movement of the compounds involved in regrowth of herbage species will permit an assessment of their relative importance not available at this time.

REFERENCES

Aberg, E. (1964). In "The Physiology and Biochemistry of Herbicides" (L. J. Audus, ed.), Academic Press, London and New York.

Archbold, H. K. (1940). New Phytol. 39, 185-219.

Arny, A. C. (1932). Minn. agric. Expn. Stn. Tech. Bull. 84.

Baker, H. K. and Garwood, E. A. (1959). J. Br. Grassld Soc. 14, 94-104.

Blacklow, W. M. and McGuire, W. S. (1971). Crop Sci. 11, 19-22.

Bray, G. A. (1960). Analyt. Biochem. 1, 279-285.

Burton, G. W. and Jackson, J. E. (1962). Agron. J. 54, 53-55.

Cugnac, A. de (1931). Annls. Sci. nat. 13 1-129.

Davidson, J. L. and Milthorpe, F. L. (1965). J. Br. Grassld Soc. 20, 15-18.

Davidson, J. L. and Milthorpe, F. L. (1966). Ann. Bot., N.S. 30, 185-198.

Dilz, K. (1966). Proc. X Int. Grassld Congr. 160-164.

Fulkerson, R. S. (1970). Proc. 11 Int. Grassld Congr. 555-559.

Graber, L. F., Nelson, N. T., Luekel, W. A. and Albert, W. B. (1927). Bull. Wis. agric. Exp. Stn 80.

Granström, B., (1954). Ann. Roy. Agric. Coll. Sweden, 21, 281-285.

Hodgkinson, K. C. and Veale, J. A. (1966). Aust. J. biol. Sci. 19, 15-21.

Hodgkinson, K. C. (1969). Aust. J. biol. Sci. 22, 1113-1123.

Hoshino, M. and Oizumi, H. (1968). Proc. Crop Sci. Soc. Japan 37, 82-86.

Hoshino, M., Nishimura, S. and Okubo, T. (1964). Proc. Crop Sci. Soc. Japan 33, 130-134.

Hoshino, M. Nishimura, S. and Okubo, T. (1967). Proc. Crop Sci. Soc. Japan 36, 269-274.

Hough, D. J. (1970). M.Sc. Thesis, Univ. of Guelph.

Hull, R. J. (1969). Weed Sci. 17, 314-320.

Hunter, R. A., McIntyre, B. L. and McIlroy, R. J. (1970). J. Sci. Fd Agric. 21, 400-405.

Jeffray, H. and Alvarez, J. (1961). Analyt. Chem. 33, 612-615.

Jewiss, O. R. and Powell, C. E. (1965). Ann. Rep. Grassld Res. Inst. Hurley, 67-72.

Julander, O. (1945). Pl. Physiol., Lancaster 20, 573-599.

Jung, G. A. and Smith, D. (1960). Pl. Physiol., Lancaster 35, 123-125.

Jung, G. A. and Smith, D. (1961). Agron. J. 53, 359-364.

Kust, C. A. and Smith, D. (1961). Crop Sci. 1, 267-269.

Langer, R. H. M. (1963). Herb. Abstr. 33, 141-148.

Langer, R. H. M. and Steinke, T. D. (1965). J. agric. Sci., Camb. 64, 291-294.

Leach, G. J. (1968). Aust. J. agric. Res. 19, 517-530.

LeBaron, H. M. (1962). Diss. Abstr. 22, 2542-2543.

Linscott, D. L. and McCarty, M. K. (1962). Weeds 10, 298-303.

Mackenzie, D. J. and Wylam, C. B. (1957). J. Sci. Fd Agric. 8, 38-45.

MacLeod, L. B. (1965). *Agron. J.* 57, 345-350.
Marshall, C. and Sager, G. R. (1968a). *Ann. Bot.* 32, 715-719.
Marshall, C. and Sager, G. R. (1968b). *J. exp. Bot.* 19, 785-794.
Matches, A. G. (1969). *Agron. J.* 61, 896-898.
May, B. H. (1960). *Herb. Abstr.* 30, 239-245.
McCarty, E. C. (1938). U.S.D.A. Tech. Bull. 598.
McIlvanie, S. K. (1942). *Pl. Physiol., Lancaster* 17, 540-557.
Milthorpe, F. L. and Davidson, J. L. (1966). *In* "The Growth of Cereals and Grasses" (F. L. Milthorpe and J. C. Ivins, eds), Butterworths, London.
Moore, R. M. and Bibbiscombe, E. F. (1964). *In* "Grasses and Grasslands" (C. Barnard, ed.), Macmillan, London.
Musgrave, R. B. and Moss, D. N. (1961). *Crop Sci.* 1, 37-41.
Nelson, C. J. and Smith, D. (1968). *Crop Sci.* 8, 25-28.
Nielsen, Hj. M. and Lysgaard, C. P. (1956). *Asskr. K. Vet.-Landbohojsk.* 77-107.
Ott, D. G., Richmond, C. R. and Trujillo, T. T. (1959). *Nucleonics* 17, 106-108.
Okajima, H. and Smith, D. (1964). *Crop Sci.* 4, 317-320.
Pearce, R. B., Fissel, G. and Carlson, G. E. (1969). *Crop Sci.* 9, 756-759.
Peters, E. J. (1956). Ph.D. Thesis, Univ. of Wisconsin, Madison.
Raese, J. T. and Decker, A. M. (1966). *Agron. J.* 58, 322-326.
Richardson, A. E. V., Trumble, H. C. and Shapter, R. E. (1932). Aust. C.S.I.R.O. Bull. 66.
Ruelke, O. C. and Smith, D. (1956). *Pl. Physiol., Lancaster* 31, 364-368.
Ryle, G. J. A. (1970). *Ann. appl. Biol.* 66, 155-167.
Sager, G. R. and Marshall, C. (1966). Proc. IX Int. Grassld Congr. 493-497.
Sheard, R. W. (1967). *J. Sci. Fd Agric.* 18, 339-342.
Sheard, R. W. (1968a). *Crop. Sci.* 8, 55-60.
Sheard, R. W. (1968b). *Crop. Sci.* 8, 658-660.
Sheard, R. W. (1970). Proc. XI Int. Grassld Congr. 570-574.
Sheard, R. W. and Winch, J. E. (1966). *J. Br. Grassld Soc.* 21, 231-237.
Small, J. G. C. and Leonard, O. A. (1969). *Am. J. Bot.* 56, 187-194.
Smith, D. (1962). *Crop Sci.* 2, 75-78.
Smith, D. (1967). *Crop Sci.* 7, 62-67.
Smith, D. (1968a). *J. Br. Grassld Soc.* 21, 231-237.
Smith, D. (1968b). *Cryobiology* 5, 148-159.
Smith, D. and Silva, J. P. (1969). *Crop Sci.* 9, 464-467.
Smith, L. H. and Marten, G. C. (1970). *Crop Sci.* 10, 146-150.
Sonneveld, A. (1962). *Neth. J. agric. Sci.* 10, 427-444.
Sprague, V. C. and Sullivan, J. T. (1950). *Pl. Physiol., Lancaster* 25, 92-102.
Stauffer, M. D. (1969). M.Sc. Thesis, Univ. of Guelph.
Steward, F. C. (1959). "Plant Physiology—a Treatise," Vol. II, Academic Press, New York and London.
Sullivan, J. T. and Sprague, V. G. (1943). *Pl. Physiol., Lancaster* 18, 656-670.
Thomas, M., Ranson, S. L. and Richardson, J. A. (1956). "Plant Physiology." J. and A. Churchill Ltd., London.
Troughton, A. (1957). *Bull. Commonw. Agric. Bur.* 44.
Ueno, M. and Smith, D. (1970). *Crop Sci.* 10, 396-399.
Wardlaw, I. F. (1968). *Bot. Rev.* 34, 79-105.
Webb, J. A. and Gorham, P. R. (1964). *Pl. Physiol., Lancaster* 39, 663-672.
Weinmann, H. (1948). *J. Br. Grassld Soc.* 3, 115-140.
Weinmann, H. (1952). Proc. 6 Int. Grassld Congr. 655-660.
Weiser, C. J. (1970). *Science, N.Y.* 169, 1269-1278.
Williams, R. D. (1964). *Ann. Bot. N.S.* 28, 419-426.
Wolf, D. D. (1967a). Conn. Agr. exp. Stn. Res. Rept. 22.
Wolf, D. D. (1967b). *Crop Sci.* 7, 317-320.

CHAPTER 26

Genetic Variation in Herbage Constituents

J. P. COOPER

Welsh Plant Breeding Station, Aberystwyth
Wales, United Kingdom

I. INTRODUCTION

Forage crops are grown to provide digestible energy and other nutrients, including protein and minerals, for livestock, particularly ruminants. The crop is thus a means of converting the seasonal inputs of light energy and soil nutrients into forage of high nutritive value, and hence into animal products. In any environment, the possible output of digestible energy will be limited primarily by the seasonal input of light energy (see Chapter 17) but can also be modified by other climatic factors such as temperature and water supply or by the availability of soil nutrients (Cooper, 1970). Similarly, the protein and mineral content of the crop are greatly influenced by the input of soil nutrients which, however, are usually under the closer control of the farmer. It is the aim of the plant breeder to develop varieties which provide the most efficient conversion of these environmental inputs in terms of the requirements of particular livestock systems.

The optimum composition of the forage will depend on the type of livestock and the system of management (A.R.C., 1965). The first pre-requisite is usually the maximum production of digestible energy, but the content of other constituents required will depend on the management system, i.e. whether the herbage is intended to provide a balanced diet, catering for all the animal requirements, or whether it is regarded as a basic energy source to be supplemented with protein and/or minerals from elsewhere (Raymond, 1969; Rogers, 1970). Where the crop itself is to be used as a protein and mineral-rich supplement, as with many legumes, its protein and mineral content will be of particular importance.

Although the total yield of digestible energy is usually of primary importance, two other factors also determine the efficient use of this energy; (i) a high voluntary intake, to ensure that the maximum amount of the forage is eaten by the animal, and (ii) the efficient conversion into economic end products of the digestible energy consumed (Raymond, 1969; Rogers, 1970). In addition, the absence of harmful constituents, including excessive levels of such minerals as selenium, or more complex compounds such as oestrogens in subterranean clover is important. In many livestock systems, the content of protein and such minerals as calcium and magnesium may also require attention, although these can often be supplemented from elsewhere. Finally, since many forages are grown for conservation either as hay, dried grass or silage, chemical composition in relation to processing requirements must also be considered.

In terms of plant breeding objectives, these nutritional characteristics can be placed in four groups.

(1) Those directly concerned with the energy content of the crop and its utilization by the animal.
   (a) Digestibility
   (b) Features which influence voluntary intake
   (c) Features which influence utilization of the energy consumed.
(2) Toxic or harmful constituents

(3) Those which are necessary for animal production but which can often be supplemented from other sources.
  (a) Protein
  (b) Minerals
(4) Physical and chemical features which influence the efficiency of conservation, either as hay, dried grass or silage.

In the future, of course, it may be necessary to consider features concerned with more direct forms of processing, such as microbial digestion or direct extraction of leaf protein (Arkcoll and Festenstein, 1971 and Chapter 29, Vol. 3).

The significance of genetic variation in these nutritional characteristics and their incorporation in a breeding programme will depend on the stage which that programme has reached. In the early stages of any forage improvement programme the most appropriate species, usually grass or legumes, are chosen on the basis of survival and seasonal production of dry matter rather than on nutritional composition. It is usually not until a more advanced stage of the programme that deliberate selection for chemical composition is attempted, in order to provide more efficient animal output from the same inputs of energy and nutrients.

Such an improvement programme usually poses the following questions:

  (i) What nutritional factors are required in the crop for a particular livestock system, and how far do existing species and varieties provide for these requirements?
  (ii) How much genetic variation exists in these characteristics, and how can they best be incorporated into an approved variety?

The present chapter therefore discusses, firstly, the variation available in these nutritional characters in herbage and, secondly, the possibilities of their genetic modification in a breeding programme for improved animal production.

II. SPECIES AND VARIETAL DIFFERENCES IN HERBAGE CONSTITUENTS

The forage species grown in a particular region are usually chosen for their climatic adaptation and seasonal dry matter production, rather than for their chemical composition in terms of animal requirements. The first criterion of a forage species is that it will grow satisfactorily and provide an adequate seasonal yield of dry matter, i.e. digestible energy, resulting in economic animal production. Once a reasonable level of production has been achieved, further improvement in plant composition in relation to animal nutrition may be attempted. Such improvement may involve the choice of different forage species, the genetic modification of current species and varieties by the plant breeder, or supplementation of protein and minerals either directly or through complementary species such as legumes.

Forage species adapted to the same climatic or agronomic environment can differ markedly in chemical composition. In both tropical and temperate environments, for instance, legumes are usually richer in protein and in many minerals, particularly calcium and magnesium and certain trace elements, than the corresponding grasses and their digestibility and intake are also higher. On the other hand, certain legumes, such as subterranean clover (*Trifolium subterraneum*), contain oestrogenic compounds which adversely affect fertility in livestock.

In general, tropical grasses are usually less digestible and have lower protein and mineral contents than temperate grasses, but it is not certain how far these differences are genetic, or how far due to low nutrient inputs under tropical grassland systems (Hutton, 1970).

Furthermore, even within the same taxonomic group, species or genera may differ in important nutrient characteristics. In the temperate grasses, for instance, cocksfoot (*Dactylis glomerata*) is usually less digestible than ryegrass (*Lolium perenne*) when cut at the same stage of maturity, and timothy (*Phleum pratense*) is invariably low in sodium compared with ryegrass and most other temperate grasses. Again, certain species have the ability to accumulate harmful minerals in excessive amounts, such as the uptake of selenium by species of *Astragalus* in parts of the United States.

From the plant breeder's point of view, however, it is the variation within species or between closely related species which is important, since this can be manipulated in a breeding programme. The usual sequence in such a breeding programme is, firstly, to see whether differences exist between varieties or between individual genotypes; secondly, to estimate the heritability of these differences, including their stability over a range of environments and seasons, and finally to use this information to set up a practical breeding and improvement programme. This present section, therefore, examines the range of variation in the important nutritional characteristics listed earlier, paying special attention to the variation between populations and genotypes of the same species.

A. Energy Content and Utilization

1. Digestibility

One of the most important characteristics of herbage is its digestibility, since this determines not only the proportion of the herbage which can be used by the animal, but also markedly affects its intake (Osbourn, 1967; Walters, 1971). There are no large regular differences in total energy content per unit of dry matter between different forage species or even between grasses and legumes (Golley, 1961; Butterworth, 1964; Hunt, 1966). The digestibility of this dry matter as measured *in vivo* or *in vitro* can, however, vary greatly (Jones and Walters, 1969; Mowat, 1969).

In most temperate grasses, including ryegrass and cocksfoot, digestibility remains high ($>$ 65-70%) during spring and early summer, but declines rapidly and regularly (to $<$ 50%) after ear emergence or, in timothy, even before. Red clover (*Trifolium pratense*) and lucerne (alfalfa, *Medicago sativa*) show a similar decline with maturity, but white clover (*Trifolium repens*) maintains a high digestibility for most of the year as only the leaves and petioles are harvested (Raymond, 1969). Varietal differences in date of heading, therefore, result in marked seasonal differences in the course of this fall in digestibility with the late flowering varieties maintaining a high *in vitro* digestibility for a longer period (Dent and Aldrich, 1963; 1968). Walters *et al.* (1967) for instance, found that during the primary growth of perennial ryegrass, cocksfoot and timothy, varietal differences in *in vitro* digestibility were related largely to stage of maturity, but that differences in the regrowth were not so related.

Similarly, in tropical grasses, which are usually high in fibre and low in crude protein (French, 1957, Marshall *et al.*, 1969), digestibility falls off very rapidly with increasing maturity compared with tropical legumes, resulting in a decrease in voluntary intake and a rapid drop in the level of animal production. An important breeding aim is thus to develop varieties which do not decline so rapidly in digestibility (Hutton, 1970). However, Minson and McLeod (1970) in a comparison of two temperate and two tropical grasses found a high negative correlation between *in vivo* digestibility and both the temperature of growth (r = -0.76) and the evaporation over the growing period (r = −0·64). Over 65% of the variation in digestibility was accounted for by this relationship, suggesting that it may be difficult to breed tropical grasses with digestibilities as high as those of the temperate species.

Even at the same stage of maturity, species and varieties can differ in digestibility, both *in vivo* and *in vitro* (Raymond, 1969). Cocksfoot, for instance, is usually less digestible than ryegrass, although certain cocksfoot varieties possess *in vitro* digestibilities within the ryegrass range (Dent and Aldrich, 1963); even within the same cocksfoot maturity group Roskilde and Scotia are more digestible than Germinal and S. 37 (Dent and Aldrich, 1968). Similarly, in perennial ryegrass, both the leaf and stem fractions of the tetraploid Reveille are more digestible than the diploid S. 24 (Dent and Aldrich, 1968), while in the subtropical Coastal Bermuda grass (*Cynodon dactylon*) Burton *et al.* (1967) were able to introduce higher digestibility from the variety Kenya 56/14 to the standard Coastal Bermuda grass.

Individual genotypes, even within the same variety, can be more digestible than others. In cocksfoot, significant differences in *in vitro* digestibility have been reported between individual plants (Cooper *et al.*, 1962; Julen and Lager, 1966; Walters *et al.*, 1967; Mowat, 1969) and between diallel progenies (Rogers, 1970). Similarly, Christie and Mowat (1968) find large differences in *in vitro* digestibility among clones of cocksfoot and brome grass (*Bromus inermis*), while marked differences in *in vitro* digestibility have been reported between single clones of Dupuits lucerne and of *Medicago glutinosa* (Allinson *et al.*, 1969), and between clones of *M. media, M. falcata* and *M. sativa* (Heinrichs *et al.*, 1969, Heinrichs, 1970).

Although the use of *in vitro* techniques (Raymond, 1969) has made it possible to assess digestibility rapidly on small herbage samples, rather less is known about the relationship of digestibility to chemical composition. The work of van Soest (1968) on the relative digestibility of various structural carbohydrate and lignin components of the cell wall, and their possible assessment *in vitro,* is beginning to throw valuable light on this topic. Sosulki *et al.* (1960) have reported varietal differences in the degree of lignification in cocksfoot, smooth bromegrass (*Bromus inermis*) and the wheat grasses (*Agropyron* spp.) while Gil *et al.* (1967) find highly significant additive genetic variation among 10 clones of *M. sativa,* not only for *in vitro* digestibility, but also for more specific components, such as cell wall constituents (CWC), acid detergent fibre (ADF) and acid detergent lignin (ADL). More recently, Muller *et al.* (1971) report that the major recessive genes (**bm**$_1$, **bm**$_3$) which determine the brown midrib character in maize (*Zea mays*) also markedly decrease lignin content (see Chapter 7, Vol. 1).

As Jones (1970a) points out, the differences between species in *in vitro* digestibility at the same stage of growth may not always be based on the same structural constituents. For instance, the low digestibility of timothy at ear emergence compared with cocksfoot is associated with a higher content of structural carbohydrate, while the lower digestibility of cocksfoot relative to ryegrass is associated with a higher lignin content. As discussed later, these differences in composition may have important effects on intake. Similarly, in a comparison of Italian ryegrass (*Lolium multiflorum*) lucerne and sainfoin (*Onobrychis* spp.), Allinson and Osbourn (1970) found that differences in digestibility within a species or variety due to stage of maturity are usually highly correlated with the digestibility of the cellulose fraction, and negatively with the content of lignin. Species and varietal differences, however, are less simply related to the cellulose/lignin complex, and may be associated with the qualitative value of the lignin fraction in the cell wall material.

Digestibility can also be influenced by other plant constituents, for instance, by tannin content in *Lespedeza* (Donnelly and Anthony, 1970) and in *Desmodium* (Hutton, 1970). In *M. falcata*, Schillinger and Elliott (1966) found that differences between clones of up to 15% in digestibility were due to the presence of anti-metabolites with an adverse effect on the cellulolytic activity of the rumen microorganisms; this effect could be counteracted by the addition of certain amino acids, such as glycine, aspartic acid or glutamine, to the *in vitro* system. Similarly, the low protein content of certain tropical grasses may limit the activity of the rumen microflora, and hence reduce digestibility, which can, however, be restored by legume or urea supplements (Minson and Milford, 1967). From the plant breeder's point of view, the recent development of reliable *in vitro* techniques has revealed useful genetic variation between varieties and between individual genotypes for overall digestibility.

2. Features Influencing Voluntary Intake

Voluntary intake is often the major factor determining animal production (Osbourn, 1967; Raymond, 1969). Ingalls *et al.* (1965) for instance, found that about 70% of the variation in the production potential of forage was determined by differences in intake, and only 30% by digestibility.

In non-ruminants, intake is usually controlled by the level of certain blood metabolites. In ruminants, however, it depends more on the capacity of the rumen and the digestive tract. Less digestible forages occupy more volume in the rumen for a longer period, resulting in a lower intake, whereas more digestible forages pass more rapidly through the rumen, resulting in a higher intake (Balch and Campling, 1962 and Chapter 31, Vol. 3).

Even so, the relationship between intake and digestibility is not the same for all species or even varieties (Walters, 1971), and Osbourn (1967) suggests that differences in digestibility account for only 40 to 60% of differences in intake. Legumes, for instance, usually show a higher intake than grasses of the same digestibility, in both temperate (Raymond, 1969; Walters, 1971) and tropical (Milford, 1967; Milford and Minson, 1968) species. Within the temperate grasses, Walters (1971) reports low intake for S.51 timothy, and high for S.170 tall fescue (*Festuca arundinacea*) and S.345 cocksfoot if harvested at the same level of digestibility. Similarly in tropical grasses, Minson (1971) finds significant differences in intake between three varieties of *Panicum coloratum* and three of *P. maximum,* and regression analysis showed that the relationship between digestibility and intake differed markedly between the varieties.

In many cases, these differences in intake appear to be related to the relative proportions of water-soluble or pepsin-soluble material, and digestible fibre (Raymond, 1969). Osbourn *et al.* (1966) and Osbourn (1967), for instance, reported marked differences in voluntary intake, at the same digestibility, in the order lucerne > ryegrass > timothy, associated with the proportion of pepsin-soluble material to digestible fibre. Similarly, van Soest (1965) found that lucerne contained a lower proportion of cell wall and higher proportion of cell contents than grass of the same digestibility, and so occupied less volume x time in the rumen. In fact, a high cell wall content usually results in a lower intake even though the digestibility remains high.

Even within the same species group, Bailey (1964) found that the probable greater voluntary intake of Italian compared to perennial ryegrass was associated with a higher content of water-soluble carbohydrates and a lower cellulose content in the Italian. Similarly, Osbourn *et al.* (1966) found a higher intake of a diploid variety of

Italian ryegrass compared to the corresponding tetraploid, again associated with a higher content of pepsin-soluble material.

The two main components of pepsin-soluble material, soluble carbohydrate and protein, have often been reported to influence intake; for instance, in cocksfoot Bland and Dent (1962) found a high correlation between intake (under free choice) and content of water-soluble carbohydrate, while in sudan grass (*Sorghum vulgare* var. *sudanensis*) Gangstad (1964, 1966) found a high correlation between grazing preference and both total sugars (+0·97) and crude protein (+0·84). On the other hand, Rabas *et al.* (1970) found the grazing preferences of sheep and cattle on sorghum (*Sorghum vulgare*) and sudan grass to be negatively related to hydrocyanic acid content rather than positively to soluble carbohydrate.

In both ryegrass and cocksfoot, marked varietal and genotypic differences in water soluble carbohydrates have been reported (Cooper, 1962), and it has been possible to develop high and low selection lines for this character (Cooper and Breese, 1971). Similarly, Dent and Aldrich (1963) report large variation in the water soluble carbohydrate content of varieties of ryegrass, timothy and cocksfoot of the same digestibility, while in *S. almum* breeding lines have been produced with 20% higher soluble carbohydrate than the standard variety (Hutton, 1970).

Differences in intake may not always be due to the same cause. The marked differences in intake between cocksfoot, ryegrass and timothy of the same level of *in vitro* digestibility (Walters, 1971) have been analysed further by Jones (1970a). He finds that the higher intake of cocksfoot was associated with a higher lignin content, but a lower content of structural carbohydrate which may promote more rapid breakdown in the rumen. The lower intake of timothy than of ryegrass or cocksfoot was associated with a high content of cell wall constituents, a slower rate of digestion *in vitro* and a lower rate of passage *in vivo*. Similarly, in a comparison of voluntary intake, digestibility and the cellulose-lignin complex in Italian ryegrass, lucerne and sainfoin, Allinson and Osbourn (1970) find that while the variation in intake with maturity within a species is positively related to cellulose digestibility, and negatively to lignin, in species or varietal comparisons the direct relation between intake and the cellulose-lignin complex is reduced, and the proportion of digesta derived from cellulose assumes major importance.

Although intake is not always associated with palatability, as estimated from animal preferences under free choice (Barnes and

Mott, 1970), certain specific constituents can decrease both palatability and intake. In *Phalaris arundinacea*, for instance, Roe and Mottershead (1962) and O'Donovan *et al.* (1967) found large differences in voluntary intake between lines, the hexaploids being much less acceptable than tetraploids. These differences were apparently based on the contents of certain alkaloids, and an extract from the low preference strain sprayed on to the high line made it unacceptable to sheep. Similarly, in *Lespedeza* (Donnelly and Anthony, 1970), tannin content markedly depresses digestibility, and Wilkins *et al.* (1953) have shown that the voluntary intake of fresh herbage by sheep is closely related to tannin content.

Palatability can however, often be improved by selection, as in *Eragrostis curvula* (Voigt *et al.*, 1970) and in tall fescue (Buckner and Fergus, 1960; Craigmiles *et al.*, 1964) even without any knowledge of its chemical basis. Intake can also be limited by the mineral or protein content of the herbage. Patil and Jones (1970) for instance, have shown that the intake by sheep of perennial ryegrass, tall fescue and timothy can be markedly increased by cobalt supplementation.

As already mentioned, in certain tropical grasses of very low protein content, the rate of digestion, and hence the voluntary intake, can be limited by lack of nitrogenous substrates for the rumen flora. The intake of Pangola grass (*Digitaria decumbens*), for instance, could be increased over 50% by protein supplementation with legumes or urea (Minson and Milford, 1967). There may, however, be more direct effects of protein or urea supplementation on the intake of low protein roughage, operating through the nitrogen status of the animal itself (Egan, 1965).

The chemical and physical basis of voluntary intake is, however, still not clear and unlike digestibility, no standard *in vitro* method has yet been developed for screening breeding material, although Evans (1964) and Wilson (1965) have shown that in the ryegrasses, leaf tensile strength provides a useful indication of cellulose content. In view of the important role of voluntary intake in determining animal production (Raymond, 1969), increased knowledge of the chemical basis of intake and the development of appropriate screening techniques is most important.

In discussing the problem of developing rapid and low cost screening methods, van Soest (1968) has stressed the value of analysing cell wall material into its cellulose, hemicellulose and lignin components, using the acid detergent method. He points out the importance in breeding work of identifying the limiting constituent for a particular forage. For instance, lucerne is high in lignin, which limits its digestibility, whereas bermudagrass is low in lignin, but high in cell wall contents and in silica. Limitations of this method, particularly for legumes, have been described by Bailey and Ulyatt (1970). Raymond (1969) has suggested that measurement of (i) the proportion of acid pepsin soluble material and (ii) the rate of cellulose digestion *in vitro* should be useful, while Thomson and Rogers (1971) also have suggested pepsin solubility (i.e. crude protein + soluble carbohydrate) as a useful indication of voluntary intake. They find that in polycross progenies of timothy, digestibility is not always correlated with pepsin solubility so that it should be possible to select for pepsin solubility among lines which are already highly digestible. However, more recently, Minson and Haydock (1971) in a survey of the relationship between pepsin solubility and intake in a range of *Panicum* varieties, concluded that pepsin solubility is only likely to be of value as a preliminary screening test in a breeding programme, rather than for detailed prediction.

3. Features Influencing Utilization of Energy

Even given the same intake of digestible energy, the efficiency with which it is used for animal production may vary. Most (70 to 80%) of the energy supply of ruminants is derived from volatile fatty acids (VFAs), mainly acetic, propionic and butyric, produced by fermentation of feeds in the rumen. The relative proportions of these acids differ with the type of feed, and in general, as digestibility of the forage decreases the proportions of propionic and butyric acid decrease, but that of acetic acid increases. Similarly, a high content of water-soluble carbohydrate seems to result in a high proportion of propionic acid. It has been suggested that the relative proportions of the different volatile fatty acids produced in the rumen influence the efficiency with which the feed is used by the ruminant. All these VFAs are used with equal efficiency for maintenance but acetic acid is used much less efficiently than butyric or propionic for the production of body fat. On the other hand, acetic acid appears to be used more efficiently than propionic for milk production, while there is little difference between their relative efficiencies for the production of body protein (Raymond, 1969).

It is, however, often difficult to relate the utilization of energy by the ruminant to any single characteristic of the herbage or even to determine whether differences in animal production are due to variation in the intake or in the utilization of energy.

In New Zealand, detailed comparisons have been made of the nutritive value to sheep of white clover and of contrasting ryegrass varieties. Rae *et al.* (1963) and Johns *et al.* (1963), for instance, found that lambs fed on clover or on ryegrass/clover swards showed higher liveweight gains than on ryegrass alone. Although there were no significant differences in digestibility or intake, the clover had a lower cellulose content and a higher ratio of readily fermentable (i.e. soluble) to structural carbohydrate. This led to a more intense fermentation and rapid physical breakdown of the clover or the clover/ryegrass mixture in the rumen and a more rapid turnover time, together with a greater concentration of VFAs than from the grass. The amount or ratios of volatile fatty acids provide one possible reason for the more efficient use of the clover or ryegrass/clover mixture in producing body fat. Furthermore, the more rapid passage of the clover or ryegrass/clover mixture through the digestive tract may mean that more digestible energy is absorbed from the small intestine.

Similarly, within the ryegrasses, Bailey (1964, 1965) found that the liveweight gains in sheep followed the sequence: Manawa ryegrass > Ariki ryegrass > perennial ryegrass; Italian ryegrass gave similar gains to Manawa. The Manawa ryegrass had a consistently lower cellulose content and usually a higher soluble sugar content than the perennial, while Ariki was similar to the perennial.

In a later comparison of sheep kept on a slight excess of herbage of white clover, Manawa reygrass and perennial ryegrass, Ulyatt (1969, 1970) also found liveweight gains in the order: clover > Manawa > perennial. There was little difference between the herbages in crude protein, lignin content or *in vitro* organic matter digestibility. Although the intake of clover was higher than that of the ryegrasses, only 20% of the variation in liveweight gain appeared to be due to differences in intake. Clover again had a higher ratio of fermentable to structural carbohydrates than the grasses. The two ryegrasses showed similar digestibility and intake, but the Manawa was higher in readily fermentable carbohydrates, but lower in structural carbohydrates than the perennial. Lignin-ratio measurements indicate that the rate of digestion and turnover in the rumen was more rapid in Manawa than in perennial, possibly resulting in greater post-ruminal utilization of organic matter in Manawa. As in the earlier work, the clover gave consistently greater concentrations of rumen VFAs than the two ryegrasses.

On the other hand, Thomson (1964) fed lambs on ryegrass of

high sugar (19%) and low crude protein (16%) content, and ryegrass of low sugar (7%) and high crude protein (29%) content, and found no difference in energy gains, while, as already mentioned, intake and energy utilization can also be influenced by the protein and mineral content of the herbage. It is clear that mechanisms other than the proportion of the different rumen fatty acids must be involved in determining the efficiency of energy conversion by the ruminant, and Raymond (1969) suggests that more emphasis should be placed on the level of feed digestibility, rather than on rumen acid pattern.

It is clear that for the same intake of digestible energy the efficiency of its utilization for meat or milk production may differ with the forage species or even variety, but our knowledge of the chemical basis of these differences is still fragmentary. As with intake, the development of appropriate selection and screening techniques in a breeding programme awaits more information on the physical or chemical basis of energy utilization by different types of livestock.

B. Toxic or Harmful Constituents

The content of digestible energy of herbage and its efficient utilization are usually the primary criteria of nutritional value, but animal production can sometimes be limited or even prevented by particular harmful constituents. A breeding programme involving appropriate screening techniques for these constituents is often the only way to overcome their effects. Such compounds include:

(1) oestrogens in subterranean clover and other legumes;
(2) coumarin in sweet clover;
(3) alkaloids as in lupins and *Phalaris*;
(4) cyanogenetic glucosides as in white clover and sorghum;
(5) nitrate and other toxic minerals;
(6) bloat-producing constituents in white clover and other legumes

1. Oestrogens in Subterranean Clover and Other Legumes

In subterranean clover certain isoflavones (formononetin, genistein, biochanin A) show high levels of oestrogenic activity, resulting in impaired reproduction in sheep. Pastures high in formononetin consistently show high oestrogenic activity, but those containing high levels of genistein and biochanin A only occasionally show high activity (Lindsay *et al.*, 1970). Rapid chemical screening methods have been developed for these compounds (Francis and Millington, 1965a) and the isoflavone contents are now known for over 150 lines of subterranean clover (Morley and Francis, 1968). Certain varieties, such as Dwalganup, Geraldton and Yarloop are high in formononetin; others, such as Clare, contain mainly genistein, while Mount Barker and Bacchus Marsh are low in both formo-nonetin and genistein but contain larger amounts of biochanin A. The differences are large enough to indicate the possibilities of breeding for low oestrogenic activity, and more recently individual genotypes with greatly reduced content of particular isoflavones have been induced by the chemical mutagen, ethyl methane sulphonate (Francis and Millington, 1965b). The biochemical relationships of these isoflavones and the effect of these single genes on the pathways of isoflavone synthesis will be discussed later (see also Chapter 6, Vol. 1).

In a survey of some 100 species of *Trifolium* (Francis *et al.*, 1967), only 14 had isoflavone contents comparable with subterranean clover, and seven of these were in the same section of the genus as *T. subterraneum* (see Chapter 6, Vol. 1, Table I). There were considerable varietal differences within the self fertilizing species, but not between the varieties of *T. pratense* examined, although individual plants of this latter species varied 3-fold in isoflavone content. In lucerne and Ladino clover (*T. repens*), on the other hand, the active oestrogenic principle appears to be a coumestrol (Todd, 1970), and large varietal differences in coumestrol content have been reported in lucerne (Hanson *et al.* 1965).

2. *Coumarin in Sweet Clover*

Coumarin, the lactone form of coumarinic acid is found in many legumes, including sweet clover (*Melilotus officinalis*). It reduces palatability and its presence in spoiled hay or silage causes "sweet clover bleeding disease" in cattle and sheep (Smith and Gorz, 1965), which is due to progressive weakening of the clotting ability of the blood resulting in internal haemorrhage. The anti-coagulant is not coumarin itself, but an oxidation product, dicumarol.

Intact plants of sweet clover contain little, if any, free coumarin, which occurs in the plant predominantly as a β-D-glucoside, which is readily hydrolyzed by a β-D-glucosidase also contained in sweet clover. Two independent gene pairs (**Cucu**, and **Bb**) determine the presence or absence of bound coumarin and of glucosidase respectively, high activity being dominant in both cases (Goplen *et al.*, 1957; Micke, 1962; Goplen, 1969). The variety Pioneer which is low in free coumarin contains ample bound coumarin but has very low enzyme activity.

M. officinalis and *M. alba* are usually high in activity, but *M. dentata* is low and has been used as a source of breeding material for low coumarin content. A number of low coumarin varieties of *M. alba* have recently been released, including Cumino (Canada), Acumar (Germany) and Denta (U.S.A.) (Smith and Gorz, 1965). However, when Goplen (1969) tested high and low coumarin isosynthetics of *M. officinalis* and *M. alba* at four different locations, he found a consistent superiority of the high coumarin iso-synthetics in both species for forage and seed yield, for height and vigour and for winter hardiness.

3. *Alkaloids in* Lupin *and* Phalaris (see also Chapter 8, Vol. 1)

One of the earliest examples of breeding for the absence of harmful constituents was the selection of alkaloid-free genotypes of lupin (*Lupinus luteus, L. angustifolius* and *L. albus*) to form the basis of improved sweet lupin varieties (Hackbarth and Troll, 1957). Such selection has continued in many parts of the world, as, for instance, in the development of the variety "Uniwhite" in western Australia (Gladstones, 1967). Most of these alkaloid-free genotypes prove to be single major gene mutants, at least 4 loci being known in *L. luteus,* 3 in *L. angustifolius* and 3 in *L. albus*.

In parts of Australia, sheep grazing on *Phalaris tuberosa* frequently suffer from an acute disorder known as "Phalaris sudden death," which is thought to be due to certain alkaloids, derivatives of methylated tryptamines (Gallagher *et al.*, 1964; Moore and Hutchings, 1967). Twenty-fold differences in alkaloid content have been reported between different lines of *P. tuberosa* (Oram and Williams, 1967). Commercial Australian varieties were consistently high while lines from Algeria were all of low alkaloid content, offering possibilities of breeding non-toxic varieties. Rendig *et al.* (1970) also found great variation between individual plants of *P. tuberosa* in the content of *NN*-dimethyltryptamine and *NN*-dimethyl-5-methoxy-tryptamine. On the other hand, in more intensive studies on three contrasting varieties, Sirocco (high), Seedmaster (intermediate) and CPI 19351 (low), Oram (1970) found no correlation between toxicity and tryptamine content.

Similar alkaloids occur in *P. arundinacea,* and Woods and Clark (1971) have recently reported a major gene controlling tryptamine content in this species.

4. Cyanogenetic Glucosides in White Clover and Sorghum

White clover and other legumes, such as *Lotus* spp., can contain cyanogenetic glucosides (*lotaustralin* and *linamarin*) and/or the appropriate glucosidase, which releases hydrocyanic acid from them (see Chapter 9, Vol. 1). Varieties of white clover differ markedly in the proportion of both glucoside and enzyme, with a regular latitudinal cline from 100% cyanogenesis in the Mediterranean region to 0% in northern Europe (Daday, 1954a), and a similar altitudinal cline from sea level to high altitudes in the Alps (Daday, 1954b). The acyanogenetic genotypes appear to have a selective advantage at low temperatures, and breeders selecting for adaptation to cold climates may automatically select acyanogenetic plants.

As discussed later, both glucoside and enzyme are controlled by single loci and can, therefore, be readily manipulated by the plant breeder (Corkill, 1942; Atwood and Sullivan, 1943). The significance to the animal of these constituents is, however, not certain. It was at one time thought that the hydrocyanic acid produced might be implicated in bloat, or that clover pastures with a high content of cyanogenetic material could actually be poisonous or have a goitrogenic effect on livestock (Davies, 1970, and Chapters 9, Vol. 1 and 33, Vol. 3).

In sorghum and Sudan grass also, hydrocyanic acid (HCN) can be released from a cyanogenetic glucoside, dhurrin (Franzke, 1945, 1948). In contrast to white clover, however, the inheritance or even the biochemistry of this character is not well known, although large varietal differences in HCN content have been reported (Carlson, 1958; Gray *et al.,* 1968; Gillingham *et al.,* 1969; Loyd and Gray, 1970). Sudan grass is usually lower in cyanoglucoside than sorghum, and low HCN content tends to be dominant (Franzke, 1948). No sorghums have yet been found which are free from cyanogenetic glucoside, so attention has been concentrated on the selection of lines of low activity. Rancher, for instance, developed in South Dakota from Dakota Amber, had only $\frac{1}{10}$ of the HCN content of most other named varieties at the time it was produced (Franzke *et al.*, 1939).

5. *Nitrate and Toxic Minerals*

The accumulation of nitrate in forages can cause acute poisoning in livestock, the nitrate being reduced to nitrite, which reacts with the haemoglobin in the blood to form methaemoglobin (Wright and Davison, 1964 and Chapter 19). Species differ in their ability to accumulate nitrate. Sudan grass appears to be a particularly active accumulator, as are orchard grass and tall fescue. Brome grass, timothy and Ladino clover are intermediate in nitrate content, and alfalfa and wheat are low (Murphy and Smith, 1967). Varietal differences in nitrate content have been reported in oats grown for forage (Gul and Kolp, 1960; Crawford *et al.*, 1961), in perennial ryegrass (ap Griffith and Johnston, 1960), and in cocksfoot (Dotzenko and Henderson, 1964). The possibility of selecting for lower nitrate content is, however, complicated by the fact that nitrate accumulation is strongly influenced by the level of nitrogen application to the crop. Free nitrate usually accumulates appreciably only at crude protein values above 12-14%.

Certain forage species are known to accumulate other deleterious minerals. In some seleniferous soils in the United States, for instance, species of *Astragalus* and *Xylorrhiza* take up selenium to toxic levels, while in South Africa, certain fodder shrubs can accumulate copper up to 4-15 times the concentration in grasses (Todd, 1970). The usual remedy here is to choose species which are non-accumulators, but it may be possible to select genetic material which takes up less of the particular mineral. The form in which the minerals are accumulated may also be important. Peterson and Butler (1962) for instance, found that the selenium accumulator, *Neptunia amplexicaule* and the crop species, perennial ryegrass, white clover, red clover and wheat take up similar amounts of ^{75}Se, but in all except *Neptunia,* it is rapidly incorporated into seleno-amino-acids (see also Chapters 1, Vol. 1 and 19).

6. *Bloat-Producing Constituents*

The problem of bloat illustrates the difficulty of dealing with an animal disorder from the breeding point of view, when its exact cause is not known. Bloat was at one time thought to be a result of HCN production from cyanogenetic legumes such as white clover, but is now regarded as due to the formation of a stable foam within the rumen which traps the fermentation gases (Reid, 1960). The main foaming agent appears to be plant cytoplasmic protein. A protein fraction with a sedimentary velocity of 18 Svedberg units (Fraction 1, see Chapter 2, Vol. 1) has been isolated and claimed to be specifically active. Bloat-producing forages, such as lucerne, contain more than 4-5% of this protein, while non-toxic forages contain less than 1% (McArthur and Miltimore, 1966, 1969). Other studies, however, suggest that other cytoplasmic proteins can also form stable foams (Jones and Lyttleton, 1969). Some legumes, such as *Lotus corniculatus* and *L. major,* which do not cause bloat, contain tannins, which may form a non-foaming complex with protein (Jones *et al.,* 1970). Little information is yet available on varietal or genotypic differences in the content of the active protein fraction or of the tannins.

7. *Miscellaneous*

A number of other plant constituents have been suggested as the cause of certain animal disorders or as factors limiting palatability, intake and digestibility.

Tannins which decrease the palatability, intake and digestibility of the herbage are found in species of *Lespedeza* (Donnelly and Anthony, 1970) and in the tropical legume *Desmodium* (Hutton, 1970). In *Lespedeza* large varietal and genotypic differences in tannin content have been reported (Stitt, 1943; Donnelly, 1954; Bates and Henson, 1955; Cope, 1962).

Many tropical legumes may contain potentially harmful constituents (Hutton, 1970). *Leucaena leucophila,* for instance, contains mimosine, which affects reproduction in monogastric animals, but is usually detoxified in the rumen. Brewbaker and Hylin (1965) found a large variation in mimosine content among species of *Leucaena,* while Gonzalez *et al.* (1967) report a three-fold variation in mimosine content in *L. leucophila.* Similarly, *Indigofera spicata* contains the hepatotoxic amino acid, indospicine, and a low content of this compound may be an important breeding objective (see Chapter 1, Vol. 1).

Clearly there is ample genetic variation for many plant constituents which are known to be responsible for specific animal disorders. Many other disorders, however, appear to be nutritional in origin, although the causal agent has not yet been identified. Once the causal agent is known, it should be possible to develop rapid and reliable assay techniques for use in a breeding programme.

C. Protein and Mineral Content

The nutritional characteristics already discussed, i.e. a high intake of digestible energy and its efficient conversion, and the absence of harmful constituents, are of primary importance in a breeding programme. Other features, however, such as the content of protein and certain minerals, may be useful secondary breeding objectives, although these constituents can often be supplemented from elsewhere, particularly in an intensive farming system.

1. Protein Content

The protein requirements of different classes of livestock are well documented (A.R.C., 1965), but the protein content required in a particular forage will depend on whether it is to be used as a complete diet for the animal, mainly as an energy source, or, as in many legumes, as a high protein supplement. The overall protein content of the diet can, of course, be adjusted by varying the relative proportions of the energy and protein feed, while in intensive systems the required protein supplement can often be given in the form of non-protein nitrogen, such as urea (Minson and Milford, 1967; Raymond, 1969). We may therefore require different protein levels for different purposes, for instance, high for green crop drying, medium for grazing systems and low for silage (Raymond, 1969). The biological value of leaf protein in terms of amino acid composition does not differ greatly between forage species and varieties (Nelson, 1969), but this is not usually important in ruminant nutrition where the necessary amino acids are synthesized afresh by the microflora of the rumen.

In the past, protein content has often been considered as an index of quality since it is usually highly correlated with digestibility, but in most intensive grassland systems in temperate environments the level of protein is rarely limiting in animal production (Raymond, 1969). In tropical and sub-tropical forages, however, protein is more likely to be limiting (Milford, 1964). Many tropical grasses decline to a very low protein level at maturity, with a corresponding reduction in digestibility and intake (French, 1957; Marshall et al., 1969).

The protein content of forages is strongly influenced by nitrogen fertilization, though legumes usually have higher values than grasses (Davies, 1970). Even so, varietal and genotypic differences in crude protein content for the same nitrogen input have been reported in ryegrass (Cooper, 1962; Vose and Breese, 1964; Lazenby and Rogers,

1965), in *Dactylis* (Cooper, 1962; Dotzenko and Henderson, 1964), in *Phalaris tuberosa* (Clements, 1969; Clements *et al.*, 1970), in *P. arundinacea* (Asay *et al.*, 1968) and in *Cynodon dactylon* (Burton *et al.*, 1967). Similarly, in lucerne, significant differences in crude protein content have been reported between genotypes of *Medicago sativa, M. media* and *M. falcata* (Heinrichs *et al.*, 1969; Heinrichs, 1970).

A high protein content is, however, often associated with lower dry matter production (Asay *et al.*, 1968; Rogers, 1970) and there is often an inverse correlation between the content of crude protein and that of soluble carbohydrate (Cooper, 1962, Rogers, 1970), which may be important for intake and silage quality. A high crude protein content may also be associated with the accumulation of free nitrate and other harmful non-protein nitrogenous compounds (Wright and Davison, 1964). Excessive protein levels should therefore be avoided. In intensive forage systems, where protein content can be increased so readily by nitrogen fertilization, there may in fact be a need to select for lower protein content, and so develop varieties which respond to nitrogen by increased dry matter and energy production, rather than by increased protein content.

2. *Mineral Content*

In view of the importance of the mineral content of herbage in animal nutrition, comprehensive analyses of most temperate forage species have been made (Fleming, 1965; Whitehead, 1966; Davies *et al.*, 1968; Reid *et al.*, 1970), and similar information is becoming available for tropical species (French, 1957; Dougall and Bogdan, 1966; Long *et al.*, 1969) (see Chapters 12, Vol. 1, and 19). In general, tropical grasses are lower in mineral content than temperate species, though it is not always clear whether this is an intrinsic property of the species or due to low mineral status of the soil (French, 1957; Long *et al.*, 1969). In both tropical and temperate environments, legumes are usually higher in minerals than grasses, particularly in calcium and magnesium (Davies *et al.*, 1968; Davies, 1970; Todd, 1970).

Deficiencies of specific minerals are often found in extensive livestock systems on soils of low nutrient status, but are also likely to be important under intensive systems of management with high nitrogen inputs and high dry matter yields. In both temperate (Davies, 1970; Mudd, 1970) and tropical (Hutton, 1970) environments it has been questioned whether the mineral content of a highly producing pasture of grass alone is adequate for maximum animal production. The calcium and magnesium levels of perennial ryegrass, for instance, may well fall below those regarded as necessary for intensive milk production (Davies *et al.*, 1968; Davies, 1970) compared with those of white and red clover.

The mineral content of herbage will often reflect the mineral status of the soil. In parts of Australia, for instance, the balance between copper and molybdenum is important in determining the copper status of the animal. On soils of low molybdenum content, both subterranean clover and grasses contain little or no molybdenum resulting in accumulation of copper in the liver and chronic copper poisoning. On soils with a very high molybdenum content, on the other hand, subterranean clover can accumulate up to ten times more molybdenum than grasses, resulting in copper deficiency in livestock (Todd, 1970). Such differential accumulation of minerals *within* a species could naturally complicate any programme of selecting genetic material for higher or lower mineral content.

Even so, regular species differences in mineral content have been detected within both grasses and legumes. Among the temperate grasses, for instance, timothy and meadow fescue (*Festuca pratensis*) are consistently low in sodium relative to ryegrass and cocksfoot (ap Griffith and Walters, 1966; de Loose and Baert, 1966), while timothy also has a lower content of cobalt and copper than have cocksfoot and ryegrass (Fleming, 1965). Similarly, Johnson and Butler (1957) reported more than 20-fold differences between forage grass species in iodine content, Yorkshire fog (*Holcus lanatus*) being particularly low, and perennial ryegrass high, while Alderman and Jones (1967) found a lower iodine content in ryegrass and tall fescue than in cocksfoot and timothy, usually below the level recommended for pregnant or lactating animals.

Varietal differences in mineral content have also been reported. Large differences in sodium and potassium content have been detected between varieties of perennial ryegrass, cocksfoot and tall fescue (ap Griffith *et al.*, 1965; ap Griffith and Walters, 1966) with an inverse relationship between potassium and sodium. Significant differences have also been found in the content of phosphorus, magnesium, calcium, potassium and sodium between contrasting varieties of Italian and perennial ryegrass (Table I). Similarly, in perennial ryegrass, Vose (1963) found marked varietal differences in the uptake of potassium and calcium, often associated with differences in cation exchange capacity, and also in the ratio of potassium to calcium plus magnesium, which has been implicated in hypo-magnesemia in dairy cows (Grunes *et al.*, 1970). Large varietal differences in iodine content in ryegrass (0·10 to 0·28 ppm) and in white clover (0·14 to 0·28 ppm) have been reported by Alderman and Jones (1967), while Davies *et al.*, (1968) found significant varietal differences for phosphorus, potassium, sodium and calcium in white clover, and for potassium, sodium and calcium in red clover. Again, in white clover, large differences in the phosphorus (Snaydon and Bradshaw, 1962) and the calcium and sodium (Snaydon and Bradshaw, 1969) contents of locally adapted ecotypes have been

TABLE I

Heritability estimates for nutritional constituents, derived from a 6 x 6 diallel analysis of contrasting ryegrass varieties[a] (Cooper, unpublished)

Constituent	Heritability (narrow-sense)	Range between extreme varieties (% of dry matter)
In vitro digestibility	0·42	72·9—76·7
Water-soluble carbohydrates	0·84	22·4—47·0
N	0·63	1·96—3·78
P	0·68	0·32—0·47
Mg	0·86	0·14—0·21
Ca	0·78	0·34—0·56
K	0·80	2·65—3·85
Na	0·55	0·07—0·15

[a] Young vegetative material in seeding year.

reported, often related to the phosphorus and calcium content in the soil of their original habitat.

Large differences in mineral content can also occur between individual genotypes, even from the same variety. For instance, significant differences were found by Robinson (1942) in the content of calcium, phosphorus and potassium between eight clones of white clover, by Seay and Henson (1958) in phosphorus and potassium content between 30 clones of Kenland red clover, and by Heinrichs *et al.* (1969) in the phosphorus, sulphur, potassium, magnesium and calcium content of genotypes of lucerne (*M. sativa*, *M. media* and *M. falcata*).

In New Zealand, 15-fold variation in iodine content has been detected between different ryegrass clones (Butler and Glenday, 1962), while Butler *et al.* (1962) reported significant differences between seven clones of long-duration ryegrass for the content of phosphorus, sodium, iron, titanium, aluminium, manganese, copper, zinc and iodine, none of these differences being related to dry matter production.

There is clearly appreciable genetic variation in the mineral content of forage grasses and legumes, though these characteristics are markedly influenced by the mineral status of the soil. It should, therefore, be technically possible to select and develop varieties with high or low levels of particular minerals. As discussed later, however, the priority given to such selection must take account of the alternative possibility of modifying the mineral content of the feed by fertilizer application to the crop or by direct supplementation to the animal (Raymond, 1969).

D. Processing Requirements

Forage conservation to provide a storage buffer between the seasonal production of the crop and the requirements of the animal plays an important part in many livestock systems, but little is yet known of the chemical and physical features of the crop which make for most efficient processing in the form of hay, dried grass or silage. Jones (1970b) in a comparison of varieties of ryegrass, timothy and cocksfoot has shown that a high content of water-soluble carbohydrates is important in providing a lactic acid type fermentation, resulting in well-preserved silage of high quality. The ryegrass varieties were regularly higher in water-soluble carbohydrates and produced silage of higher quality than cocksfoot when grown under the same environment and management, but individual varieties of both ryegrass and cocksfoot differed in water-soluble carbohydrate content and correspondingly in silage quality.

Clearly, as more information becomes available on the chemical and physical features of the crop which make for efficient conservation, the breeder will need to consider such features for incorporation in his improvement programme.

Throughout the present discussion, herbage composition has been considered in relation to processing through the ruminant, either directly or via some form of conservation. In the future, however, more direct methods of processing, such as microbial digestion on an industrial scale, or the direct extraction of leaf protein, may impose rather different requirements in terms of chemical composition of herbage.

In conclusion, there appears to be appreciable genetic variation for many of the herbage constituents which are important in animal nutrition. Before developing an operational plant breeding programme, however, the breeder needs to decide which of these characteristics are likely to be limiting in the livestock systems with which he is concerned. In many cases, such a decision is complicated by lack of information on the chemical or physical basis of such important nutritional features as digestibility, the intake of digestible energy and its efficient utilization by the animal.

III. GENETIC CONTROL

Although ample variation exists for many nutritional characteristics, their ease of manipulation in a breeding programme and eventual incorporation into an improved variety will depend on the type of genetic control, the amount of genetic variation available, and its heritability (i.e. the proportion of genetic to total variation). Genetic correlations between desirable and undesirable characters may also be important limitations in a breeding programme.

A. Type of Genetic Control

An important distinction is whether the character is controlled by one or a few major genes, which can be easily isolated and manipulated in a breeding programme, or whether it is under the control of many genes and consequently must be handled by more complex biometrical techniques.

1. Major Gene Control

Certain well-defined plant constituents are under the control of single major genes. These are often constituents which are not essential to the survival of the plant, i.e. whose presence or absence has no over-riding selective advantage. Examples are the oestrogenic isoflavones in subterranean clover, cyanogenetic glucosides in white clover, coumarin in sweet clover, and alkaloid content in lupin.

In subterranean clover, mutations at individual loci which control various steps in the synthesis of the oestrogenic isoflavones have been produced from the variety Geraldton by the chemical mutagen, ethyl methane sulphonate (EMS) (Francis and Millington, 1965b), and their biochemical effects have been analysed by Wong and Francis (1968a) (Fig. 1).

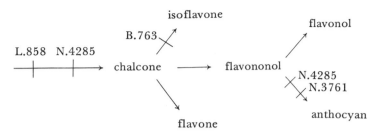

Fig. 1. Probable sites of action controlling isoflavone synthesis in some mutants of subterranean clover (Wong and Francis, 1968a).

The four main isoflavones found in subterranean clover are genistein and biochanin A, its methyl derivative, and diadzein and formononetin, its methyl derivative. In the variety Geraldton, genistein, biochanin A and formononetin form the main isoflavones. In intact leaves these occur in the "bound" form, probably as glucosides, but the free form is released on crushing.

Treatment with EMS resulted in four main types of mutant:

Group 1 (hh) e.g. 972 which shows only traces of isoflavones after crushing, but normal quantities are released after adding Geraldton leaf extract, i.e. these mutants lack the enzymic capacity to release isoflavone.

Group 2 (mm) e.g. A. 258 which possesses very large amounts of the isoflavones diadzein and genistein, but only small quantities of their methyl derivatives, suggesting that these mutants lack an enzyme for 4'O methylation of isoflavones. The other flavonoids are little changed.

Group 3 (a'a') e.g. B. 763 which has very low but detectable amounts of each isoflavone but all other flavonoids are greatly enhanced. An extract of B. 763 can release isoflavone from 972, so it contains the isoflavone-releasing enzyme.

Group 4 (b'b') e.g. L. 858 which also has a very low content of isoflavones, though it can release isoflavone from 972. It is deficient in anthocyanin, but the flavonoids, kaempferol, quercetin and isorhamnetin are only slightly affected.

Each of these mutants proved to be controlled by a single recessive gene.

Two further anthocyan mutants, N. 4285 (white seed) and N. 3761 (pale seed) were also studied. In N. 4285, the isoflavones were reduced, but to a lesser extent than in L. 858, and other constituents including flavonols were also reduced. In addition, there was complete inhibition of anthocyanin suggesting a double effect of this mutant. In N. 3761, there was little change in any of the flavonoids, but a visible reduction of anthocyanin in seeds and leaves.

Similar biochemical analyses of subterranean clover varieties of contrasting isoflavone content (Wong and Francis, 1968b) revealed a more complex situation. Yarloop, for instance, is similar to Geraldton but with genistein as the most abundant component. Clare, on the other hand, has a very high genistein content, with less biochanin A, and formononetin is greatly reduced, together with complete absence of flavonols. Mount Barker has a very low formononetin content with the absence of other 5-deoxy compounds, and a much higher content of genistein and biochanin A than in Geraldton.

These varietal differences clearly require genetic changes at more than one point in the biosynthetic pathway. The pattern shown in Clare, for instance, which is fairly typical of the subspecies *brachycalycinum* would involve:

(i) inhibition of the 5-deoxy pathway at an early stage,
(ii) blockage of the route to flavonols and
(iii) decrease in the capacity for $4'O$ methylation of isoflavones.

Production of these single gene mutations not only throws light on the possible pathways of biosynthesis of these oestrogenic compounds (Fig. 1), but also provides most useful isogenic material to examine the oestrogenic effects of individual isoflavones (Millington *et al.*, 1966). Feeding sheep on the mutant 972, which is deficient in enzyme, had the same oestrogenic effect as the original Geraldton, indicating that hydrolysis can occur in the sheep in the absence of the plant enzyme. Similarly, feeding the non-methylated mutant A. 258 showed that diadzein was just as active as formononetin, suggesting that formononetin may in fact be demethylated in the rumen; while feeding mutants B. 763 and L. 858 indicated that the precursors of isoflavones were not oestrogenically active.

In white clover the cyanogenetic glucosides (lotaustralin and linamarin) and the corresponding glucosidase are controlled by single dominants (**Acac** and **Lili** respectively) (Corkill, 1942; Atwood and Sullivan, 1943). These genes can be easily screened in a breeding programme, but there appears to be a selective advantage of acyanogenetic plants at low temperatures (Daday, 1954a, b).

In sweet clover (Smith and Gorz, 1965), both the presence of bound coumarin and of the appropriate glucosidase are controlled by single genes (**Cucu,** and **Bb**) (Goplen *et al.*, 1957; Micke, 1962;

Goplen, 1969). **Cucu** is possibly concerned with the *o*-hydroxylation of *trans*-cinnamic acid to form *o*-cinnamic acid. It has proved possible to induce mutations for low coumarin content by the chemical mutagen, ethyl urethane (Scheibe and Hülsmann, 1958).

In fodder lupins, the production of varieties containing little or no alkaloid has been made possible by the development of rapid screening methods to detect natural single gene mutations in existing bitter varieties (Hackbarth and Troll, 1957). In *L. luteus,* for instance, at least four loci have been detected (**du, am, lib,** and **v**), in *L. angustifolius* at least three (**iuc, esc,** and **dep**), and in *L. albus* at least three (**nut, pan** and **mit**). Similarly, in *Phalaris arundinacea* Woods and Clark (1971) report a single major gene controlling the presence of the alkaloid tryptamine.

Even single genes which are not specific for chemical components can have important effects on nutritive value, as with the single dwarfing gene (d_2) in pearl millet (Burton *et al.,* 1969), which markedly decreases stem percentage and increases *in vitro* digestibility. Similarly, in maize, the major genes (bm_1) and (bm_3), which produce the brown midrib phenotype, also greatly reduce lignin content (Muller *et al.,* 1971).

Such single major genes are comparatively easy to manipulate in a breeding programme, though they may be correlated with other less desirable characteristics. In sweet clover, for instance, low coumarin isosynthetics seem to be also low in dry matter production (Goplen, 1969). On the other hand, in white clover there appears to be an advantage in acyanogenetic types at low temperatures (Daday, 1954a, b).

2. *Polygenic Control*

Most nutritional characteristics, however, such as digestibility and protein content, show continuous variation and are under the control of many genes, while even those showing major gene control may be modified quantitatively (Corkill, 1942). In selecting for such quantitative characters it is important to know both the amount of genetic variation available and its heritability (i.e. the proportion of genetic to total variation), since this indicates the precision with which the variation can be identified and used in a selection programme (Falconer, 1960).

Heritability, defined in general terms as that proportion of the total variation which is genetic in origin, can be estimated in several ways. At the simplest level, clonal replication over one or more environments provides a measure of the overall genetic variation compared to the total variation, i.e. of heritability "in the broad sense". For the more precise assessment of heritability "in the narrow sense", breeding tests of increasing complexity, which can distinguish between the different components of genetic variation, are required (Latter, 1964; Breese and Hayward, 1970). These techniques range from the polycross design, in which heritability is estimated from the regression of progeny performance on the female parent, through pair crosses, in which both parents are known, to the more critical diallel cross which allows of the estimation of additive and non-additive genetic variation (corresponding to "general" and "specific" combining ability) and also of maternal and reciprocal effects. In general, it is only the additive portion of the genetic variation which can be fixed in a breeding programme, although dominance and interaction effects can be made use of in hybrid varieties. It must be stressed that a particular estimate of heritability refers strictly to the population or range of genetic material from which it was derived and does not necessarily apply to other populations, or even to the same population grown in a different environment. In most breeding work, the stability of a character over a range of seasons or locations is· also important, and appropriate biometrical techniques have been developed for measuring any genotype-environment interactions (Breese and Hayward, 1970).

For any character, an estimate of the genetic variation, together with its heritability, makes it possible to predict the likely response to selection, since $R = h^2 S$, where R is the response to selection, h^2 is the heritability and S is the possible selection differential, which is proportional to the amount of variation present. In general, a heritability value of over 0·30 indicates that useful response to selection should be possible. Furthermore, if two or more characters are measured on the same genetic material, it is possible to estimate not only the heritability of each character, but also any genetic correlations between them, i.e. the extent to which selection for one character would be expected to result in correlated response in the other (Falconer, 1960).

Strictly speaking, an estimate of heritability predicts response for the first generation of selection only, and for largely self-fertilizing species, such as subterranean clover, in which most plants are rather homozygous, little or no response to selection might be expected after the first generation. Most forage species, on the other hand, are cross-fertilizing, and each variety is built up of many genetically different individuals. In addition, most plants are very heterozygous and contain a large store of potential genetic variation, some of which can be released in each generation by segregation and re-combination and so become available for selection to act upon. In these cross-fertilizing forage species, therefore, as has been shown for flowering time in ryegrass (Cooper, 1963), steady response to selection can continue for several generations, often extending well outside the range of the original population.

B. Estimates of Heritability and Response to Selection

In spite of the extensive varietal or genotypic differences reported for many nutritional characteristics, comparatively few estimates of heritability and subsequent response to selection are available.

1. Digestibility and Intake Characteristics

Encouraging heritability estimates for digestibility and intake characteristics are available from both clonal replication and progeny tests. In cocksfoot, for instance, Christie and Mowat (1968) found that most (73%) of the variation in digestibility between 444 clones was genetic in origin, while heritabilities of over 0·5 have been reported for digestibility in this species (Cooper, 1962; Cooper et al., 1962; Rogers, 1970). Similarly, in a 6 x 6 diallel analysis of bromegrass, Ross et al. (1970) found highly significant general combining ability for this character.

In many cases, useful variation both in digestibility and in other associated nutritive characteristics has been reported. Rogers (1970), for instance, found significant differences between diallel progenies of cocksfoot for both in vitro digestibility and pepsin solubility, while Gil et al. (1967) reported highly significant general combining ability in Medicago sativa for in vitro digestibility, cell wall consti-tuents, acid detergent fibre, and acid detergent lignin, as well as protein. There was, however, a negative genetic correlation between digestibility and yield, and these authors conclude that an increase in total digestible energy can best be achieved by increasing dry matter.

Other characteristics associated with nutritive value also show useful genetic variation. Wilson (1965) examined the variation in leaf tensile strength and cellulose content in a range of Lolium genotypes. He found high heritabilities (about 0·8) for both these character-istics, with a genetic correlation of 0·93 between them. He suggests that these characteristics may serve as a useful screening technique for nutritive value in ryegrass.

Even where there are no differences in digestibility there may still be variation in other important nutritional characters. In perennial and Italian ryegrass (Table I), a 6 x 6 diallel analysis shows high heritability (0·84) for water soluble carbohydrate but not for digestibility, while even within a single variety of perennial ryegrass, S. 23, useful variation in water soluble carbohydrates occurs (Cooper, 1962). In both cases there was a strong negative genetic correlation between WSC and crude protein content. Useful heritability values for palatability (animal preference) also have been

TABLE II

Response to selection for *in vitro* digestibility in cocksfoot, and correlated responses in other characters (Breese, 1970)

| | % DOMD[a] (conservation cut) | Seedling vigour[b] | Date of ear emergence | Leaf size | |
				length (cm)	width (cm)
High selection lines (mean of 3)	72·3	11·1	11·2 May	27·7	1·09
Control variety (S.37)	65·3	6·8	18·8 May	22·8	0·83
Low selection lines (mean of 3)	60·5	4·9	16·5 May	21·8	0·85

[a] DOMD = Digestible organic matter in the dry matter.
[b] g dry wt/100 seedlings, 6 weeks after sowing.

reported for tall fescue (Craigmiles *et al.*, 1964) and for *Phalaris arundinacea* (Barnes *et al.*, 1970).

Year to year variation in digestibility and other characteristics may, however, complicate the task of the breeder. Rogers and Thomson (1970) for instance, in a clonal 5 x 5 diallel of *Lolium perenne* found significant general combining ability for *in vitro* digestibility and acid pepsin solubility in one year, but not in the next, although significant specific combining ability could be detected for all characters in both years.

The above heritability values indicate the possibilities of rapid change under selection. In cocksfoot, for instance, selection for increased digestibility has resulted in marked response over three generations (Table II), the high selection lines reaching values well within the ryegrass range. Such selection has also increased the content of water-soluble carbohydrates and resulted in other correlated changes, including earlier flowering, larger leaves and thicker and more succulent leaf sheaths (Breese, 1970).

Similarly, in ryegrass, it has proved possible to select lines with high and low soluble carbohydrate content from within a single variety (S. 23), (Table III). The high carbohydrate lines had a lower crude protein content and showed improved winter-hardiness both in the field and in freezing tests, but unfortunately were also more susceptible to crown rust (*Puccinia coronata*) (Breese, 1970) (see also Chapter 3, Vol. 1).

TABLE III

Response to selection for water-soluble carbohydrate content in S.23 perennial ryegrass, and correlated responses in other characters (Breese, 1970)

	Water-soluble carbohydrate (%)	Crude protein (%)	Susceptibility to rust (*Puccinia coronata*)[a]	Relative cold tolerance[b] %
High selection lines (mean of 3)	26·0	13·3	2·7	159
Low selection lines (mean of 3)	16·1	15·6	1·2	71

[a] eye score: 0 = resistant; 5 = most susceptible.
[b] relative to original S.23 as 100%.

2. Harmful Constituents

Although single gene differences have been isolated which control such harmful constituents as the isoflavone content of subterranean clover, even here modifying effects of polygenic systems have also been found, and intervarietal crosses indicate useful additive genetic variation (Morley and Francis, 1968).

Many other harmful constituents which are apparently under polygenic control prove to be highly heritable. In *Phalaris tuberosa*, for instance, the content of tryptamine alkaloids shows a high heritability of over 0·9, and 20-fold differences in alkaloid content have been produced between high and low selection lines (Oram, 1970). Similarly, heritabilities of about 0·4 (Bates and Henson, 1955) and about 0·7 (Cope, 1962), have been reported for tannin content in *Lespedeza*, with no adverse genetic correlation with yield. These estimates indicate that selection for low tannin content should be quite effective, and in fact, in *L. cuneata* Donnelly and Anthony (1970) have been able to select low tannin lines with a content of 2-3% and a digestibility of 65%, compared to high tannin lines with a content of 5-6% and a digestibility of only 58%. Again, "broad sense" heritabilities of 0·4-0·7 have been reported for the content of hydrocyanic acid in crosses between sorghum and Sudan grass (Barnett and Caviness, 1968).

3. *Content of Protein and Minerals*

Useful heritability values have been found for crude protein content (i.e. total nitrogen) in both ryegrass (Cooper, 1962; Thomson and Rogers, 1970) and in cocksfoot (Cooper, 1962), with usually a strong negative genetic correlation between crude protein and water-soluble carbohydrate. In *Phalaris arundinacea,* Asay *et al.* (1968) reported "broad sense" heritabilities for crude protein of 0·44-0·70, and "narrow sense" heritabilities of 0·19-0·41. About 70% of the total genetic variation was additive indicating possibilities of rapid response to selection.

Similarly, in a highly variable breeding population of *P. tuberosa,* Clements (1969) found a marked response to selection for high and low crude protein content at heading time, with a realized heritability of 0·25 and 0·20 in the high and low lines respectively. This selection produced positive correlated changes in *in vitro* digestibility, but as in the results of Asay *et al.* (1968) from *P. arundinacea* high nitrogen content was associated with a low relative growth rate and low dry matter yield.

Seasonal or environmental effects can complicate the expression of this genetic variation. Thomson and Rogers (1970) found significant general and specific combining ability for crude protein content in a 5 x 5 diallel cross between clones of perennial ryegrass, but the exact effects varied greatly between nitrogen rates and between years.

For mineral content, the results of Butler *et al.* (1962) on seven clones of long rotation ryegrass indicate encouraging "broad sense" heritabilities for phosphorus, nitrogen, sulphur, sodium, iron, titanium, aluminium, manganese, copper and zinc, while Butler and Glenday (1962) found iodine content to be strongly inherited. As shown in Table I, the heritability estimates derived from a diallel cross between six contrasting ryegrass varieties indicate useful additive genetic variation for nitrogen, phosphorus, magnesium, calcium, potassium and sodium in this range of material.

These results, although somewhat fragmentary, suggest that much of the variation for nutritional characteristics is genetic and additive and thus can be used in a selection and breeding programme. In fact, in those forage grasses and legumes which are cross-fertilizing, and hence rather heterozygous, useful additive genetic variation can often be detected even within a single variety. In such material, continued response to selection for several generations would be expected, and has already been demonstrated for such characters as digestibility and content of water-soluble carbohydrates. Unfavourable correlated responses to selection may, however, occur, as in the increased susceptibility to fungal attack of the high carbohydrate lines of ryegrass.

IV. USE OF GENETIC VARIATION IN A BREEDING PROGRAMME

It is clear that useful genetic variation exists within forage species for many nutritionally important constituents. A few of these are controlled by single major genes, but most show polygenic control, often with reasonably high heritability, which allows of useful response to selection, although correlated responses in undesirable characters may sometimes prove a problem.

However, before an attempt is made to incorporate these features in an operational breeding programme, the economic justification of breeding for nutritional characters needs to be carefully considered. Where animal production is limited by one or a few well-defined constituents, such as oestrogens in subterranean clover or coumarin in sweet clover, the effort is clearly well justified, but a breeding programme for such characteristics as digestibility or mineral content may be more difficult to justify. In principle, selection for digestibility and intake characteristics should be valuable by virtue of it resulting in a greater utilizable output of forage, and hence more efficient animal production, for the same environment inputs of light energy and soil nutrients. Even so, the primary criterion of a breeding programme is likely to be the total annual or seasonal production of digestible energy, which is usually highly correlated with the total dry matter (Raymond, 1969; Rogers, 1970). In South Australia, for instance, Knight and Yates (1968) have questioned the value of selecting directly for digestibility in cocksfoot, since this characteristic is modified so much by differences in maturity date and degree of summer dormancy.

Similarly, in many livestock systems, deficiencies in protein or in certain minerals may be more easily remedied by fertilizer application, by supplementation with a legume, or by direct mineral supplementation of the feed, than by a breeding programme.

Before embarking on a breeding programme for nutritional characteristics, the breeder also needs to consider the purpose for which the forage is to be used, i.e. whether it is to be regarded as a complete and balanced diet for livestock, or primarily as an energy source, to be supplemented with protein and/or minerals from elsewhere. Any management system is likely to require the maximum content of digestible energy, a high voluntary intake and efficient utilization of the energy consumed, together with the absence of harmful constituents (Raymond, 1969).

Having decided on the nutritional requirements for a particular livestock system, the breeder then needs to assess the deficiencies which exist in currently used species and varieties, together with the relative economic importance of these deficiencies in terms of animal production. He can then examine possible sources of variation for these characters. For certain characters, it may be possible to select within already established and adapted varieties, as for digestibility in cocksfoot or low alkaloid content in lupins, but often the breeder may have to try to introduce the required character from less adapted or less productive varieties, as in the introduction of low coumarin content from *Melilotus dentata* to *M. alba* and *M. officinalis,* or of high digestibility from Kenya 56/14 into Coastal bermudagrass (Burton *et al.,* 1967). A more sophisticated source of variation is the production of induced mutants, either by irradiation or by chemical means, as has been possible for isoflavone content in subterranean clover (Francis *et al.,* 1970). Such induction of mutation is likely to be of most value for single major gene differences in self-fertilizing species.

Finally, having selected his breeding material and produced a potentially improved variety with the required nutritional characters, the breeder must test the effect of his selection in terms of actual animal production. There are comparatively few cases in which such a breeding programme has been followed through to the stage at which improved animal production can be demonstrated. In pearl millet (*Pennisetum glaucum*) however, Burton *et al.* (1969) have assessed the effect of introducing the dwarfing gene (d_2) on *in vitro* digestibility, intake and animal production. This gene increases the percentage of leaf and decreases that of stem, but also decreases the total dry matter production to 70-90% of the original variety. It also, however, increases the digestibility and protein content of the stems. Dairy heifers ate 21% more of the dwarf type than of the original variety and gained 49% faster, while under rotational grazing, they showed a 20% increase in daily gains in spite of the lowered dry matter production. Similarly, Burton *et al.* (1967) have demonstrated that selection for increased digestibility in Coastal bermudagrass resulted in greater daily liveweight gains.

In conclusion, the incorporation of nutritional characters in a breeding programme is likely to be of value in two main types of livestock enterprise. Firstly, it will be important in the comparatively simple case where one particular constituent has a major limiting effect on animal production, as with oestrogens in subterranean clover and alkaloids in *Phalaris tuberosa*, or possibly in certain tropical grasses, where low digestibility and low protein are major limiting factors (Clements, 1970). Such major limitations are likely to be particularly important in tropical species, which have not yet been studied so intensively (Hutton, 1970).

Secondly, it may well be valuable in more complex and well developed intensive livestock systems, where forage varieties of good adaptation and high seasonal dry matter production are already available (Cooper and Breese, 1971). In such high input-output systems, it may well be economic to pay attention to further improvement of digestibility, of intake and of the efficient conversion of digestible energy. Rogers (1970) suggests that for this purpose the assessment of *in vitro* digestibility, crude protein content, pepsin-soluble components and sugars should be most useful. Such an advanced programme needs a considerable understanding of the nutritional requirements of the particular livestock system, together with the development of appropriate chemical and physiological techniques to screen genetic material rapidly and reliably in the breeding programme.

REFERENCES

Agricultural Research Council (1965). "The Nutrient Requirements of Farm Livestock. No. 2. Ruminants." Agricultural Research Council, London.
Alderman, G. and Jones, D. I. H. (1967). *J. Sci. Fd Agric.* 18, 197-200.
Allinson, D. W., Elliott, F. C. and Tesar, M. B. (1969). *Crop Sci.* 9, 634-637.
Allinson, D. W. and Osbourn, D. F. (1970). *J. agric. Sci., Camb.* 74, 23-36.
ap Griffith, G. and Johnston, T. D. (1960). *J. Sci. Fd Agric.* 11, 623-626.
ap Griffith, G. and Walters, R. J. K. (1966). *J. agric. Sci., Camb.* 67, 81-89.
ap Griffith, G., Jones, D. I. H. and Walters, R. J. K. (1965). *J. Sci. Fd Agric.* 16, 94-98.

Arkcoll, D. B. and Festenstein, G. N. (1971). *J. Sci. Fd Agric.* **22**, 49-56.
Asay, K. H., Carlson, I. T. and Wilsie, C. P. (1968). *Crop Sci.* **8**, 568-571.
Atwood, S. S. and Sullivan, J. T. (1943). *J. Hered.* **34**, 311-320.
Bailey, R. W. (1964). *N.Z. Jl agric. Res.* **7**, 496-507.
Bailey, R. W. (1965). *Proc. N.Z. Grassld Ass.* **1964**, 164-172.
Bailey, R. W. and Ulyatt, M. J. (1970). *N.Z. Jl agric. Res.* **13**, 591-604.
Balch, C. C. and Campling, R. C. (1962). *Nutr. Abstr. Rev.* **32**, 669-686.
Barnes, R. F. and Mott, G. O. (1970). *Agron. J.* **62**, 719-721.
Barnes, R. F., Nyquist, W. E. and Pickett, R. C. (1970). *Proc. XI int. Grassld Congr.* 202-206.
Barnett, R. D. and Caviness, C. E. (1968). *Crop Sci.* **8**, 89-91.
Bates, R. P. and Henson, P. R. (1955). *Agron. J.* **47**, 503-507.
Bland, B. F. and Dent, J. W. (1962). *J. Brit. Grassld Soc.* **17**, 157-158.
Breese, E. L. (1970). *Rep. Welsh Pl. Breed Stn* **1969**, 33-37.
Breese, E. L. and Hayward, M. D. (1970). Proc. Eucarpia, Forage Crop Section, Lusignan, France.
Brewbaker, J. L. and Hylin, J. W. (1965). *Crop Sci.* **5**, 348-349.
Buckner, R. C. and Fergus, E. N. (1960). *Agron. J.* **52**, 173-177.
Burton, G. W., Hart, R. H. and Lowrey, R. S. (1967). *Crop Sci.* **7**, 329-332.
Burton, G. W., Mouson, W. G., Johnson, J. C., Lowrey, R. S., Chapman, H. D. and Marchant, W. H. (1969). *Agron. J.* **61**, 607-612.
Butler, G. W. and Glenday, A. C. (1962). *Aust. J. biol. Sci.* **15**, 183-187.
Butler, G. W., Barclay, P. C. and Glenday, A. C. (1962). *Pl. Soil* **16**, 214-228.
Butterworth, M. H. (1964). *J. agric. Sci., Camb.* **3**, 319-321.
Carlson, I. T. (1958). *Agron. J.* **50**, 302-306.
Christie, B. R. and Mowat, D. N. (1968). *Can. J. Pl. Sci.* **48**, 67-73.
Clements, R. J. (1969). *Aust. J. agric. Res.* **20**, 643-652.
Clements, R. J. (1970). *Proc. XI int. Grassld Congr.* 251-254.
Clements, R. J., Oram, R. N. and Scowcroft, W. R. (1970). *Aust. J. agric. Res.* **21**, 661-676.
Cooper, J. P. (1962). *Rep. Welsh Pl. Breed. Stn.* 1961, 145-156.
Cooper, J. P. (1963). *In* "Environmental Control of Plant Growth". (L. T. Evans, ed.), pp. 381-400, Academic Press, New York and London.
Cooper, J. P. (1970). *Herb. Abstr.* **40**, 1-15.
Cooper, J. P. and Breese, E. L. (1971). *In* "Potential Crop Production (P. F. Wareing and J. P. Cooper, eds), pp. 295-318, Heinemann Educational Books, London.
Cooper, J. P., Tilley, J. M. A., Raymond, W. F. and Terry, R. A. (1962). *Nature, Lond.* **195**, 1276-1277.
Cope, W. A. (1962). *Crop Sci.* **2**, 10-12.
Corkill, L. (1942). *N.Z. Jl Sci. Technol.* **23B**, 178-193.
Craigmiles, J. P., Crowder, L. V. and Newton, J. P. (1964). *Crop Sci.* **4**, 658-661.
Crawford, R. F., Kennedy, W. K. and Johnson, W. C. (1961). *Agron. J.* **53**, 159-162.
Daday, H. (1954a). *Heredity* **8**, 61-78.
Daday, H. (1954b). *Heredity,* **8**, 377-384.
Davies, W. E. (1970). *In* "White Clover Research" (J. Lowe, ed.), Occasional Symposium 6, pp. 99-122. British Grassland Society, Hurley, Berks.
Davies, W. E., Thomas, T. A. and Young, N. R. (1968). *J. agric. Sci., Camb.* **71**, 261-281.
de Loose, R. and Baert, L. (1966). *Pl. Soil.* **24**, 343-350.
Dent, J. W. and Aldrich, D. T. A. (1963). *J. natn. Inst. agric. Bot.* **9**, 261-281.
Dent, J. W. and Aldrich, D. T. A. (1968). *J. Br. Grassld Soc.* **23**, 13-19.

Donnelly, E. D. (1954). *Agron. J.* **46**, 96-97.
Donnelly, E. D. and Anthony, W. B. (1970). *Crop Sci.* **9**, 200-202.
Dotzenko, A. D. and Henderson, K. E. (1964). *Agron. J.* **56**, 152-155.
Dougall, H. W. and Bogdan, A. V. (1966). *E. Agr. agric. For. J.* **32**, 45-49.
Egan, A. R. (1965). *Aust. J. agric. Res.* **16**, 463-472.
Evans, P. S. (1964). *N.Z. Jl agric. Res.* **7**, 508-513.
Falconer, D. S. (1960). "Introduction to quantitative genetics," Oliver and Boyd, Edinburgh.
Fleming, G. A. (1965). *Outl. Agric.* **4**, 270-285.
Francis, C. M. and Millington, A. J. (1965a). *Aust. J. agric. Res.* **16**, 557-564.
Francis, C. M. and Millington, A. J. (1965b). *Aust. J. agric. Res.* **16**, 565-573.
Francis, C. M., Gladstones, J. S. and Stern, J. S. (1970). *Proc. XI int. Grassld Congr.* 214-218.
Francis, C. M., Millington, A. J. and Bailey, R. W. (1967). *Aust. J. agric. Res.* **18**, 47-54.
Franzke, C. J. (1945). Circ. S. Dak. agric. Exp. Stn 57, pp. 8.
Franzke, C. J. (1948). *J. Amer. Soc. Agron.* **40**, 396-406.
Franzke, C. J., Puhr, L. F. and Hume, A. N. (1939). *Tech. Bull. S. Dak. agric. Exp. Stn* 1, pp. 51.
French, M. H. (1957). *Herb. Abstr.* **27**, 1-9.
Gallagher, C. H., Koch, J. H., Moore, R. M. and Steel, J. D. (1964). *Nature, Lond.* **204**, 542-545.
Gangstad, E. O. (1964). *Crop Sci.* **4**, 269-270.
Gangstad, E. O. (1966). *Crop Sci.* **6**, 334-336.
Gil, H. C., Davis, R. L. and Barnes, R. F. (1967). *Crop Sci.* **7**, 19-21.
Gillingham, J. T., Shirer, M. M., Starnes, J. J., Page, N. R. and McClain, E. F. (1969). *Agron. J.* **61**, 727-730.
Gladstones, J. S. (1967). *J. Agric. West. Aust.* **8**, 190-196.
Golley, F. B. (1961). *Ecology* **42**, 581-584.
Gonzalez, V., Brewbaker, J. L. and Hamill, D. E. (1967). *Crop Sci.* **7**, 140-143.
Goplen, B. P. (1969). *Crop Sci.* **9**, 477-480.
Goplen, B. P., Greenshields, J. E. R. and Baenziger, H. (1957). *Can. J. Bot.* **35**, 583-593.
Gray, E., Rice, J. S., Wattenbauger, D., Benson, J. A., Hester, A. J., Loyd, R. C. and Green, B. M. (1968). *Bull. Tenn. agric. Exp. Stn* 445, pp. 48.
Grunes, D. L., Stout, P. R. and Brownell, J. R. (1970). *Adv. Agron.* **22**, 331-374.
Gul, A. and Kolp, B. J. (1960). *Agron. J.* **52**, 504-506.
Hackbarth, J. and Troll, H. J. (1957). In "Handbuch der Pflanzenzuchtung" (H. Kappert and W. Rudorf, eds), Band IV, pp. 1-51, Paul Parey, Berlin.
Hanson, C. H. Loper, G. M., Kohler, G. O., Bichoff, E. M. and Taylor, K. W. (1965). *Tech. Bull. U.S. Dep. Agric.* 1333, pp. 72.
Heinrichs, D. H. (1970). *Proc. XI int. Grassld Congr.* 267-270.
Heinrichs, D. H., Troelsen, J. E. and Worder, F. G. (1969). *Can. J. Pl. Sci.* **49**, 293-305.
Hunt, L. A. (1966). *Crop Sci.* **6**, 507-509.
Hutton, E. M. (1970). *Adv. Agron.* **22**, 1-74.
Ingalls, J. R., Thomas, J. W., Benne, E. J. and Tesar, M. (1965). *J. Anim. Sci.* **24**, 1159-1164.
Johns, A. T., Ulyatt, M. J. and Glenday, A. C. (1963). *J. agric. Sci., Camb.* **61**, 201-207.
Johnson, J. M. and Butler, G. W. (1957). *Physiologia Pl.* **10**, 100-111.
Jones, D. I. H. (1970a). *J. Sci. Fd Agric.* **21**, 559-562.
Jones, D. I. H. (1970b). *J. agric. Sci., Camb.* **75**, 293-300.

Jones, D. I. H. and Walters, R. J. K. (1969). *In* "Grass and Forage Breeding" (L. Phillips and R. Hughes, eds), Occasional Symposium 5, pp. 37-43. British Grassland Society, Hurley, Berks.

Jones, W. T. and Lyttleton, J. W. (1969). *N.Z. Jl agric. Res.* **12**, 31-46.

Jones, W. T., Lyttleton, J. W. and Clarke, R. T. J. (1970). *N.Z. Jl agric. Res.* **13**, 149-156.

Julen, G. and Lager, A. (1966). *Proc. X int. Grassld Congr.* 654-657.

Knight, R. and Yates, N. G. (1968). *Aust. J. agric. Res.* **19**, 373-380.

Latter, B. D. H. (1964). *In* "Grasses and Grasslands" (C. Barnard, ed.), pp. 168-181, MacMillan, London.

Lazenby, A. and Rogers, H. H. (1965). *J. agric. Sci., Camb.* **65**, 79-90.

Lindsay, D. R., Francis, C. M. and Kelly, R. W. (1970). *Proc. XI int. Grassld Congr.* 782-784.

Long, M. I. E., Ndyanabo, W. K., Marshall, B. and Thornton, D. D. (1969). *Trop. Agric., Trin.* **46**, 201-209.

Loyd, R. C. and Gray, E. (1970). *Agron. J.* **62**, 394-397.

Marshall, B., Long, M. I. E. and Thornton, D. D. (1969). *Trop. Agric. Trin.* **46**, 43-46.

McArthur, J. M. and Miltimore, J. E. (1966). *Proc. X int. Grassld Congr.* 518-521.

McArthur, J. M. and Miltimore, J. E. (1969). *Can. J. anim. Sci.* **49**, 69-75.

Micke, A. (1962). *Z. PflZucht.* **48**, 1-13.

Milford, R. (1964). *Bull. Commonw. Bur. Past. Fld Crops* **47**, 144-153.

Milford, R. (1967). *Aust. J. exp. Agric. Anim. Husb.* **1**, 540-545.

Milford, R. and Minson, D. J. (1968). *Aust. J. Exp. Agric. Anim. Husb.* **8**, 409-412.

Millington, A. J., Francis, C. M., and Lloyd-Davies, H. (1966). *Aust. J. agric. Res.* **17**, 901-906.

Minson, D. J. (1971). *Aust. J. expt. Agric. Anim. Husb.* **11**, 18-25.

Minson, D. J. and Haydock, K. P. (1971). *Aust. J. expt. Agric. Anim. Husb.* **11**, 181-185.

Minson, D. J. and McLeod, M. N. (1970). *Proc. XI inter. Grassld Congr.* 719-722.

Minson, D. J. and Milford, R. (1967). *Aust. J. expt. Agric. Anim. Husb.* **7**, 546-551.

Moore, R. M. and Hutchings, R. J. (1967). *Aust. J. expt. Agric. Anim. Husb.* **7**, 17-21.

Morley, F. H. W. and Francis, C. M. (1968). *Aust. J. agric. Res.* **19**, 15-26.

Mowat, D. N. (1968). *In* "Forage Economics-Quality" (C. M. Harrison, ed.), ASA, special publication; 13, pp. 85-95. American Society of Agronomy, Madison, Wisconsin, U.S.A.

Mudd, A. J. (1970). *Br. vet. J.* **126**, 38-44.

Muller, L. D., Barnes, R. F., Bauman, L. F. and Colenbrander, V. F. (1971). *Crop Sci.* **11**, 413-415.

Murphy, L. S. and Smith, G. E. (1967). *Agron. J.* **59**, 171-175.

Nelson, O. E. (1969). *Adv. Agron.* **21**, 171-194.

O'Donovan, P. B., Barnes, R. F., Plumlee, M. P., Mott, G. O. and Packett, L. V. (1967). *J. Anim. Sci.* **26**, 1144-1152.

Oram, R. N. (1970). *Proc. XI int. Grassld Congr.* 785-788.

Oram, R. N. and Williams, J. D. (1967). Nature, Lond. **213**, 946-947.

Osbourn, D. F. (1967). *In* "Fodder Conservation" (R. J. Wilkins, ed.), Occasional Symposium 3, pp. 20-28. British Grassland Society, Hurley, Berks.

Osbourn, D. F., Thomson, D. J. and Terry, R. A. (1966). *Proc. X int. Grassld Congr.* 363-366.
Patil, B. D. and Jones, D. I. H. (1970). *Proc. XI int. Grassld Congr.* 726-730.
Peterson, P. J. and Butler, G. W. (1962). *Aust. J. biol. Sci.* 15, 126-146.
Rabas, D. L., Schmid, A. R. and Marten, G. C. (1970). *Agron. J.* 62, 762-764.
Rae, A. L., Brougham, R. W., Glenday, A. C. and Butler, G. W. (1963). *J. agric. Sci., Camb.* 61, 187-190.
Raymond, W. F. (1969). *Adv. Agron.* 21, 2-108.
Reid, C. S. W. (1960). *Proc. VIII int. Grassld Congr.* 668-671.
Reid, R. L., Post, A. J. and Jung, G. A. (1970). *Bull. W. Va. Univ. agric. Exp. Stn* 589T, pp. 35.
Rendig, V. Y., Welsh, R. M. and McComb, E. A. (1970). *Crop Sci.* 10, 682-684.
Robinson, R. R. (1942). *J. Amer. Soc. Agron.* 34, 933-939.
Roe, R. and Mottershead, B. E. (1962). *Nature, Lond.* 193, 255-286.
Rogers, H. H. (1970). *Proc. XI int. Grassld Congr.* A70-A78.
Rogers, H. H. and Thomson, A. J. (1970). *J. agric. Sci., Camb.* 75, 145-158.
Ross, J. C., Bullis, S. S. and Lin, K. C. (1970). *Crop Sci.* 10, 672-674.
Scheibe, A. and Hulsmann, G. (1958). *Z. PflZücht.* 39, 299-324.
Schillinger, J. A. and Elliott, F. C. (1966). *Q. Bull. Mich. St. Univ. agric. Exp. Stn* 48, 570-579.
Seay, W. A. and Henson, L. (1958). *Agron. J.* 50, 165-169.
Smith, W. K. and Gorz, H. J. (1965). *Adv. Agron.* 17, 164-233.
Snaydon, R. W. and Bradshaw, A. D. (1962). *J. exp. Bot.* 13, 422-434.
Snaydon, R. W. and Bradshaw, A. D. (1969). *J. appl. Ecol.* 6, 185-202.
Sosulski, F. W., Patterson, J. K. and Low, A. G. (1960). *Agron. J.* 52, 130-134.
Stitt, R. E. (1943). *J. Amer. Soc. Agron.* 35, 944-954.
Thomson, A. J. and Rogers, H. H. (1970). *J. agric. Sci., Camb.* 75, 159-168.
Thomson, A. J. and Rogers, H. H. (1971). *J. agric. Sci., Camb.* 76, 283-294.
Thomson, D. J. (1964). *Expt. Prog. Grassld Res. Inst.* 16, 67-68.
Todd, J. R. (1970). *In* "White Clover Research" (J. Lowe, ed.), Occasional Symposium 6, pp. 297-307. British Grassland Society, Hurley, Berks.
Ulyatt, M. J. (1969). *Proc. N.Z. Soc. Anim. Prod.* 29, 114-123.
Ulyatt, M. J. (1970). *Proc. XI int. Grassld Congr.* 709-713.
van Soest, P. J. (1965). *J. Anim. Sci.*, 24, 834-843.
van Soest, P. J. (1968). *In* "Forage Economics—Quality" (C. M. Harrison, ed.), A.S.A. special publication 13, pp. 63-76. American Society of Agronomy, Madison, Wisconsin, U.S.A.
Voigt, P. W., Kneebone, W. R., McIlvain, E. H., Shoop, M. C. and Webster, J. E. (1970). *Agron. J.* 62, 673-676.
Vose, P. B. (1963). *Pl. Soil* 19, 49-64.
Vose, P. B. and Breese, E. L. (1964). *Ann. Bot.* N.S. 28, 251-270.
Walters, R. J. K. (1971). *J. agric. Sci., Camb.* 76, 243-252.
Walters, R. J. K., ap Griffith, G., Hughes, R. and Jones, D. I. H. (1967). *J. Br. Grassld Soc.* 22, 112-116.
Whitehead, D. C. (1966). Mimeogrd Publs Commonw. Bur. Past. Fld Crops 1/1966, pp. 83.
Wilkins, H. L., Bates, R. P., Henson, P. R., Lindahl, I. L. and Davies, R. E. (1953). *Agron. J.* 45, 335-336.
Wilson, D. (1965). *J. agric. Sci., Camb.* 65, 285-292.
Wong, E. and Francis, C. M. (1968a). *Phytochemistry* 7, 2131-2137.
Wong, E. and Francis, C. M. (1968b). *Phytochemistry* 7, 2139-2142.
Woods, D. L. and Clark, K. W. (1971). *Can. J. Pl. Sci.* 51, 323-329.
Wright, M. J. and Davison, K. L. (1964). *Adv. Agron.* 16, 197-247.

Author Index

Numbers followed by an asterisk are those pages on which References are listed

Bach, M. K., 209, 222*
Bagi, G., 334, 349*
Baig, M. W., 6, 12*
Bailey, R. W., 16, 23*, 84, 101*, 196, 222*, 386, 389, 390, 392, 414*, 415*
Baker, C. J. L., 16, 21, 23*
Baker, D. A., 121, 125*
Baker, D. N., 95, 98*
Baker, E., 118, 125*
Baker, H. K., 363, 376*
Balch, C. C., 386, 414*
Baldwin, E., 237, 244*
Baldwin, I. L., 205, 223*
Balz, H. P., 334, 349*
Bar-Akiva, A., 140, 158*, 161*
Barber, D. A., 111, 117, 118, 121, 123*, 125*, 280, 281, 309*
Barber, S. A., 119, 123*, 267, 279, 297, 309*, 311*
Barclay, P. C., 19, 23*, 120, 124*, 400, 410, 414*
Barkoczy, M., 330, 351*
Barnes, R. F., 384, 385, 387, 388, 405, 407, 408, 414*, 415*, 416*
Barnes, R. J., 178, 179, 187*
Barnett, A. J. G., 147, 158*
Barnett, N. M., 336, 349*
Barnett, R. D., 410, 414*
Barrios, S., 204, 225*
Barrow, N. J., 266, 267, 271, 272, 289, 290, 291, 306, 310*, 311*
Barshad, I., 153, 158*
Bartholomew, W. V., 229, 238, 239, 244*, 246*
Barton, R. A., 329, 349*
Bassham, J. A., 27, 39, 40, 53*
Bates, R. P., 388, 396, 409, 414*, 417*
Bauman, L. F., 385, 405, 416*
Baumgardner, M. F., 278, 285, 310*
Bear, F. E., 265, 266, 283, 310*
Beath, O. A., 155, 158*
Beaton, G. H., 2, 11*
Beaton, J. D., 289, 310*
Bedri, A. A., 167, 187*
Beeson, W. M., 157, 161*
Beevers, H., 39, 53*
Beevers, L., 130, 136, 137, 138, 158*
Beidleman, R. G., 236, 244*
Belkengren, R. P., 197, 225*
Bell, F., 201, 223*
Benacchio, S. S., 278, 285, 310*
Benemann, J. R., 220, 222*, 226*

Benne, E. J., 385, 415*
Benson, J. A., 394, 415*
Bentley, L. E., 177, 187*
Ben-Zioni, A., 336, 349*
Berg, C. C., 20, 24*
Bergersen, F. J., 191, 194, 196, 197, 201, 202, 203, 205, 208, 210, 211, 212, 213, 214, 215, 216, 217, 218, 219, 221, 222*, 223*, 224*, 225*
Bernstein, L., 140, 158*, 331, 340, 349*, 350*
Berry, J., 52, 53*
Berry, J. A., 35, 53*
Bertramson, B. R., 287, 310*
Bertrand, A. R., 277, 313*
Berzborn, R., 30, 54*
Bethke, R. M., 147, 159*
Bibbiscombe, E. F., 371, 377*
Bichoff, E. M., 392, 415*
Biddulph, O., 83, 98*, 122, 123*, 272, 289, 310*, 331, 349*
Biddulph, S. F., 272, 310*, 331, 349*
Bidwell, R. G. S., 325, 351*
Bieleski, R. L., 132, 133, 138, 158*, 161*
Bierhuizen, J. F., 70, 71, 72, 76, 98*, 100*
Birch, H. F., 174, 187*
Birnstel, M. L., 133, 159*
Bishop, I., 130, 158*
Bisset, K. A., 193, 223*
Bixby, D. W., 289, 310*
Bjorkman, O., 45, 47, 50, 52, 53*, 67, 76, 79, 82, 98*
Blachère, H., 196, 222*
Black, A. L., 149, 160*
Black, C. A., 237, 244*, 267, 268, 269, 272, 279, 280, 297, 310*, 313*
Black, C. C., 32, 35, 38, 49, 52, 53*, 54*, 65, 68, 98*
Black, J. N., 89, 98*
Blacklow, W. M., 362, 376*
Blackman, G. E., 93, 98*
Bland, B. F., 387, 414*
Blaser, R. E., 88, 89, 90, 96, 98*, 230, 244*, 281, 310*
Blaxter, K. L., 156, 158*
Blue, W. G., 277, 285, 310*
Blumenfeld, A., 348, 350*
Boardman, N. K., 27, 29, 30, 31, 33, 35, 36, 53*, 55*
Bogdan, A. V., 398, 415*
Böhning, K., 167, 183, 187*

Subject Index

A

Abscisic acid, 348
Abscisins, 348
Acetylene reduction assay of nitrogen
    fixation, 212–213
Acid production in rhizobia, 195
Acids, organic, 163–186
    and carboxylation, 167
    and ionic balance, 173
    degradation, 170–173
    fractionation, 174
    in herbage, generation by cation
      uptake, 166–168
    in relation to anion mechanism, 108
    levels in herbage, 175–183
    measurement by ash alkalinity,
      174
    partition between plant organs, 177
    synthesis in roots, 172
    synthesis sites, 170–173
    time course of accumulation,
      183–185
    transport, 170–173
Ageing of plant tissue, 323–335
Agmatine, and potassium deficiency,
    338
Alkali disease, 154
    blind staggers, 153
Alkaloids
    lupin, 393
    phalaris, 393–394
Aluminium
    pH tolerance, in plants, 141
    tolerance to in herbage, 141, 143
Amino acids
    in nitrogen fixation, 210
    in senescence, 330
    seleno-, 142
    seleno-, in herbage, 134
    translocation of, 330
    utilization by rhizobia, 194

Ammonium
    and organic acids in ryegrass, 185
    uptake and organic acids, 164–166
    uptake in roots, 172
Animal nutrition, minerals in, 143–151
Anion absorption, 114–115
Anions, metabolism in herbage,
    168–169
Ascorbic acid, 6
Aspartate, in C_4 photosynthesis
    pathway, 42–46
Assimilates
    distribution, 85–87
    transport, 83–85
Auxins, 343–344
    and *Rhizobium*, 204

B

B-vitamins and *Rhizobium* growth,
    194
Bacteroids, 194
    in nodules, 205–207
    metabolism, 215–216
    respiration, 216
    suspensions, 214–217
6-Benzylaminopurine, 346–348
Bicarbonate
    and organic acid accumulation in
      ryegrass, 185
    uptake in roots, 172–173
Biochanin A, 392
Biotin, 11
Blind staggers, 154
Bloat, 395–396
Bone chewing, 145
Boron
    excretion, 266–267
    form in plants, 134
    function in plants, 130
    loss from leaves, 122
Brassicas, organic acid levels in, 181

Genera and Species Index

A

Agropyron repens, photosynthate transport, 85
Agrostis spp., Zn in, 134
A. stolonifera, Zn tolerance, 142
A. tenuis
  organic acid levels in, 179
  Zn tolerance, 142
Andropogon caricosus, oxalic acid in, 176
A. gayanus, water in, 20
Atriplex spongiosa, chloroplasts, 35
Axonopus compressus, oxalic acid in, 176

B

Brassica napus, water in, 16
B. oleracea, water in, 16
Bromus inermis
  regrowth and reserves, 361
  reserves in, 357
B. unioloides
  organic acid levels in, 179
  water in, 16

C

Chloris inflata, oxalic acid in, 176
Cynodon dactylon
  digestibility, 384
  genetic variation of, 384
  oxalic acid in, 176
  photosynthesis in, 65
  protein content, 398
  reserves in, 374
  water in, 20
C. plectostachyus, oxalic acid in, 176

D

Dactylis glomerata
  digestibility, 382, 384
  genetic variation of, 384
  Mo in, 153
  organic acid levels in, 178–180
  photosynthate transport, 84
  photosynthesis, 65, 69, 72, 77
  protein content, 398
  regrowth and reserves, 361
  soluble sugars, 387
  water in, 16, 20
Desmodium uncinatum, water in, 20
Digitaria decumbens
  intake of, 388
  nitrogen levels in, 229
  oxalic acid in, 176
  photosynthesis, 69
  water in, 20
D. smutsii, water in, 16

E

Equisetum spp., thiaminase in, 7
Eragrostis curvula, palatability of, 388
Eriochloa polystachya, oxalic acid in, 176

F

Festuca arundinacea
  intake of, 386
  palatability inheritability, 408
  palatability of, 388
  photosynthesis, 77
  regrowth and reserves, 362
  water in, 16, 20
  Zn in, 134
F. pratensis, organic acid levels in, 178–180
F. rubra, organic acid levels in, 179

Glossary of Common Names

Alfalfa	*Medicago sativa*
Altar fescue	*Festuca arundinacea*
Alsike clover	*Trifolium hybridum*
Bahia grass	*Paspalum notatum*
Bajra grass	*Pennisetum typhoides*
Barley	*Hordeum vulgare*
Barley grass	*Hordeum murinum*
Beach grass	*Sporobolus virginicus*
Bent grass	*Agrostis tenuis*
Bent grass	*Agrostis stolonifera*
Bermuda grass	*Cynodon dactylon*
Berseem	*Trifolium alexandrinium*
Birdsfoot trefoil	*Lotus corniculatus*
Blue grass	*Bouteloua gracilis*
Blue grass	*Poa pratensis*
Brome grass	*Bromus inermis*
Brown top	*Agrostis tenuis*
Buckwheat	*Polygonum fagopyrum*
Buffel grass	*Cenchrus ciliaris (Pennisetum ciliare)*
Carpet grass	*Axonopus compressus*
Centipede grass	*Eremochloa ophiuroides*
Cerrillo	*Sporobolus indicus*
Chewings fescue	*Festuca rubra* var. *commutata*
Chou moellier	*Brassica oleracea*
Coastal bermuda grass	*Cynodon dactylon*
Cocksfoot	*Dactylis glomerata*
Common spear grass	*Stipa comata*
Cotton grass	*Imperata cylindrica*
Couch grass	*Cynodon dactylon*
Crab grass	*Digitaria sanguinalis*
Creeping soft grass	*Holcus mollis*
Crested dogstail	*Cynosurus cristatus*
Cuban grass	*Andropogon carricosus*
Elephant grass	*Pennisetum purpureum*
Esparto grass	*Stipa tenacissima*
Fescue	*Festuca glauca*
Field pea	*Pisum sativum*
Fog	*Holcus lanatus*
Giant star grass	*Cynodon plectostachyus*
Guatemala grass	*Tripsacum laxum*

Guinea grass	*Panicum maximum*
Italian ryegrass	*Lolium multiflorum*
Johnson grass	*Sorghum halepense*
Jowar	*Andropogon sorghum*
Kentucky bluegrass	*Poa pratensis*
Kenya sheep grass	*Brachiaria decumbens*
Kikuyu	*Pennisetum clandestinum*
Ladino clover	*Trifolium repens*
Lucerne	*Medicago sativa*
Maize	*Zea mays*
Malojilla	*Eriochloa polystachya*
Marrow stem kale	*Brassica oleracea*
Meadow fescue	*Festuca pratensis*
Merker grass	*Pennisetum purpureum* var. *merkerii*
Mexican blue grass	*Chloris inflata*
Millet	*Setaria italica*
Molasses grass	*Melinis minutiflora*
Napier grass	*Pennisetum purpureum*
Nashiva star grass	*Cynodon plectostachyus*
New Zealand brown top	*Agrostis tenuis*
Oat	*Avena sativa*
Orchard grass	*Dactylis glomerata*
Pangola grass	*Digitaria decumbens*
Pangola river grass	*Digitaria pentzii*
Para grass	*Brachiaria mutica (Panicum purpuraceum)*
Pearl millet	*Pennisetum typhoides*
Perennial ryegrass	*Lolium perenne*
Prairie grass	*Bromus unioloides*
Pusa giant Napier grass	*Pennisetum purpureum* x *P. typhoides*
Quack grass	*Agropyron repens*
Rape	*Brassica napus*
Red clover	*Trifolium pratense*
Red fescue	*Festuca rubra*
Red top	*Agrostis gigantea*
Reed canary grass	*Phalaris arundinacea*
Rhodes grass	*Chloris gayana*
Rough fescue	*Festuca sabrella*
Rye	*Secale cereale*
Sainfoin	*Onobrychis viciifolia*
San Augustine	*Stenotaphrum secondatum*
Sempre verde grass	*Panicum maximum*
Short rotation ryegrass	*Lolium perenne* x *L. multiflorum*
Smooth brome grass	*Bromus inermis*
Snow tussock	*Chionochloa* sp.
Sorghum	*Sorghum bicolor*
Soya bean	*Glycine max*
Spear grass	*Heteropogon contortus*
Star grass	*Cynodon plectostachyus*
Subterranean clover	*Trifolium subterraneum*
Sudan grass	*Sorghum vulgare* var. *sudanensis*
Sunflower	*Helianthus annuus*
Suwanee bermuda grass	*Cynodon dactylon*
Sweet clover	*Melilotus officinalis*

Sweet grass	*Paspalum plicatulum*
Sweet maize	*Zea mays* var. *saccharata*
Sweet vernal grass	*Anthoxanthum odoratum*
Switch grass	*Panicum virgatum*
Tall fescue	*Festuca arundinacea*
Tall oat grass	*Arrhenatherum elatius*
Thousand-headed kale	*Brassica oleracea*
Timothy	*Phleum pratense*
Townsville stylo	*Stylosanthes humilis*
Trefoil	*Lotus* sp.
Trefoil	*Medicago lupulina*
Vetch	*Vicia villosa*
Venezuela grass	*Melinis minutiflora*
Venezuela grass	*Paspalum fasciculatum*
Wheat	*Triticum vulgare*
White clover	*Trifolium repens*
Yorkshire fog	*Holcus lanatus*

Sweet grass	*Paspalum plicatulum*
Sweet maize	*Zea mays* var. *saccharata*
Sweet vernal grass	*Anthoxanthum odoratum*
Switch grass	*Panicum virgatum*
Tall fescue	*Festuca arundinacea*
Tall oat grass	*Arrhenatherum elatius*
Thousand-headed kale	*Brassica oleracea*
Timothy	*Phleum pratense*
Townsville stylo	*Stylosanthes humilis*
Trefoil	*Lotus* sp.
Trefoil	*Medicago lupulina*
Vetch	*Vicia villosa*
Venezuela grass	*Melinis minutiflora*
Venezuela grass	*Paspalum fasciculatum*
Wheat	*Triticum vulgare*
White clover	*Trifolium repens*
Yorkshire fog	*Holcus lanatus*